CHIP ON BOARD TECHNOLOGIES FOR MULTICHIP MODULES

CHIP ON BOARD TECHNOLOGIES FOR MULTICHIP MODULES

Edited by John H. Lau

VAN NOSTRAND REINHOLD
An International Thomson Publishing Company

New York • London • Bonn • Boston • Detroit • Madrid • Melbourne • Mexico City
Paris • Singapore • Tokyo • Albany NY • Belmont CA • Cincinnati OH

Copyright © 1994 by Van Nostrand Reinhold

Library of Congress Catalog Card Number 93-40440
ISBN 0-442-01441-4

I(T)P Van Nostrand Reinhold, an International Thomson Publishing company.
       ITP logo is a trademark under license.

Printed in the United States of America

Van Nostrand Reinhold
115 Fifth Avenue
New York, NY 10003

International Thomson Publishing
Berkshire House
168-173 High Holborn
London WC1V 7AA, England

Thomas Nelson Australia
102 Dodds Street
South Melbourne, Victoria 3205
Australia

Nelson Canada
1120 Birchmount Road
Scarborough, Ontario
M1K 5G4, Canada

International Thomson Publishing GmbH
Königswinterer Strasse 418
53227 Bonn
Germany

International Thomson Publishing Asia
221 Henderson Road
#05 10 Henderson Building
Singapore 0315

International Thomson Publishing Japan
Hirakawacho Kyowa Building, 3F
2-2-1 Hirakawa-cho, Chiyoda-ku
Tokyo 102
Japan

ARCFF 16 15 14 13 12 11 10 9 8 7 6 5 4 3 2

Library of Congress Cataloging-in-Publication Data

Chip on board technologies for multichip modules / edited by John H. Lau.
    p.   cm.
    Includes bibliographical references and index.
    ISBN 0–442–01441–4
    1. Electronic packaging.   2. Surface mount technology.   I. Lau,
John H.
TK7870.15.C55   1994
621.381′046—dc20   93–40440
                           CIP

CONTENTS

Preface   *xiii*
Acknowledgments   *xvii*

1. **A Brief Introduction to Wire Bonding, Tape Automated
   Bonding, and Flip Chip on Board for Multichip Module
   Applications**   **1**
   1.1 Introduction   1
   1.2 The Chip   3
        1.2.1 The Wafer   3
        1.2.2 Number of Chips per Wafer   3
        1.2.3 Number of Yielded Chips per Wafer   4
        1.2.4 Performance   6
        1.2.5 Number of Possible I/Os (Pads) per Chip   12
        1.2.6 Number of Required I/Os (Pads) per Chip   13
        1.2.7 COB and MCM Yield   15
        1.2.8 Known Good Dies (KGD)   19
        1.2.9 Unknown Bad Dies (UBD)   21
        1.2.10 Fracture Toughness Measurement of Chip   22
   1.3 Chip Level Interconnects   25
        1.3.1 Wire Bonding   25
        1.3.2 Tape Automated Bonding   27
        1.3.3 Solder Bumped Flip Chip   27
   1.4 First Level Packages   29
        1.4.1 Single-Chip Modules   29

1.4.2 Multichip Modules (MCM)   31
1.4.3 Multichip Carrier Modules (MCCM or MC^{2}M)   33
1.4.4 Distinction Between COB and MCM-L   35
1.5 The Board   37
    1.5.1 Wireability   37
    1.5.2 Propagation Delay   40
    1.5.3 Characteristic Impedance   42
    1.5.4 Known Good Substrates (KGS) and Unknown Bad
        Substrates (UBS)   47
    1.5.5 Material Requirements   47
    1.5.6 Flexural Rigidity and Young's Modulus Measurement of
        Board   55
1.6 Chip on Board   57
    1.6.1 Wire Bonding Chips on Board   58
    1.6.2 Tape Automated Bonding Chips on Board   61
    1.6.3 Cost Comparison Between Wire Bonding and TAB   63
    1.6.4 Solder Bumped Flip Chip on Board   65
1.7 Packaging Alternatives for Miniature Products   80
    1.7.1 PCMCIA PC Cards by COB Technology   81
    1.7.2 LCD Modules by COG Technology   84
1.8 Summary   86
    Acknowledgments   87
    References   87

2. Making COB Testing Tractable: Chip Pretest and System
   Diagnostics   101
2.1 Introduction   101
2.2 Known Good Die   102
2.3 COB Design for Test and Diagnostics   107
    2.3.1 Diagnostics   108
    2.3.2 Design for Testability Tools   109
    2.3.3 Defect Analysis and Test Strategies   118
    2.3.4 Concurrent Engineering   119
    Acknowledgment   120
    References   120

3. Chip Level Interconnect: Wire Bonding for Multichip Modules   124
3.1 Introduction   124
3.2 Wire Bonding Principles   125
    3.2.1 Thermocompression Bonding   128
    3.2.2 Ultrasonic Bonding   131
    3.2.3 Thermosonic Bonding   132
    3.2.4 Automatic Wire Bonding   134

3.2.5 Other Wire-Like Interconnects   134
3.3 Bonding Process   137
   3.3.1 Bonding Parameters   137
   3.3.2 Process Optimization   141
3.4 Materials   153
   3.4.1 Chip Metallizations and Dielectrics   153
   3.4.2 Bonding Wire   155
   3.4.3 Lead Frames and Packages   157
   3.4.4 Substrate Metallizations and Materials   159
3.5 Wire Bond Testing   161
   3.5.1 Wire Bond Testing Equipment   162
   3.5.2 Wire Bond Pull Tests   164
   3.5.3 Ball Shear Testing   170
3.6 Wire Bond Quality Assurance/Reliability   171
   3.6.1 Quality Assurance (QA)   171
   3.6.2 Reliability   174
3.7 MCM Wire Bonding Design   180
3.8 Summary   181
   Acknowledgments   181
   References   181

4. Chip Level Interconnect: Wafer Bumping and Inner Lead Bonding   **186**
4.1 Introduction   186
4.2 TAB Inner Lead Bond Bump Options   188
4.3 Wafer Bumping   189
   4.3.1 Gold Bump   189
   4.3.2 Copper Bump   193
   4.3.3 Aluminum Bump   193
   4.3.4 Solder Bump   195
4.4 Tape Bumping   197
   4.4.1 Bumped-TAB Tape   197
   4.4.2 Transfer-Bumped Tape   198
   4.4.3 Balltape   200
4.5 TAB ILB Processes and Tools   201
4.6 TAB Inner Lead Thermocompression Bonding Process   205
   4.6.1 Thermocompression Bonding Considerations   205
   4.6.2 The Mechanics of the Bonding Cycle   206
   4.6.3 Thermocompression Bonding Mechanical Process Window   214
4.7 ILB Solder Joining Tool – Hot Air Thermode   216
4.8 ILB Solder Attach Processes   218
4.9 Future Trends in TAB Technology   222

References   223

5. Chip Level Interconnect: Solder Bumped Flip Chip   **228**
    5.1 Introduction   228
    5.2 Basic Process Description   230
    5.3 Chip Passivation   231
        5.3.1 Inorganics   231
        5.3.2 Polymers   232
    5.4 Terminal Metals and Solders – Materials   232
        5.4.1 Terminal Metals   233
        5.4.2 Solders   234
        5.4.3 Reliability Considerations   236
    5.5 Terminal Metals and Solder – Processes   240
        5.5.1 Sputter Cleaning   240
        5.5.2 Masking   240
        5.5.3 Deposition Processes   243
        5.5.4 Representative Process Flows   245
    5.6 Conclusion   245
        References   246

6. Chip Attachment   **251**
    6.1 Introduction   251
    6.2 PWB Materials and Chip and Pad Preparation   252
    6.3 Chip Attach   255
        6.3.1 Types of Chip Attach   255
        6.3.2 Polymer Bonding   256
        6.3.3 Glass Adhesives   259
        6.3.4 Soft Solders   259
        6.3.5 Hard Solders   261
        6.3.6 Bonding Below the Eutectic Point of Solders   262
    6.4 Attachment Quality   267
        6.4.1 Thermal Stress Considerations   267
        6.4.2 Attachment Quality   267
    6.5 Summary   268
        References   269

7. Wire Bonding Chip on Board   **275**
    7.1 Introduction   275
    7.2 Applications of Wire Bonded COB   278
    7.3 Development of the COB Design, Materials, and Process
        Sets   281
    7.4 Substrates   286
    7.5 Circuitization   287

7.6 Layouts   288
7.7 Substrate Plating   291
   7.7.1 Electroplated Gold/Nickel (Full Additive)   291
   7.7.2 Electroplated Gold/Nickel (Semi-Additive)   292
   7.7.3 Electroless Nickel/Immersion Gold   294
   7.7.4 Nickel and Alloys   295
   7.7.5 Platings – Miscellaneous   296
7.8 Chip Metallurgy   296
7.9 Substrate Cleanliness   298
7.10 Die Attachment   301
   7.10.1 Materials   301
   7.10.2 Process   303
   7.10.3 Quality Controls   307
7.11 Wire Bonding   307
   7.11.1 General   307
   7.11.2 Wire Bonding Options   308
   7.11.3 Wire Metallurgy/Controls   313
   7.11.4 Thermal   314
   7.11.5 Electrical Current Capabilities   314
   7.11.6 Wire Bond Process   315
   7.11.7 Wire Bond Process Set-Up   319
   7.11.8 In-Process Monitoring   320
   7.11.9 Visual Inspection of Bonds   321
   7.11.10 Rework   324
7.12 Electrical Testing   328
   7.12.1 Wafer Testing   328
   7.12.2 Raw Card Testing   329
   7.12.3 Board Assembly Testing   329
7.13 Encapsulation   330
   7.13.1 Materials   330
   7.13.2 Silicone   330
   7.13.3 Epoxies   331
   7.13.4 Other COB Encapsulation Systems   334
   7.13.5 Process   336
   7.13.6 Equipment   337
7.14 Environmental Stress Testing   338
7.15 Manufacturing Process Flow Sequence   338
7.16 Summary   338
   Acknowledgments   339
   References   340

8. Tape Automated Bonding Chip on Board and on MCM-D   **343**
8.1 Introduction   343

8.2 Classification of TAB Packaging Configurations   345
8.3 Design Considerations for TAB on Board and on
     MCM-D   346
     8.3.1 OLB Window and Leadform Design   349
     8.3.2 Keeper Bar Design   353
     8.3.3 Corner Tie Bar to Hot Bar Clearance   354
     8.3.4 Hot Bar Selection   354
     8.3.5 Hot Bar Location   356
     8.3.6 Carrier and Test/Burn-in Socket Selection   357
     8.3.7 Substrate Pad Design   357
     8.3.8 Substrate Keep Clear Areas   358
8.4 Material Considerations for TAB on Board and on
     MCM-D   358
     8.4.1 Substrate   359
     8.4.2 TAB Tape   361
     8.4.3 Die-Attach Material   363
     8.4.4 Flux and Cleaning Solvent   364
8.5 Assembly Considerations for TAB on Board and on
     MCM-D   365
     8.5.1 TAB Mount Options   365
     8.5.2 Assembly Process Overview   366
     8.5.3 Outer Lead Bonding Options   371
     8.5.4 Flux Cleaning   382
     8.5.5 Inspection   383
     8.5.6 Repair and Rework   386
     8.5.7 Assembly Equipment Selection   393
8.6 How to Determine if TAB is the Right Technology for a COB
     Application   395
8.7 Examples of TAB Applications   396
     8.7.1 TAB on Board   398
     8.7.2 TAB on MCM-D   401
8.8 Future Trends of TAB Technology   402
     References   405

9. Solder Bumped Flip Chip Attach on SLC Board and Multichip
   Module   410
   9.1 Requirements for Packaging   410
   9.2 SLC Technology   411
       9.2.1 Structure of SLC   411
       9.2.2 Manufacturing Process of SLC   413
       9.2.3 Quasi Module   417
   9.3 Flip Chip Attach (FCA) Technology   418
       9.3.1 Flip Chip Bonding   418

9.3.2 Stress Analysis of FCA Joint   422
9.3.3 Manufacturing Process of FCA   427
9.4 Advantages of SLC/FCA Package   430
9.5 SLC/FCA Applications   434
    9.5.1 Coverage of Applications   434
    9.5.2 Token Ring LAN Adapter Card   435
    9.5.3 Extension to MCM-L   435
9.6 Summary   442
    References   442

10. Micron Bump Bonding Chip on Board   **444**
10.1 Introduction   444
10.2 Outline and Futures of New Technology   444
    10.2.1 Method of Process   446
    10.2.2 Principle of this Technology   447
    10.2.3 Electrical Characteristics   449
    10.2.4 Reliability   451
10.3 Typical Applications of MBB Method   453
    10.3.1 LED Array Module   453
    10.3.2 Thermal Head Module   457
    10.3.3 MCM Module   463
10.4 Summary   468
    References   468

11. Chip on Board Encapsulation   **470**
11.1 Introduction   470
11.2 Encapsulation of COB   472
    11.2.1 Purposes and Goals for Encapsulation   472
    11.2.2 Material Requirements   474
    11.2.3 Potential Encapsulants   474
11.3 Manufacturing Process for Encapsulation of COB   491
    11.3.1 Pre-encapsulation Cleaning   492
    11.3.2 Application of Encapsulants   494
    11.3.3 Curing of Encapsulants   495
11.4 Summary   500
    References   500

12. Underfill Encapsulation for Flip Chip Applications   **504**
12.1 Introduction   504
    12.1.1 Packaging Strategies for Flip Chips   505
    12.1.2 Candidate Underfill Encapsulant Materials   505
    12.1.3 The Underfill Encapsulation Process   508
12.2 Material Requirements for Underfill Encapsulants   508

12.2.1 Properties of the Uncured Encapsulant   509
12.2.2 Properties of the Cured Encapsulant   514
12.3 Reliability Gains and Underlying Mechanisms   518
12.3.1 Mechanics of C4 Fatigue Enhancement   519
12.3.2 Reliability Testing and Case Studies   523
12.4 Recommendations for Manufacturing   526
12.5 Summary and Future Directions   528
Acknowledgments   529
References   529

Authors' Biographies   533

Index   545

Preface

The major trend in the electronics industry today is to make products more *personal* by making them lighter, smaller, thinner, shorter, and faster, while at the same time making them more friendly, functional, powerful, reliable, and less expensive. As the trend toward miniature and compact products continues, the introduction of more user friendly products and a wider variety of functions will provide growth in the market. One of the key technologies that is helping to make these product design goals possible is electronic packaging and assembly technology, especially the wire bonding, tape automated bonding, and flip bare Chip on Board (COB) technology.

The last few years witnessed an explosive growth in the research and development efforts devoted to COB and multichip module (MCM) as a direct result of the rapid growth of surface mount technology and miniaturization. Some examples are: video camcorders, electronic organizers, personal computers (PCs), notebook PCs, subnotebook PCs, laptop PCs, palmtop PCs, PCs with built-in portable phones, PCMCIA (Personal Computer Memory Card International Association) PC cards, cellular phones, wireless phones, pagers, portable electronics products, audiovisual products, multimedia products, etc.

We are now beginning to obtain useful insight into and understanding of the economic, design, material, process, equipment, quality, and reliability issues of COB and MCM. The important COB/MCM parameters such as the chip (e.g., performance, test, known good die, unknown bad die), chip level interconnects (e.g., wire bonding, TAB, and flip chip), first level packages (e.g., single chip modules and MCM), and the board (e.g., wireability, propagation delay, characteristic impedance, and material requirements), the chip attachment, and the encapsulant have been studied by many experts. Their results already have been disclosed in diverse journals or, more incidentally, in the proceedings of many conferences, symposia, and workshops. There is no single source of information devoted to the state of the art of COB technology. This book aims to remedy this deficiency and to present, in one volume, a timely summary of progress in all aspects of this fascinating field.

The sequence of this book is basically following the COB and MCM processes. After a brief introduction to COB and MCM in Chapter 1, Charles Hawkins, David Palmer, and Keith Treece discuss the bare wafer/chip test standards ranging from the IEEE 1149.1 boundary scan standard to the proposed system standard P1149.5 (Chapter 2).

The book's next three chapters present the most common chip level interconnects. In Chapter 3, Harry Charles, Jr. describes the wire bonding technology and its applications to MCMs. The wafer bumping and inner lead bonding for TAB is discussed by Sung Kang, William Chen, Richard Hammer, and Frank Andros (Chapter 4). In Chapter 5, Lewis Goldmann and Paul Totta present the well-known Controlled Collapse Chip Connection (C4) technology.

For wire bonding and TAB bare chip on board, chip attachment is usually needed. In Chapter 6, Goran Matijasevic, Chin Lee, and Chen Wang examine various chip attachment materials, methods, and reliability.

The next four chapters of this book present the wire bonding, TAB, and flip bare chip on board technology. In Chapter 7, Robert Christiansen defines the key parameters in design, materials, and processing of wire bonding chip on board. Key considerations of design, materials, assembly, and equipment for TAB chip on board and on MCM-D related applications are discussed in Chapter 8 by Tom Chung, James Chang, and Ali Emamjomeh. In Chapter 9, Yutaka Tsukada presents the solder bumped flip chip attach on Surface Laminar Circuit (SLC) board and on SLC MCM. Chapter 10, by Kenzo Hatada, provides the design, material, process, and reliability of the Micron Bump Bonding (MBB) flip chip on board technology.

The last two chapters of this book present two very important subjects in COB and MCM applications. In Chapter 11, C. P. Wong, John Segelken, and Courtland Robinson define the material requirements, screening tests and procedures, and appropriate manufacturing processes of potential encapsulants for COB applications. Darbha Suryanarayana and Donald Farquhar (Chapter 12) examine the role, candidate materials and their requirements, process, and reliability of underfill encapsulants for solder bumped flip chip applications.

For whom is this book intended? Undoubtedly it will be of interest to three groups of specialists: (1) those who are active or intend to become active in research and development of COB and MCM; (2) those who have encountered practical COB and MCM problems and wish to understand and learn more methods of solving such problems; and (3) those who have to choose a high performance and cost effective packaging technique for their interconnect system.

I hope this book will serve as a valuable source of reference to all those faced with the challenging problems created by the ever more expanding use of COB and MCM in electronics packaging and interconnection. I also hope that it will aid in stimulating further research and development on bare wafer/chip testing and burn-in, wafer bumping, wire bonding, TAB, fine line/space high density low cost printed circuit board, chip attachment, encapsulant, underfill epoxy, flux, printing, placement, mass reflow, cleaning, inspection, testing, equipment, material, process, and design, and more sound use of COB technology in either single-chip or multichip packaging applications.

The organizations that learn how to design and manufacture COB and MCM in their interconnect systems have the potential to make major advances in electronics packaging and to gain great benefits in cost, performance, quality, size, and weight. It is my hope that the information presented in this book may assist in removing road blocks, avoiding unnecessary false starts, and accelerate design, material, and process development of these technologies. The COB and MCM technologies are limited only by the business constraints, ingenuity and imagination of engineers, visionless managements, and infrastructure.

John H. Lau, PhD, PE
Hewlett-Packard Company

Acknowledgments

Development and preparation of *Chip on Board Technologies for Multichip Modules* was facilitated by the efforts of a number of dedicated people at Van Nostrand Reinhold (VNR) and Keyword Publishing Services. I would like to thank them all, with special mention to Alan Chesterton of Keyword for his effective coordination of the publication process and to Mr. Steve Chapman, Ms. Marjorie Spencer, and Ms. Dianne Littwin of VNR for their unswerving support and for solving many problems that arose during the book's preparation. It has been a great pleasure and fruitful experience to work with them.

The material in this book has clearly been derived from many sources including individuals, companies, and organizations, and the various contributing authors have attempted to acknowledge, in the appropriate parts of the book, the assistance that they have been given. It would be quite impossible for them to express their thanks to everyone concerned for their cooperation in producing this book, but on their behalf, I would like to extend due gratitude, especially to the American Society of Mechanical Engineers (ASME) Electronic Packaging & Interconnection (EP&I) Conferences, Proceedings, and Transactions, e.g., *Journal of Electronic Packaging*, the Institute of Electrical and Electronic Engineers (IEEE) EP&I Conferences, Proceedings, and Transactions, e.g., *Hybrids, Packaging, and Manufacturing Technology*, the International Society of Hybrid Microelectronics (ISHM) and the International Electronic Packaging Society (IEPS) EP&I Conferences, Proceedings, and Transactions, e.g., *Microcircuits & Electronic Packaging*, and American Society of Metals (ASM) EP&I

Conferences, Proceedings, and books, e.g., *Electronic Materials Handbook, Volume 1, Packaging*.

I express my deep appreciation to the 24 contributing authors, all experts in their respective fields, for their many helpful suggestions and cooperation in responding to requests for revisions. Their depth of knowledge, dedication, and patience have been demonstrated throughout the process of preparing this book. Working with them has been an adventure and a privilege, and I learned a lot about life and electronic packaging from them. Their brief technical biographies are presented at the end of this book.

Lastly, I want to thank my employer, Hewlett-Packard, for providing an excellent environment in which completing this book was possible. I also want to thank my managers, Steve Erasmus and Anita Danford, for their trust, respect, and support of my real work at HP. Finally, I want to thank my daughter (Judy, who also helped design the book cover) and my wife (Teresa) for their love, consideration, and patience by allowing me to work on many weekends for this private project. Their simple belief that I am making my small contribution to mankind was strong motivation for me, and to them I have dedicated my efforts on this book.

John H. Lau, PhD, PE
Palo Alto, California

CHIP ON BOARD TECHNOLOGIES FOR MULTICHIP MODULES

1

A Brief Introduction to Wire Bonding, Tape Automated Bonding, and Flip Chip on Board for Multichip Module Applications

John H. Lau

1.1 INTRODUCTION

The major trend in the electronics industry today is to make products more personal by making them lighter, smaller, thinner, shorter, and faster, while at the same time making them more friendly, functional, powerful, reliable, and less expensive.[1-42] Some of the examples are video camcorders, electronic organizers, personal computers (PC), notebook PC, sub-notebook PC, laptop PC, palmtop PC, PC with built-in portable phones, PCMCIA (Personal Computer Memory Card International Association) PC cards, cellular phones, wireless phones, pagers, portable electronics products, audiovisual products, multimedia products, etc.

As the trend toward miniature and compact products continues, the introduction of more user-friendly and wider variety of functions will provide growth in the market. One of the key technologies that is helping to make these product design goals possible is electronic packaging and assembly technology.[1-42]

Figure 1-1 shows schematically the hierarchy of an electronic package.[43] It should be noted that the wafer is not in the packaging hierarchy. It is included in Fig. 1-1 to show where the integrated circuit (IC) chip originates. Packaging focuses on how chips are packaged efficiently and reliably.

The chip is not isolated. It must communicate with other chips in a circuit through an input/output (I/O) system of interconnects. Furthermore, the chip and its embedded circuitry are delicate, requiring the package to carry, connect, and protect it. Consequently, the major functions of electronic

1

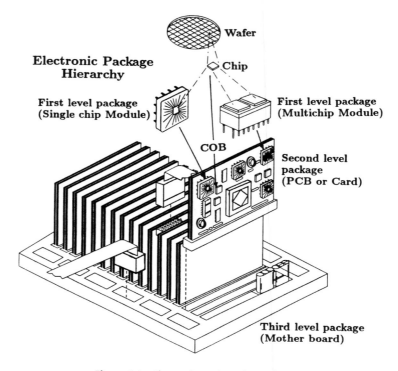

Figure 1-1 Electronic package hierarchy.

packaging are:[1-42] (1) to provide a path for the electrical current that powers the circuits on the chip; (2) to distribute the signals onto and off the chip; (3) to remove the heat generated by the circuit; and (4) to support the chip and protect it from hostile environments.

Packaging is an art based on the science of establishing interconnections ranging from zero level packages (chip level connections), to first level packages (either single chip or multichip modules), second level packages (e.g., printed circuit boards (PCB)), and third level packages (e.g., mother boards) (Fig. 1-1). Because of the recent trend in wire bonding, tape automated bonding, and flip bare chips on board (COB) technology, the distinction between the first and second levels of packages is blurred. Usually, COB is called the one-and-half (or 1.5) level package. Figure 1-2 schematically shows the three most common COBs.

In this chapter, all the necessary information for COB is briefly introduced, including chips, chip level interconnects, first level packages, boards, encapsulants, and wire bonding, tape automated bonding, and flip bare chips on board. For more information, read Chapters 2 through 12.

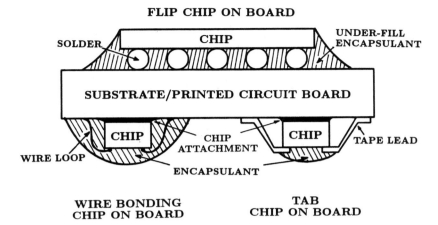

Figure 1-2 Wire bonding, TAB, and flip bare chips on board (COB).

1.2 THE CHIP

1.2.1 The Wafer

Figure 1-3 shows schematically a typical silicon wafer.[44] The overall orientation of wafer cells is called the wafer step plan. The wafer cells and their orientation are the focus of wafer design and have great impact on bump mask design for wafer bumping. The wafer diameter is, usually, 3 in. (7.3 cm), 4 in. (10.2 cm), 5 in. (12.7 cm), 6 in. (15.2 cm), and 8 in. (20.3 cm). The x and y are the dimensions of the chips on the wafer. A wafer usually consists of many product unit cells (PC), a few test unit cells (TC), and a few combined mask alignment (CMA) unit cells (Fig. 1-3). For more information see ref. 44.

1.2.2 Number of Chips per Wafer

The possible number of undamaged chips (N_c) stepped from a wafer may be given[45] by

$$N_c = \pi \, \frac{(D - \alpha)^2}{4x^2} \tag{1-1}$$

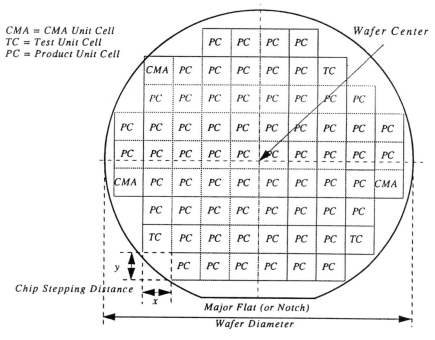

Figure 1-3 A wafer.[44]

where x is the dimension of one side of a square chip (mm), D is the wafer diameter (mm), α is the useless scrap edge distance of a wafer (mm) and is assumed equal to $2x$. Then, Eq. (1-1) becomes

$$N_c = \pi \, \frac{(D - 2x)^2}{4x^2} \tag{1-2}$$

Figures 1-4 and 1-5 show the number of chips for different sizes of wafers. It can be seen from Fig. 1-5 that for a 5 in. (12.7 cm) wafer there are about 100 of the 9 mm square chips.

1.2.3 Number of Yielded Chips per Wafer

Murphy's law[46] states the finite probability that the possible number of yielded chips after processing is given by

$$F = (1 + A\delta/c)^{-c} \tag{1-3}$$

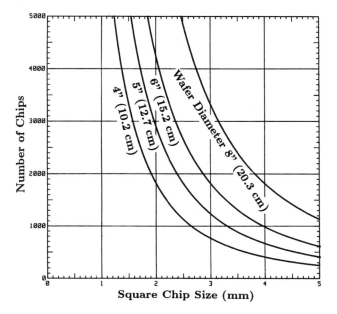

Figure 1-4   Number of chips on a wafer (small chip sizes).

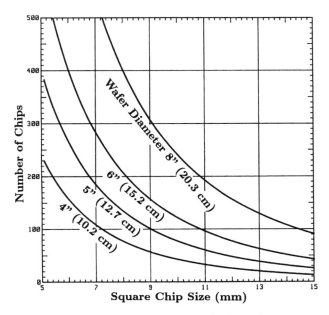

Figure 1-5   Number of chips on a wafer (large chip sizes).

where F is the yield probability, A is the area of a single chip (mm²), δ is the defect density (defects/mm²), and c is a constant that indicates defect clustering (for $c = 1$ we have Seeds' formula,[47] and for $c = \infty$ we are left with Poisson's yield formula). Thus, the total number of yielded unassembled chips per wafer ($N_y = FN_c$) is given by

$$N_y = \pi\, \frac{(D - 2x)^2}{4x^2}\, (1 + A\delta/c)^{-c} \qquad (1\text{-}4)$$

Figures 1-6 to 1-9 show the number of yielded chips ($c = 2.5^{45}$) for different wafer sizes (4 in. or 10.2 cm and 6 in. or 15.2 cm) and defect densities ($\delta = 0.01, 0.05, 0.1, 0.2$).

1.2.4 Performance

Figure 1-10 shows the processing performance of computer engineering.[48] High-end applications include super, mainframe and mid-range computers. Low-end applications usually include consumer, audiovisual, multimedia products, all kinds of PCs, and workstation computers. However, with the

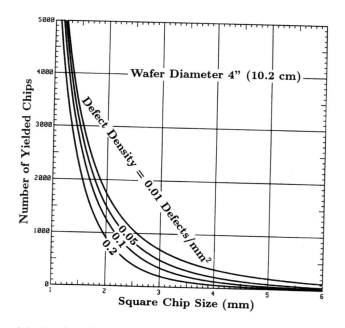

Figure 1-6   Number of yielded chips on a 4 in. (10.2 cm) wafer (small chip sizes).

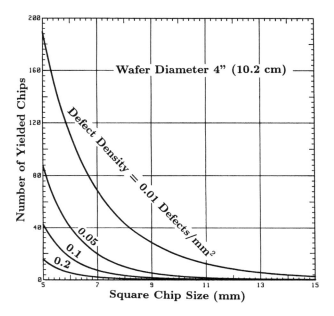

Figure 1-7 Number of yielded chips on a 4 in. (10.2 cm) wafer (large chip sizes).

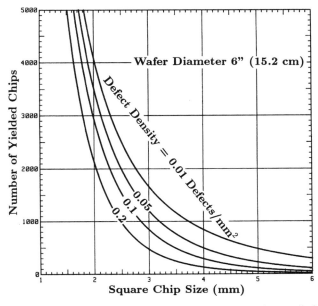

Figure 1-8 Number of yielded chips on a 6 in. (15.2 cm) wafer (small chip sizes).

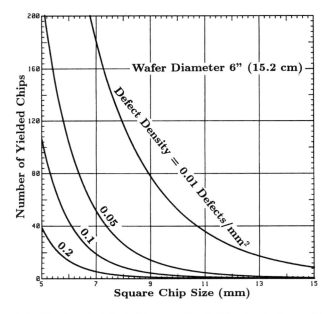

Figure 1-9   Number of yielded chips on a 6 in. (15.2 cm) wafer (large chip sizes).

advance of semiconductor and packaging technologies, some workstations are designed and manufactured to run at very high speed. For example, one of the 1991 HP workstations executes approximately 75 million instructions per second (MIPS). Thus, according to Fig. 1-10, it should be considered as a high-end application.

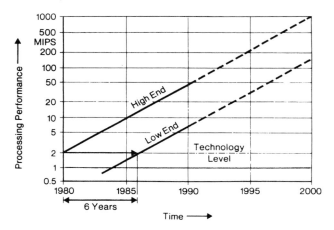

Figure 1-10   Processing performance (MIPS) (conservative estimate).[48]

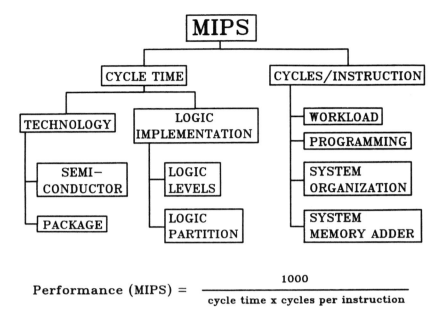

$$\text{Performance (MIPS)} = \frac{1000}{\text{cycle time x cycles per instruction}}$$

Figure 1-11   Factors affecting the performance (MIPS).

Figure 1-11 shows the factors affecting the performance (MIPS) of a computer system. It can be seen that packaging technology plays one of the key roles in performance. Proper choice, design, and manufacture of a packaging system can reduce the cycle time and thus improve the performance (Fig. 1-11).

The other technology which affects the performance of a computer system (Fig. 1-11) is the semiconductor. Figure 1-12 shows schematically the key elements on a silicon chip. These elements are devices (e.g., transistor or capacitor), gates (made up of on average 3 or 4 devices and performing a function), and lines (to link blocks of gates).

There are two popular integrated circuit (IC) designs, namely, the complementary metal oxide semiconductor (CMOS) and emitter coupled logic (ECL). The gate integration of these two technologies is shown in Fig. 1-13. The gate delays of these two circuit designs are shown in Fig. 1-14. It can be seen that the gate delays of CMOS techniques are about 10 times that of ECL technology. Also, a CMOS device tends to be more susceptible to corrosion and contamination, and thus, has a much higher early-life fail.

The line delay of high-end and low-end computer applications is shown in Fig. 1-15. It can be seen that the line delay of low-end applications is about twice of that of high-end applications.

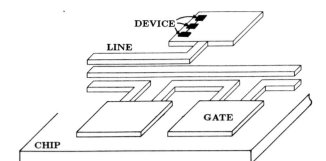

DEVICE: transistor, capacitor, etc.

GATE: make up of several devices and perform a function

LINE: "long" lines linking blocks of gates

Figure 1-12   Typical key elements on a chip.

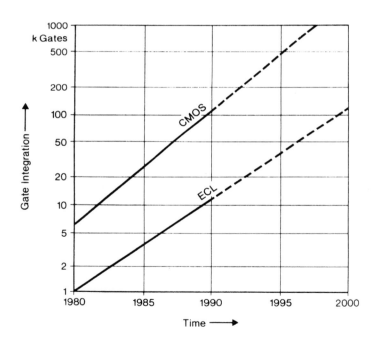

Gate integration of VLSI arrays (k=kilo=10^3)
ECL (Emitter Coupled Logic)
CMOS (Complementary Metal-Oxide-Semiconductor)

Figure 1-13   ECL and CMOS gate integration (conservative estimate).[48]

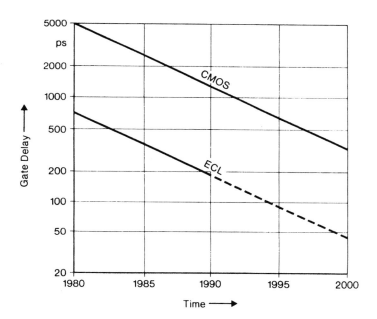

Gate delay in VLSI arrays $(p=pico=10^{-12})$
ECL (Emitter Coupled Logic)
CMOS (Complementary Metal-Oxide-Semiconductor)

Figure 1-14   ECL and CMOS gate delay (conservative estimate).[48]

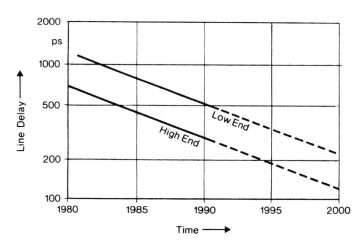

Line delay $(p=pico=10^{-12})$

Figure 1-15   Line delay for high- and low-end applications (conservative estimate).[48]

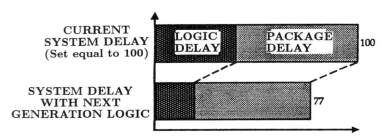

50% IMPROVEMENT IN LOGIC PERFORMANCE FOR A SYSTEM RESULTS IN MUCH SMALLER OVERALL GAINS, UNLESS PACKAGE DELAY IS ADDRESSED

Figure 1-16   Logic and package delay (Amdohl's law).

In the past few years, the on-chip delay in semiconductor devices has reduced much more than the signal delay in packaged ICs. Packages have become a great loss function[43] of a computer system. It can be seen from Fig. 1-16 that a 50% improvement in logic performance for a computing system only results in less than 23% overall gains if the packaging does not change. In order to keep pace with the advance of semiconductor technology, packaging engineers have to minimize the loss function by increasing the packaging density (see Section 1.5.1) and reducing the packaging delay (see Section 1.5.2).

1.2.5 Number of Possible I/Os (Pads) per Chip

For a given chip size, the physically possible number of I/Os (pads) depends on the pad size, pad pitch, and pad arrangement. The physically possible number of peripheral array pads P_p on a chip surface is given by

$$P_p = 4\left(\frac{x}{p} - 1\right) \qquad (1\text{-}5)$$

The physically possible number of area array pads P_a on a chip surface is given by

$$P_a = \left(\frac{x}{p} - 1\right)^2 \qquad (1\text{-}6)$$

In Eqs. (1-5) and (1-6), p is the pad to pad pitch (mm) on the chip surface, and x is a side dimension of a square chip (mm). Equation (1-5) is plotted in

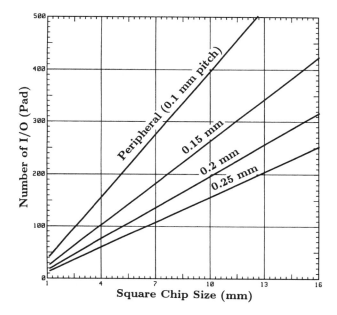

Figure 1-17 Number of possible I/Os on a peripheral array chip.

Fig. 1-17 for the peripheral array pads with $p = 0.1$ mm, 0.15 mm, 0.2 mm, and 0.25 mm. Equation (1-6) is plotted in Fig. 1-18 for the area array pads with $p = 0.2$ mm, 0.25 mm, 0.3 mm, and 0.35 mm. It can be seen that for a 10 mm square chip, the number of pads with 0.25 mm pitch for the area array is about 1500 and for the peripheral array is only about 155.

1.2.6 Number of Required I/Os (Pads) per Chip

With the advance of semiconductor technology (e.g., $0.5\,\mu$m line width), more and more circuits are crowded onto a chip. It can be seen from Fig. 1-13 that the number of gates on a common IC technology will soon grow past the one million mark. This growth in gate count affects the number of I/Os on the IC chips and, consequently, the number of package I/Os.

The number of I/Os or pads ($P_{I/O}$) required by a chip containing a given number of logic gates (G) was first proposed by E. Rent of IBM[2] and is given by

$$P_{I/O} = \beta G^{\xi} \tag{1-7}$$

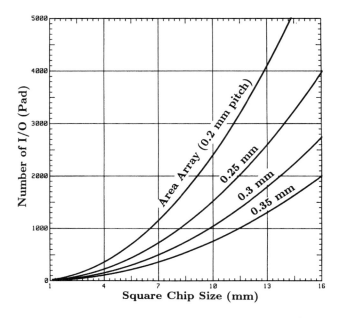

Figure 1-18   Number of possible I/Os on an area array chip.

where β and ξ are constants. Since then, many values of the constants have been proposed for different types of circuits, e.g., refs. 49, 50. For CMOS gate arrays, we have

$$P_{I/O} = 1.9G^{0.5} \qquad (1\text{-}8)$$

For group of logic cells requiring more than one IC to create a complete function, we have

$$P_{I/O} = 3.2G^{0.434} \qquad (1\text{-}9)$$

For microprocessors, we have

$$P_{I/O} = 0.82G^{0.45} \qquad (1\text{-}10)$$

For the functionally complete chip, we have

$$P_{I/O} = 7G^{0.21} \qquad (1\text{-}11)$$

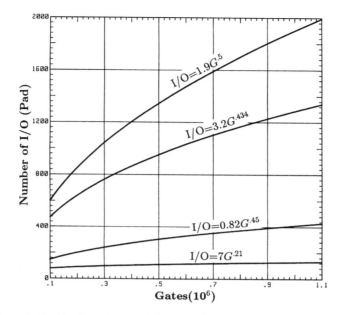

Figure 1-19   Number of required I/Os on a chip with different circuit designs.

Equations (1-8) to (1-11) are shown in Fig. 1-19. It can be seen that, even with the same number of gates, the required number of I/Os is very different. Among these four types of circuits, the functionally complete chip requires the least number of I/Os and thus could be the driver for low cost COB and laminate multichip module (MCM) applications.

1.2.7 COB and MCM Yield

With COB technology or MCM technology, there are usually more than one chip on the substrate. The COB or MCM yield depends on the chip yield and the number of chips on the substrate.

$$Y_m(\%) = 100\,Y_c^{N_c} \qquad (1\text{-}12a)$$

where Y_m is the COB or MCM yield, Y_c is the chip yield, and N_c is the number of chips on the substrate. Equation (1-12a) is plotted in Figs. 1-20a and 1-20b for different values of chip yield. Of course "good" chip yield is very important for COB and MCM applications. If one chip on a COB or MCM is bad, the entire substrate becomes worthless unless reworked and repaired. In general, for COB and laminate MCM, the cost of scrapping

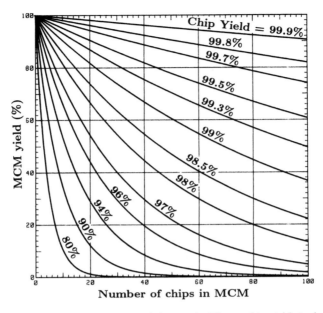

Figure 1-20a   MCM yield vs number of chips with different chip yields in the MCM.

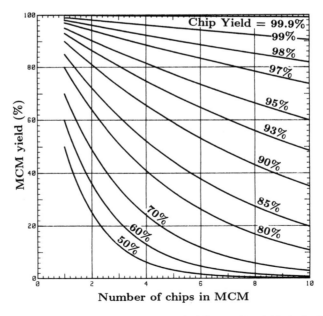

Figure 1-20b   MCM yield vs number of chips with different chip yields in the MCM (small number of chips).

them because of bad chips may be more cost-effective than troubleshooting and fixing them.

The MCM yield, Y_m, in Eq. (1-12a) will become a shipped defect if the test method used cannot identify the defect for successful rework. The resultant shipped MCM yield, Y_{ms}, is given by

$$Y_{ms}(\%) = 100\, Y_c^{(1-\eta)N_c} \qquad (1\text{-}12b)$$

where η is the test fault coverage (%) and is defined as the ability of the tester to identify defects. Thus, the probable MCM defect level, D_m, is

$$D_m(\%) = 100[1 - Y_c^{(1-\eta)N_c}] \qquad (1\text{-}12c)$$

Equations (1-12b) and (1-12c) are plotted in Figs. 1-20c, 1-20d, and 1-20e, 20f, respectively. With these figures, the trade-off between the number of chips, chip yield, MCM yield, MCM rework, resultant shipped MCM yield, and probable MCM defect level can be made.

For example, for a 10-chip MCM with chip yield $Y_c = 90\%$, and test fault coverage $\eta = 95\%$, the MCM yield is $Y_m = 35\%$ (Figs. 1-20a and 1-20b), the resultant shipped MCM yield is $Y_{ms} = 95\%$ (Figs. 1-20c and 1-20d), and the

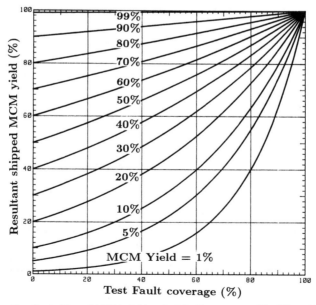

Figure 1-20c   Resultant shipped MCM yield vs test fault coverage with different MCM yields.

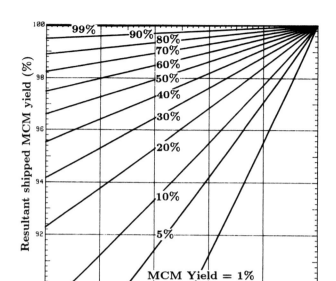

Figure 1-20d   Resultant shipped MCM yield vs test fault coverage with different MCM yields (high test fault coverage).

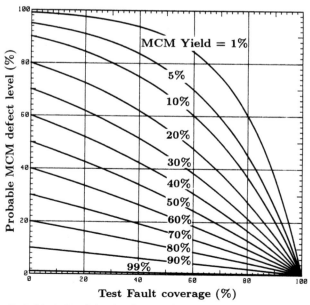

Figure 1-20e   Probable MCM defect level vs test fault coverage with different MCM yields.

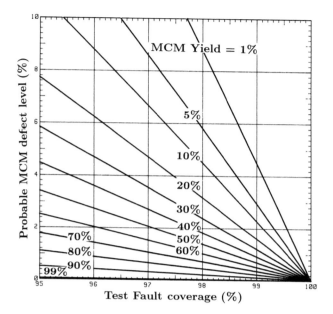

Figure 1-20f Probable MCM defect level vs fault coverage with different MCM yields (high test fault coverage).

probable MCM defect level is $D_m = 5\%$ or 50 000 ppm (Figs. 1-20e and 1-20f). In this case, 65% of the MCM will require at least one rework.

On the other hand, if the chip yield Y_c is increased from 90% to 99% and the test fault coverage remains the same, $\eta = 95\%$, then $Y_m = 90\%$, $Y_{ms} = 99.5\%$, $D_m = 0.5\%$ or 5000 ppm, and only 10% of the MCM will require rework. This shows the importance of known good dies and the logic that the less bad chips (higher chip yield) there are, the lower the chance that they will escape into MCM assembly.

1.2.8 Known Good Dies (KGD)

KGD is a very complex issue, e.g., refs. 51-56. Each company has its own methodology and philosophy. Even within the same company, there will be different strategies to deal with different customers. Even in vertically integrated companies (i.e., who make their own wafer, chip, design, process, test, burn-in . . .) there are different KGD requirements for different circuits. For example, Table 1-1 shows the different KGD requirements for IBM's bipolar and CMOS circuits. In the case of bipolar, burn-in was done at the MCM level. However, in the case of CMOS, burn-in was done at the

Table 1-1   Known Good Die (KGD) Requirements, Bipolar and CMOS Comparison[51]

SPQL Category	Bipolar	CMOS	Comments
DC untested	Controlled, extremely high test coverage achieved	Because of circuit density, 10× of bipolar	High circuit density on CMOS gives challenges even with LSSD
AC tested	Very key, DFQ and high AC test effectiveness required	Smaller by 10×, high performance application makes a key factor	Delay and AC test likely needed
Tested	Remains a factor	Remains a factor	Thorough test needed at next step
ELF	Minor problem except early in program	Key factor, over 10× of bipolar	Burn-in assumed for CMOS at die level

Note: SPQL (Shipped product quality level); ELF (Early life fail)

chip level. Also, for the CMOS technology, chip burn-in appears to be a necessity, and some form of temporary chip carrier is a requirement.[51]

The heart of the KGD issue is not so much technology as business, and thus, there is always a trade-off between the objective of KGD and the business needs. An electronic packaging system usually consists of high- and low-density interconnect features. The high-density interconnects need high band width and I/Os on the 32 and emerging 64 bit microprocessors and their supporting applications specific IC (ASIC). The low density interconnects support the on module cache memories which may range from 64 kB to 4 MB or more. The cost of the high-density ICs can be many times more than the low-density ICs.

For example, a COB (or MCM) has one ASIC chip and several static random access memory (SRAM) chips. These chips were not burned-in prior to assembly on the substrate. The COB or MCM assembly is burned-in and tested and is diagnosed to be bad. If the ASIC chip is bad and the substrate is not expensive, then the entire package is discarded since the rework would cost more than the low-cost SRAM chips and substrate. If one of the SRAM chips is bad, then it is reworked. This makes business sense, because the cost of reworking the bad SRAM chip is less than the total cost of the ASIC chip, the other SRAM chips, and the substrate.

Rework of COB or MCM can be very expensive though. Specific design rules must be followed to isolate the failure to a single chip and very good diagnostics are required. Furthermore, replacing the failure chip could be costly. Thus, burn-in at the COB or MCM level can result in an unacceptable product cost.

The ideal KGD solution is to burn-in and test the chip (either at wafer level or individual chip) at low cost (including the handling cost), to assembly it on the substrate, to minimize or even eliminate rework and repair, and to reduce yield losses after early assembly and field failures. It should be pointed out that, due to the power distribution and cooling problems and wafer contact difficulties, burn-in of chip at wafer level poses a technical challenge. Also, connections to individual chips for burn-in are very difficult. Built-in self testing (BIST) chips and boundary scan test of inter-chip connections are under active investigation. For more information about chip pretest and system diagnostics, see Chapter 2.

1.2.9 Unknown Bad Dies (UBD)

Just like KGD, UBD (including the escapes) are a very complex issue. Shipping KGD at a reasonable price will make the customers very happy and keep them. However, shipping UBD will create many problems for the customers and eventually lose them. Thus, most of the IC companies are not providing the bare wafers or chips to their customers because they do not want to be liable for the UBD. Also, they do not want to disclose their confidential wafer yields and reduce their profits because of non-packaging of chips.

The research and development efforts of COB and MCM in the past few years were great.[57-125] Unfortunately, not many products use these technologies yet. As a matter of fact, because of the infrastructure issues (e.g., design, wafer/chip availability, wafer bumping, test and burn-in, KGD, UBD), application requirements, and the resources available for each company, many designers found that COB and MCM are too expensive for their next products. Unless there are some compelling reasons (e.g., size, weight, performance) for them to use these technologies, they are just standing by and watching very closely the developments and solutions of these critical issues. It is interesting to note that, today, most of the leaders using COB and MCM technologies in their products are vertically integrated companies. One of the reasons is that they are subjected to fewer infrastructure constraints.

1.2.10 Fracture Toughness Measurement of Chip

Silicon is a brittle material. With small flaws (e.g., minute cracks, voids, inclusions) silicon chip may crack. One of the ways to prevent chip cracks is to develop tough silicon chips that are not so sensitive to flaws.

How can the toughness of silicon chips be measured? The answer is to determine experimentally the fracture toughness (K_{IC})[126-131] of the silicon. Fracture toughness is a material property in the same sense that yield strength is a material property. It is independent of crack length, geometry, or loading system. Physically, fracture toughness can be considered a material property which describes the inherent resistance of the material to failure in the presence of a crack-like defect (flaw). The unit for K_{IC} is MPa$\sqrt{m}$.

A variety of methods have been proposed for measuring fracture toughness of engineering materials. The three methods presented in this book may not be the most common specimen designs, but they are very easy to use for silicon chips.

Double Torsion Specimen

Figure 1-21 shows a double torsion specimen which is very popular for high-temperature testing. This test determines the fracture toughness of a silicon chip under compressive loads in an elevated-temperature chamber without even the crack length being known.

$$K_{IC} = Pb \sqrt{\frac{3(1 + v)}{Wdh^3}} \qquad (1\text{-}13)$$

where P is the applied compressive load, v is the Poisson's ratio, h is the thickness, W is the width of the specimen, b and d are shown in Fig. 1-21. A subject of debate about this method is that the crack grows further on the lower surface (in tension) than on the upper surface (in compression) of the specimen.

Indentation Crack Specimen

The indentation crack method is fast and easy and can be applied to any size of specimen. This method involves making an impression on the surface (Fig. 1-22) with a Vickers hardness indenter. Cracks grow from the corners

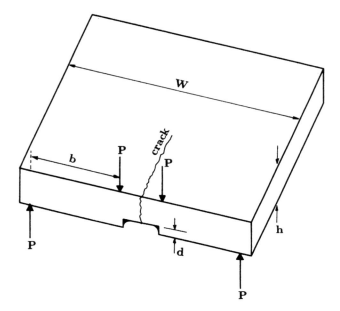

Figure 1-21 Fracture toughness, double torsion specimen.

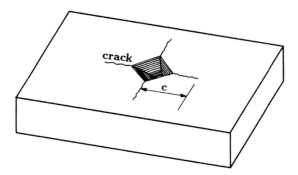

Figure 1-22 Fracture toughness, indentation crack specimen.

of the indentation and the crack lengths (c) can be measured under a microscope. The fracture toughness is given by

$$K_{IC} = \lambda P \sqrt{E/H} c^{-3/2} \qquad (1\text{-}14)$$

where P is the indentation force, E is the Young's modulus, H is the hardness of the silicon chip, and λ is a calibrated constant. Unlike other methods, the indentation method cannot determine the fracture toughness directly, but rather has been calibrated against other methods. It is noted that this method applies to a small area of the silicon chip and so more measurements at different locations of the chip are necessary.

Three-Point Loaded Bend Specimen

The three-point loaded bend specimen is shown in Fig. 1-23. It is subjected to a force (P) at the mid point of the top surface and is supported by two equal forces $(P/2)$ at the bottom with a distance (S) apart. The dimensions of the specimen should be such that the width (W) is twice the thickness (B). The American Society for Testing and Materials (ASTM) recommends that $B = 0.125S$. The fracture toughness is given by

$$K_{IC} = \frac{PS}{BW^{3/2}} \left[2.9\left(\frac{a}{W}\right)^{1/2} - 4.6\left(\frac{a}{W}\right)^{3/2} + 21.8\left(\frac{a}{W}\right)^{5/2} \right.$$
$$\left. - 37.6\left(\frac{a}{W}\right)^{7/2} + 38.7\left(\frac{a}{W}\right)^{9/2} \right]$$

$$(1\text{-}15)$$

In Eq. (1-15), the crack length a is measured after fracture. It is noted that Eq. (1-15) is valid if $2.5(K_{IC}/\sigma_y)^2$ is less than both the thickness and crack length of the specimen. Otherwise, it is necessary to use a thicker specimen to determine K_{IC}. σ_y is the 0.2% offset yield strength of the material.

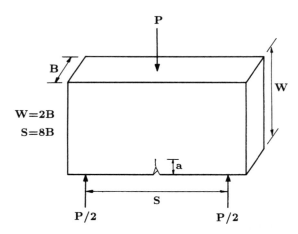

Figure 1-23 Fracture toughness, three-point loaded bend specimen.

1.3 CHIP LEVEL INTERCONNECTS

The purpose of chip level connections (zero level packages) is to provide the required connections between the chip and the package. There are at least three popular chip level connections, namely wire bonding using thermocompression or ultrasonic bonding (Fig. 1-24(a)), tape automated bonding (TAB) using thermocompression bonding or pulse-thermode reflow (Fig. 1-24(b)), and flip chip solder bumps (Fig. 1-24(c)).

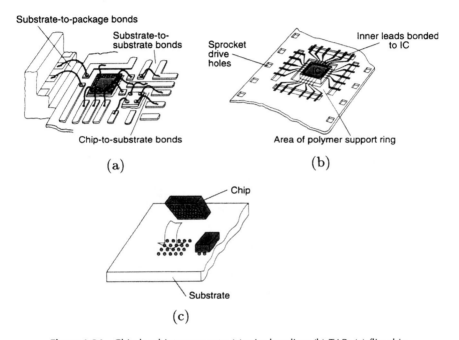

Figure 1-24   Chip level interconnects: (a) wire bonding; (b) TAB; (c) flip chip.

1.3.1 Wire Bonding

Figure 1-25 shows schematically an example of 0.1 mm pitch wire bonding on a silicon chip. Today, however, 0.076 mm to 0.889 mm pitches are more and more used in prototypes,[132,133] and the speed of wire bonding is already 4 bonds/sec. The diameter of gold ball wire bonding ranges between 20 μm and 33 μm. The corresponding ball diameters are determined to be 2.5 to 4 times the wire diameter. In general, the wire bond sites are along the periphery of the chip (Fig. 1-26). More than 90% of the chips used today are

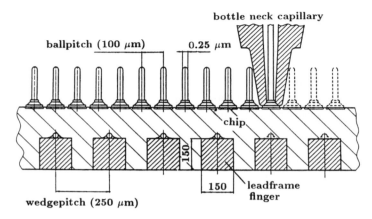

Figure 1-25   Wire bonding on chip.

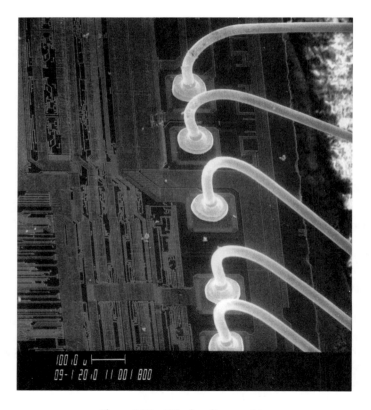

Figure 1-26   Wire bonding on chip.

used by wire bonding technology. For more information about wire bonding, see Chapters 3 and 7.

1.3.2 Tape Automated Bonding

Figure 1-27 shows a scanning electron microscopy (SEM) micrograph of a TAB inner lead bonding. The beam lead is usually made of copper with electroplated tin or gold, or electroless tin. A single inner lead with bump is shown in Fig. 1-28. It can be seen that thick metal bumps are made on the chip bond pads for inner lead bonding. These bumps are made of various metals of which the most common are gold and solder. For TAB the bump pitch can be as small as 0.076 mm and the bump sites on the chip can be either in peripheral array or area array.[14] Today, less than 7% of the chips used are by TAB technology. For more information about TAB, see Chapters 4 and 8.

Figure 1-27   Tape automated bonding.

1.3.3 Solder Bumped Flip Chip

Figure 1-29 shows schematically a solder bumped flip chip made by Sharp Corporation.[97,98] The 60wt%Sn/40wt%Pb was electroplated on the TiW-Cu interface metallurgy. The bump height was controlled so as to be

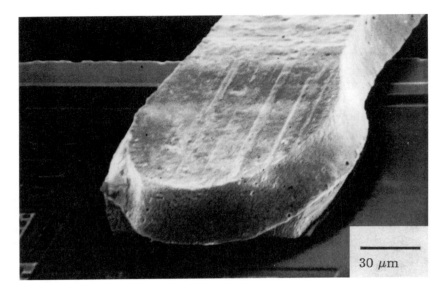

Figure 1-28 Single inner lead bonded bumped pad.

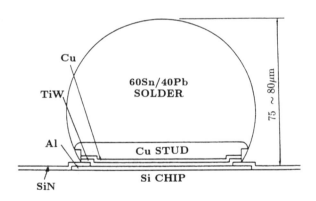

Figure 1-29 Solder bump on chip.[98]

$75 \sim 80\,\mu m$. The shear strength per bump was about 17 g and breakage always occurred at the solder ball.

The most mature method to make flip chip solder bumps is by evaporation.[134-141] Because the evaporation rate and pressure of the Pb are higher than those of the Sn, this method is usually applied to high Pb solder bumps, e.g., 95wt%Pb/5wt%Sn. The most common under bump metal (UBM) of high Pb solder bumps is Cr-Cu-Au,[134-141] which has been

Figure 1-30   Solder bump on chip.

used for most of IBM's controlled collapse chip connection (C4) technology. Figure 1-30 shows an example of flip chip solder bumps. It can be seen that the solder bumps are all over the chip. A $10\,\mu m$ pitch with a $5\,\mu m$ diameter and $20\,\mu m$ height solder bump can be made.[142] Among these three chip level interconnect technologies, solder bumped flip chip provides the highest packaging density with less packaging delay. Less than 3% of the chips used are now solder bumped. For more information about solder bumping and flip chip applications, see Chapters 5 and 9.

1.4 FIRST LEVEL PACKAGES

After the silicon chip has been wire bonded, tape bonded or solder bumped it is ready to be placed on a second level PCB (COB technology or 1.5 level packages) or packaged into either a single-chip or multichip module (first level packages), Figs. 1-1 and 1-2. In this section, only the first level packages will be considered. The function of the first level packages is to provide all the necessary wiring, interconnection, power distribution, and cooling path. It is noted that, for wire bonding technology, the chip level and the next level connections are performed simultaneously.

1.4.1 Single-Chip Modules

The definition of single-chip module is that the package (or chip carrier) has only one chip in it. Some examples are: thin quad flat pack (TQFP), thin small outline package (TSOP),[143-145] thin tape carrier package (TCP),[146] plastic ball grid array (PBGA),[108,110,147,148] ceramic ball grid array (CBGA),[63,103,109] plastic quad flat pack (PQFP),[149-155] square quad flat pack (SQFP), rectangular quad flat pack (RQFP), very small outline package (VSOP), very small quad flat pack (VSQP), flatpack, leadless chip carrier (LCC), small outline integrated circuit (SOIC),[156] pin grid array (PGA),[157,158] plastic leaded chip carrier (PLCC),[159] SOIC with J-Leads (SOJ), dual in-line package (DIP), etc.

The trends in single-chip carriers are finer pitch, higher I/O, larger horizontal body size, thinner vertical body size, and area array.[12] For example, Fig. 1-31 shows schematically a cross-section of an ultra TCP.[146] The thickness of the Si chip and package are 0.2 mm and 0.4 mm, respectively. In order to achieve this thin package, TAB technology is used. The upper and side surface of the chip are encapsulated with resin, but the bottom surface of the chip is exposed.

Table 1-2 shows a few common packages provided by Kyocera. It can be seen that a bare chip typically occupies much less than 50% area in various single-chip packages. This is due to the bonding wires, fan-out lines, and the pitch limitation of external pins. As a matter of fact, many of the ASIC chips are restricted to package carriers with fewer terminations than the total number available on the chips. This problem is caused by a chip size and package cavity mismatch, that is, the chip size is too small for a given package cavity resulting in wire bond lengths exceeding the maximum allowable. Thus, by directly attaching the bare chips on the second level PCBs could save all the available pads on the bare chips, material, weight, size, and costs.

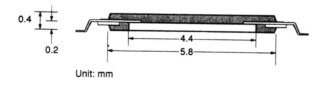

Unit: mm

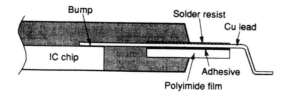

Bump   Solder resist   Cu lead   IC chip   Adhesive   Polyimide film

Figure 1-31   Ultra thin TCP.[146]

Table 1-2   Area of Bare Chip in Ceramic Single Chip Packages, Kyocera Product Catalogs

Package Type	Pin Count (Pitch in mils)	Outside Package Dimensions (in.)	Max. Cavity for Bare Chip (in.)	Bare Chip Area in Package (%)
PGA	68 (100)	1.100 × 1.100	0.550 × 0.550	25.0
	84 (100)	1.100 × 1.100	0.470 × 0.470	18.4
	100 (100)	1.320 × 1.320	0.500 × 0.500	14.4
	132 (100)	1.400 × 1.400	0.450 × 0.450	10.3
	208 (100)	1.770 × 1.770	0.450 × 0.450	6.5
	224 (100)	1.750 × 1.750	0.480 × 0.480	7.5
	257 (100)	2.000 × 2.000	0.630 × 0.630	9.9
LCC	24 (50)	0.400 × 0.400	0.270 × 0.270	45.6
	32 (50)	0.550 × 0.450	0.390 × 0.290	45.7
	68 (50)	0.950 × 0.950	0.630 × 0.630	44.0
	84 (50)	1.150 × 1.150	0.700 × 0.700	37.1
	100 (50)	1.350 × 1.350	0.390 × 0.390	8.4
	124 (50)	1.650 × 1.650	0.520 × 0.520	9.9
QFP	24 (50)	0.400 × 0.400	0.280 × 0.280	49.0
	32 (50)	0.400 × 0.400	0.265 × 0.265	43.9
	68 (50)	0.950 × 0.950	0.500 × 0.500	27.7
	84 (50)	1.150 × 1.150	0.470 × 0.470	16.7
	132 (25)	0.950 × 0.950	0.400 × 0.400	17.7
	164 (25)	1.130 × 1.130	0.570 × 0.570	25.4
SOJ	28 (50)	0.725 × 0.432	0.601 × 0.349	59.6
	32 (50)	0.830 × 0.423	0.650 × 0.270	50.0
DIP	24 (100)	1.200 × 0.610	0.650 × 0.430	38.2
	32 (100)	1.600 × 0.310	0.560 × 0.220	24.8
	48 (100)	2.400 × 0.610	0.490 × 0.400	13.4
	64 (100)	3.200 × 0.910	0.551 × 0.433	8.2

1.4.2 Multichip Modules (MCM)

The most simple definition of MCM is that the package (or chip carrier) has more than one chip in it. The past few years have witnessed an explosive growth in the research and development efforts devoted to MCM as a direct result of the density and performance limitations of single chip modules. It combines many high-performance silicon ICs (e.g., Fujitsu's MLG-MCM has 144 ICs) with a custom designed substrate structure which takes full advantage of the IC performance. This complex substrate structure is the heart of the MCM technology. It can be fabricated on multilayer ceramics, polymers, silicons, metals, glass ceramics, PCB, etc., using thin films, thick films, cofired, and layered methods.

A formal definition of MCM has been given by IPC (Institute for Interconnecting and Packaging Electronic Circuits). They define three main categories of MCMs.

1. MCM-C are multichip modules which use thick film technology such as fireable metals to form the conductive patterns, and are constructed entirely from ceramic or glass-ceramic materials, or possibly, other materials having a dielectric constant above 5. In short, a MCM-C are constructed on ceramic (C) or glass-ceramic substrates.
2. MCM-L are multichip modules which use laminate structures and employ PCB technology to form predominantly copper conductors and vias. These structures may sometimes contain thermal expansion controlling metal layers. In short, MCM-L utilize PCB technology of reinforced plastic laminates (L).
3. MCM-D are multichip modules on which the multilayered signal conductors are formed by the deposition of thin-film metals on unreinforced dielectric materials with a dielectric constant below 5 over a support structure of silicon, ceramic, or metal. In short, MCM-D uses deposited (D) metals and unreinforced dielectrics on a variety of rigid bases.

A more detailed definition of MCM is provided in Table 3-1 of Chapter 3. For silicon chips, one of the best choices of substrate material is silicon, because the silicon substrate matches the thermal expansion of the chip (even though the chip is always hotter than the substrate and a thermal gradient does exist) and has very high thermal conductivity. Furthermore, the dielectric can be either SiO_2 or polyimide as both are standard silicon IC processes. For example, Fig. 1-32 shows schematically a cross section of AT&T Bell Laboratories' very high-density thin-film structure on silicon substrate.

Which MCM technology is the right choice for a particular application? The answer is not easy. There is always a compromise between cost, performance, volume, applications requirement, technology availability, resources availability, infrastructure, business need, etc. In general, MCM-L appears to be the lowest risk (because of availability) and cost (when wiring density requirements are less than $200 \, cm/cm^2$), MCM-D leads to the highest performance, cost, and wiring density ($\geq 1000 \, cm/cm^2$), and MCM-C shows excellent power distribution and cost-effectiveness if the wiring

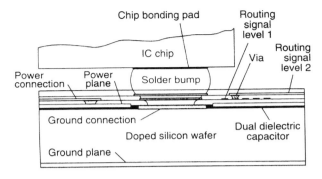

Figure 1-32   AT&T's high-density thin film package.

density is greater than 200 and less than $1000 \, \text{cm/cm}^2$ (ref. 160). A series of criteria to use when making initial evaluations of MCM technology for a given application is shown in Table 1-3.[84,87,119] Since the major driving force in electronic products today is cost and volume reduction, the COB and MCM-L technologies could be the most used.

1.4.3 Multichip Carrier Modules (MCCM or MC²M)

The MCCM technology combines the advantages of COB/MCM and the well-defined and established standard single-chip module. MCCM integrates a few chips with some discretes on a small common substrate (e.g., silicon, ceramic, FR-4 epoxy, bismalcimide triazine (BT) resin, etc.), and is then assembled into a standard single-chip module (e.g., PBGA, CBGA, PQFP, PLCC, etc.). It is a welcome package for board level manufacturers because it is surface mount technology (SMT) compatible and they have the know-how to assemble the standard single-chip module on the second-level PCB with high yield and low cost.

Figure 1-33a shows schematically an MCCM proposed by AMKOR.[108,110] It incorporates the two lowest-cost electronic interconnect technologies, PCB and BT material (MCM-L) and standard PQFP package with gull-wing leads, which have tremendous industry infrastructure and ongoing research and development (R&D).

Figure 1-33b shows schematically an IBM's MCCM.[63,103,109] It incorporates their ceramic substrate and CBGA (with composite solders) technologies for higher performance applications. Also, the solder bumped flip chips on the ceramic substrate are assembled by their C4 flip chip on substrate technology.[134-137] They have their own infrastructure and R&D.

The first level package shown in Fig. 1-33a is a peripheral package, while that in Fig. 1-33b is an area array package. Usually, the area array packages

Table 1-3 Suggested Criteria for Choosing MCM Substrates[84,87]

MCM-L	MCM-D	MCM-C
• A generous size or form factor budget exists	• Primary goal is form factor reduction	• MIL STD hermeticity is required
• CPU clock speed 50 MHz or less	• Clock speed is 25 MHz or greater	• Large numbers of embeddable passives
• A component count beyond 20 active ICs	• High level of component integration	• Size and height not critical on overall module size
• "Throwaway" test strategy (limited or no rework)	• Limited or no passives	• Clock speed is 25 MHz or greater
• Relatively low-value ICs (under $50)	• High-value silicon	
• Low level of IC integration	• Potential thermal issue	
• Very limited design budget	• Desire to increase overall silicon performance, or design cannot be functional using other than MCM-D geometries and die spacing	
• Cost-effective line geometries for PCB substrate can support the application requirements	• Possible/partial silicon control	

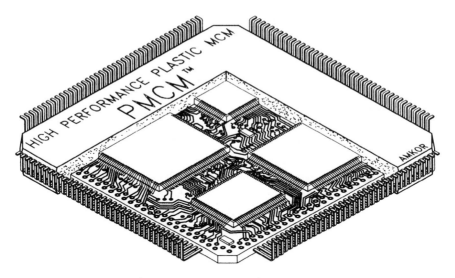

Figure 1-33a   AMKOR's plastic MCCM.

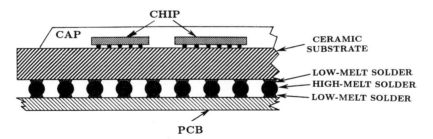

Figure 1-33b   IBM's ceramic MCCM.

have more I/Os than the peripheral packages (Fig. 1-34).[147] For example, the 40 mm body size, 0.5 mm pitch PQFP has 308 I/Os. However, the same body size with even larger pitch (1.27 mm) area array package would have more than 900 I/Os.

1.4.4 Distinction between COB and MCM-L

Since both COB and MCM-L technologies utilize the interconnections of bare chips (either by wire bonding, TAB, or flip chip methods) on a substrate (manufactured using PCB materials and processes), how can they be distinguished?

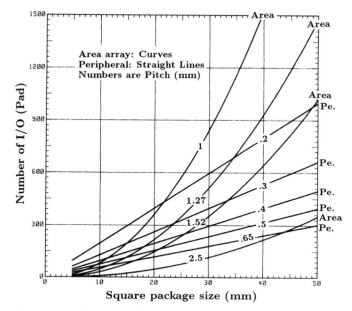

Figure 1-34 Number of I/Os, area array vs peripheral packages.

The "old" school of thought was to include all the COB applications (regardless of the number of chips or the substrate density) in MCM-L. Unfortunately, this created confusion.

Another school of thought is to distinguish the COB and MCM-L by packaging efficiency (area of silicon/area of package).[96] Messner suggested the achieved packaging efficiency of over 30% as a useful criterion for distinguishing MCM-L from COB.

The final school of thought is to follow the hierarchy of electronic packaging. It can be seen from Fig. 1-1 and Section 1.4.2 that MCM-L is a first level package. Usually, MCM-L technology uses COB technology to assemble a few bare chips and discretes on a small multilayer PCB. (In order to increase density, for example, polyimide and adhesive foils were used as interdielectric layers to permit excimer laser structuring,[48] surface laminar layers were used for signal wiring,[138-140] and photosensitive polyimide was used on top of an ordinary FR-4 PCB to make very small vias.[161]) After chips/discretes assembly, this board is either encapsulated into an MCCM or constructed as a stand-alone module with various terminations (e.g., solder balls, gull-wing leads, pins, J-leads, edge connectors). Finally, this MCCM or stand-alone module is assembled on the second level package (PCB or card). It should be noted that the PCB in the MCM-L (first level package) is different from the second level PCB packages, which have all kinds of components, such as SMT components, bare chips, chip resistors, chip capacitors, MCM-L, MCM-D, MCM-C, SMT connectors, plated-through-hole connectors, and power suppliers. Directly attaching bare chips on the second level PCB is called COB or the 1.5 level packages.

1.5 THE BOARD

1.5.1 Wireability

In substrate assessment, one of the most important factors is the interconnection density, i.e., wireability. It is a measure of package capability to provide interconnections between chips. As mentioned in Section 1.2.4, in order to keep pace with the advance of semiconductor technology, it is necessary to increase the packaging density. This can be achieved by placing individual chips closer together and increasing the I/Os of a chip. The average wire length l required for an application of N_c chips with an average distance between chips of p_c (pitch) is given by Mikhail.[1,2]

$$\frac{l}{p_c} = 0.75[1 + 0.1\ln(N_c)]N_c^{1/6} \qquad (1\text{-}16)$$

which can be approximated as

$$\frac{l}{p_c} \approx 0.77N_c^{0.245} \qquad (1\text{-}17)$$

where l and p_c should have the same unit. Equations (1-16) and (1-17) are plotted in Figs. 1-35 and 1-36. It can be seen that Eq. (1-17) is a very good approximation. Also, from Figs. 1-35 and 1-36, for a given number of chips (N_c) and the average pitch between them (p_c), the average wire length (l) can be readily determined.

The total wireability (W_i) needed to provide the required level of interconnection of N_c chips with the average of I/Os per chip (P_i) is given by:[27]

$$W_i \approx 1.25N_c^{0.245}P_i/p_c \qquad (1\text{-}18)$$

which is plotted in Figs. 1-37 and 1-38. The unit for W_i is mm/mm^2 (a total conductor length per substrate area), and for p_c is mm. Thus, for a given number of chips (N_c), average of I/Os per chip (P_i), and average pitch of chips (p_c), the wireability can be obtained from Figs. 1-37 and 1-38. There are four parameters to work on to achieve the required level of

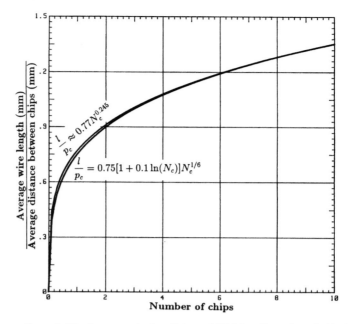

Figure 1-35   Average wire length in an MCM (small number of chips).

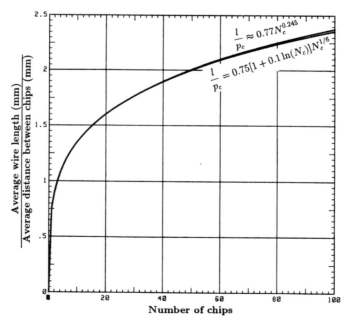

Figure 1-36   Average wire length in an MCM (large number of chips).

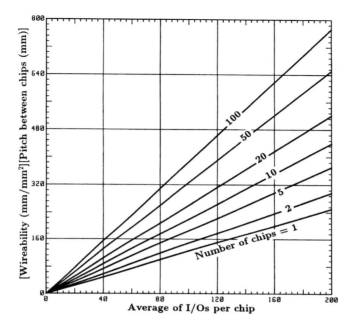

Figure 1-37 Wireability in an MCM (small number of I/Os).

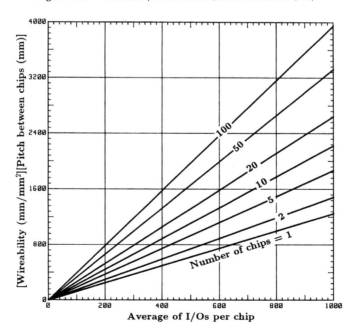

Figure 1-38 Wireability in an MCM (large number of I/Os).

interconnections: number of lines per channel, number of channels per centimeter, number of signal layers, and total area of the package. For more information on these four parameters, see ref. 27.

1.5.2 Propagation Delay

From the transmission line theory, we have

$$\tau = \sqrt{LC} = 3.33\sqrt{\varepsilon_r}, \qquad (\text{ps/mm}) \qquad (1\text{-}19)$$

where τ is the propagation time constant, ε_r is the relative dielectric constant, $p = 10^{-12}$, and C and L satisfy the following equations

$$\frac{\partial V}{\partial x} = -L \frac{\partial I}{\partial t} \qquad (1\text{-}20)$$

and

$$\frac{\partial I}{\partial x} = -C \frac{\partial V}{\partial t} \qquad (1\text{-}21)$$

In Eqs. (1-20) and (1-21), V is the voltage (volts), I is the current (amperes), L is the inductance (henries) per unit length, and C is the capacitance (farads) per unit length.

In Eq. (1-19), when $\varepsilon_r = 1$ (in a vacuum), a signal will propagate 1 mm in 3.33 ps. When $\varepsilon_r = 4$ (in a common FR-4 PCB), the same signal will require 6.66 ps to cover the same distance. Equation (1-19) is plotted in Fig. 1-39 showing the effects of the relative dielectric constant on propagation delay time per unit wire length. It can be seen that by choosing a lower ε_r the signal delay can be reduced and thus increase the package performance. As a matter of fact, the relative dielectric constant is one of the most important design trade-offs between cost, delay time, and wiring.

The signal delay time, S_t (ps), of a package with a substrate supporting N_c chips with an average distance between chips of p_c (mm) is given by

$$\frac{S_t}{p_c} = 2.56\sqrt{\varepsilon_r} N_c^{0.245} \qquad (1\text{-}22)$$

Equation (1-22) is plotted in Figs. 1-40 and 1-41 for a wide range of values of the number of chips and the relative dielectric constant (ε_r) of the substrates. For a given number of chips and performance requirements, these figures are very useful for choosing materials and arranging the chips.

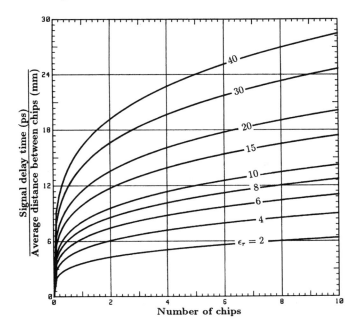

Figure 1-39 Signal delay vs relative dielectric constant.

Figure 1-40 Signal delay in an MCM (small number of chips).

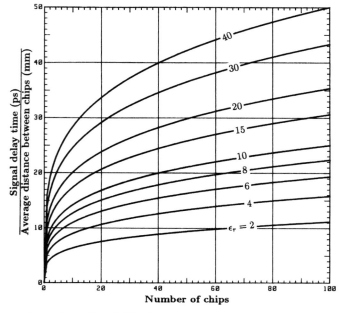

Figure 1-41   Signal delay in an MCM (large number of chips).

1.5.3 Characteristic Impedance

The general form for the characteristic impedance of a signal line is given by

$$Z_0 = \sqrt{\frac{R + i\omega L}{G + i\omega C}} \quad \text{(ohms, } \Omega\text{)} \tag{1-23}$$

where Z_0 = characteristic impedance of circuit (ohms), R = resistance per unit length of line (ohms), G = conductance per unit length of line (ohms), $i = \sqrt{-1}$, $\omega = 2\pi f$, and f = frequency (Hz). At high frequency applications, it is assumed that $\omega L \gg R$ and $\omega C \gg G$, and Eq. (1-23) becomes

$$Z_0 = \sqrt{\frac{L}{C}} \quad \text{(ohms, } \Omega\text{)} \tag{1-24}$$

For most of the PCBs and chip carriers in electronics packaging,

$$30\,\Omega \leq Z_0 \leq 100\,\Omega \tag{1-25}$$

Combining Eqs. (1-19) and (1-24), we have

$$C = \frac{3.33\sqrt{\varepsilon_r}}{Z_0} \quad (\text{pF/mm}) \tag{1-26a}$$

or

$$\sqrt{\varepsilon_r}Z_0 = \frac{3.33}{C/\varepsilon_r} \quad (\text{ohms}, \Omega) \tag{1-26b}$$

The capacitance C in Eq. (1-26) can be determined by solving an electrostatic boundary value problem.[167,168] It should be noted that C/ε_r depends on the structural geometry only and not on the relative dielectric constant.

Stripline in a Homogeneous Medium

Figure 1-42 schematically shows a cross-section of a stripline in a homogeneous medium with a relative dielectric constant ε_r above a voltage reference plane (ground or power). A rectangular cross-section signal line with width W and thickness T is at a height H above the reference plane (W, T, and H have the same unit). It can be shown[162-166] that the capacitance (C) and the characteristic impedance of this signal line are given by, respectively,

$$C = \varepsilon_r\varepsilon_0[(W_e/H) + 2.62(W_e/H)^{1/4}] \tag{1-27}$$

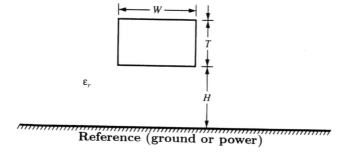

Figure 1-42   Stripline cross-section in a homogeneous medium. [162-166]

and

$$\sqrt{\varepsilon_r} Z_0 \approx \frac{377}{(W_e/H) + 2.62(W_e/H)^{1/4}} \qquad (1\text{-}28)$$

where

$$W_e = (W + T)/1.5 \qquad (1\text{-}29)$$

Most of the PCB structures satisfy the following:

$$0.3 \leq T/W \leq 0.6 \qquad (1\text{-}30)$$

and

$$0.5 \leq H/W_e \leq 5 \qquad (1\text{-}31)$$

In that case, the error of Z_0 in Eq. (1-28) is less than 3%.[162-166] Equation (1-28) is plotted in Fig. 1-43 for engineering convenience. In Eq. (1-27), the first term (W_e/H) represents the parallel plate contribution and the second term $2.26(W_e/H)^{1/4}$ yields the approximate fringe field contribution to the

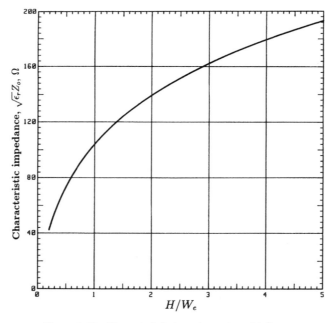

Figure 1-43   Characteristic impedance in a stripline.

signal line capacitance and impedance. Also, in Eq. (1-27), ε_0 is the permittivity of free space and is equal to 8.85×10^{-12} farads/m.

Shielded Stripline in a Homogeneous Medium

In order to increase the electrostatic capacitance of the signal line and reduce the characteristic impedance in PCB designs, it is usual to add a second reference (either ground or power) plane (Fig. 1-44). In this case, the signal line is sandwiched between two reference planes and is called shielded stripline or triplate structure. In Fig. 1-44, ε_r is the relative dielectric constant of the medium, W is the width and T is the thickness of the rectangular signal line, H_1 is the height from one of the reference planes and H_2 is the distance below the other reference plane, and $H_2 \geq H_1$.

For $H_2 > H_1$, the capacitance (C) and the characteristic impedance of this signal line is given by refs. 162-166 respectively, as:

$$C = \varepsilon_r \varepsilon_0 [(W_e/H_1) + (W_e/H_2) + 2.62(W_e/H_1)^{1/4}] \qquad (1\text{-}32)$$

$$\sqrt{\varepsilon_r} Z_0 \approx \frac{377}{(W_e/H_1) + W_e/H_2 + 2.62(W_e/H_1)^{1/4}} \qquad (1\text{-}33)$$

where

$$W_e = (W + T)/1.5 \qquad (1\text{-}34)$$

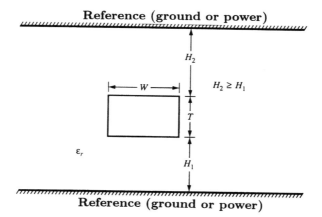

Figure 1-44 Shielded stripline (triplate) cross-section.[162-166]

Again, if

$$0.3 \leq T/W \leq 0.6 \qquad (1\text{-}35)$$

and

$$0.5 \leq H_1/W_e \leq 5 \qquad (1\text{-}36)$$

then the error of Z_0 in Eq. (1-33) is less than 3%.[162-166]

If $H_2 = H_1$, then the capacitance and characteristic impedance of this signal line (Eqs. (1-32) and (1-33)), become[162-166]

$$C = \varepsilon_r \varepsilon_0 [(W_e/H_1) + (W_e/H_2) + 2.8(W_e/H_1)^{1/4}] \qquad (1\text{-}37)$$

$$\sqrt{\varepsilon_r} Z_0 \approx \frac{377}{(W_e/H_1) + W_e/H_2 + 2.8(W_e/H_1)^{1/4}} \qquad (1\text{-}38)$$

Equations (1-33) and (1-38) are plotted in Fig. 1-45 for $H_1/H_2 = 0$, 0.5, and 1. In Eqs. (1-32) and (1-37), the first two terms (W_e/H_1) and (W_e/H_2),

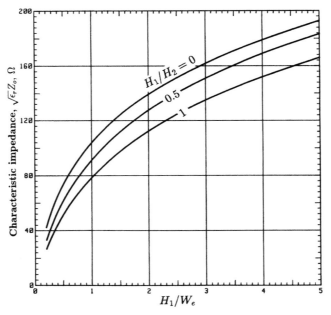

Figure 1-45   Characteristic impedance in a shielded stripline.

represent the parallel plate contribution and the last term gives the approximate fringe field contribution.

With Eqs. (1-28), (1-33), and (1-38), the geometries of the transmission signal lines can be established as soon as the required characteristic impedance values for PCB backplanes are determined by (e.g., SPICE) simulation.

1.5.4 Known Good Substrates (KGS) and Unknown Bad Substrates (UBS)

Even though KGS and UBS (including escapes) are not as important as KGD and UBD, nevertheless for COB and MCM applications, assembling some "good" chips on a bad substrate could lead to either discarding the whole assembly or very expensive remove/rework/replace/repair. In both cases, it will increase the material and manufacturing costs of the product. On the other hand, extensive electrical tests of substrates prior to board assembly will also increase the product costs. Thus, there is a trade-off between the objective of KGS, UBS, and the business needs.

Testing of COB and MCM substrates is very complex and costly. The test equipment alone will cost three to five times as much as the standard SMT equipment. A test fixture that is unique to the board must be constructed to interface the selected product test locations to specific test points of the test system. Sophisticated software that specifies the manner in which these test points should be observed must be interconnected by a properly fabricated product. Better qualified test and fixture personnel are required to operate the more complex equipment technologies. A major commitment must be made before implementing this process for bare board testing.

Table 1-4 summarizes two of the published electrical tests and requirements of PCBs for COB and MCM-L applications. It can be seen that the bare board is first subjected to the continuity test (for "open"), and during the continuity test, a current is allowed to pass through the conductor under test. The isolation test (for "short" or "leak") will verify that each network is well isolated from the rest of the board, and during the isolation test of each network, a small amount of current is injected into the network. For some products, if the characteristic impedance, Z_0, is required, it can be tested off-line using specialized radio frequency (RF) equipment.

1.5.5 Material Requirements

Less than 10 years ago, the standard PCB features were 10 mils (0.25 mm) lines and spaces with 35 and 43 mils (0.89 and 1.09 mm) holes for component

Table 1-4 Electrical Test of PCB

Electrical Test	IPC-ET-652	MIL-55110D (production)
Maximum continuity resistance pass/fail threshold test	Class 1: general electronic = 50 V Class 2: dedicated service = 20 V Class 3: high reliability = 20 V	= 10 V
Maximum continuity current test	Undefined	Per MIL-STD-275
Isolation resistance pass/fail threshold test	Class 1: general electronic = 500 kV Class 2: dedicated service > 2 MV Class 3: high reliability > 2 MV	> 2 MV
Apply voltage to passing networks during isolation test	High enough to provide sufficient current for the measurement in question, but low enough to prevent arc-over	40 V, or twice the maximum rated voltage on the board, whichever is greater

attachment. Today, 4 mils (0.1 mm) lines and spaces with 8 mils (0.2 mm) drilled holes are common. The 3 mils (0.075 mm) and emerging 2 mils (0.05 mm) lines and spaces are possible at a higher price.

Requirements on mechanical, electrical, thermal, physical, and chemical properties and manufacturability of COB and MCM substrate materials are continually increasing. The most common and cost-effective materials for COB and MCM-L applications are laminate materials. The most important requirements for improved laminate materials are: low relative dielectric constant, low dissipation factor, low thermal coefficient of expansion, low moisture absorption, high thermal conductivity, high glass transition temperature, high copper ductility, high copper peel strength, good dimensional stability, good solvent resistance, and good chemical resistance.

Low Relative Dielectric Constant and Dissipation Factor Laminate Materials

As discussed in Sections 1.5.2 and 1.5.3, low relative dielectric constant (ε_r) materials can reduce signal delay and thus increase package performance. Also, low dissipation factor Tan δ materials can reduce

Table 1-5 Low Relative Dielectric Constant Laminate Materials

Materials	Dielectric Constant (at 1 MHz)	Process Deviations	Dissipation Factor	Material Cost (ratio to FR-4)
FR-4/E-glass	4.6	None	0.025	1
FR-4/Quartz	3.8	Lamination and drilling	0.023	5
Epoxy and polyphenylene oxide/E-glass	4.0	Lamination and drilling	0.01	1.1
Epoxy/expanded PTFE	3.3	Lamination and drilling	0.02	12
BT/E-glass	4.3	Lamination and drilling	0.02	1.6
Sycar	3.6	Lamination	0.0003	3.2
Cyanate ester/ E-glass	3.5	Lamination, drilling, electroless, and copper oxide reduction	0.01	2.7
Cyanate ester/ expanded PTFE	2.7	Lamination, drilling, electroless, and copper oxide reduction	0.003	17

Note: Process deviation and material cost are compared with the standard FR-4 PCB.

signal loss at high frequency.[162-166] Thus, these two material properties are very important in high-speed and low-noise circuit design.

Attempts have been made to achieve the low ε_r materials by combining different materials with low ε_r, Table 1-5. It can be seen that the ε_r of laminate materials depend on both the resins (e.g., BT-resin, cyanate ester, poly-tetrafluoroethylene (PTFE)) and reinforcements (e.g., D-glass, E-glass (the standard reinforcement for FR-4 laminates), S-glass, quartz-glass, Aramid fiber, expanded PTFE).[169] Most of the resins have similar ε_r which are lower than the ε_r of E-glass. Thus, replacing E-glass with other lower ε_r reinforcement materials (e.g., expanded PTFE) usually leads to better results.

High Glass Transition Temperature and Thermal Conductivity Laminate Materials

Other important material properties of COB and MCM-L are the glass transition temperature (T_g) and thermal conductivity (k). In polymer or glass chemistry, transformation from a plastic (or glassy) to a rubbery behavior (or liquid) is termed the glass transition. The temperature at which glass transition takes place is called glass transition temperature, T_g (°C). The thermal conductivity, k, is a material constant (possibly temperature dependent) relating the rate of heat flow by conduction to the area of the material normal to the heat flow path and to the temperature gradient along the heat flow path ($W/m-K$).

Figures 1-46 and 1-47 show the typical plots of the elongations versus temperature of an FR-4 laminated PCB by thermomechanical analysis (TMA). It can be seen that the elongations in both in-plane (X- and Y-directions, Fig. 1-46), and out-of-plane (Z-direction, Fig. 1-47), are a nonlinear function of temperatures. The T_g for this particular sample is about 115°C, which is estimated by the intersection of the tangent lines of the glass (lower temperature) and liquid (higher temperature) regions.

High T_g materials are desirable for PCB assembly because they can be reflowed, tested, burned-in, and operated at higher temperatures. High T_g PCBs can be achieved by the development of higher cured resin systems such as multifunctional epoxy resin, BT, BT-epoxy blends, polyimide, polyimide-epoxy blends and cyanate ester resin.[169] The state of the art is given in Table 1-6.

In both COB and MCM-L technologies, the bare chips are directly attached to the laminate PCBs which are the primary source for conducting

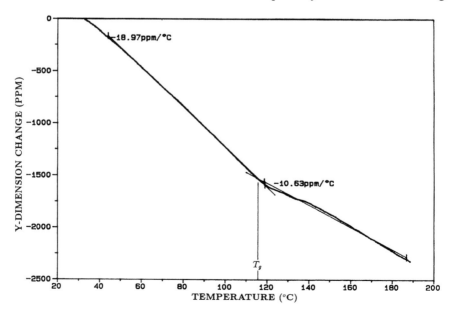

Figure 1-46 In-plane elongation of a PCB.

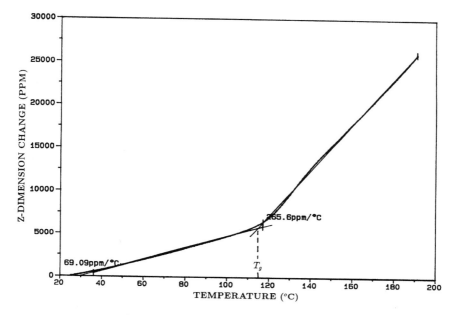

Figure 1-47 Out-of-plane elongation of a PCB.

Table 1-6 High T_g Laminate Materials

Resin Materials	T_g °C	Process Deviations	Material Cost (Ratio to FR-4)
FR-4	140	None	1
High T_g epoxy	190	Lamination and drilling	1.2
Epoxy and polyphenylene oxide	180	Lamination and drilling	1.1
BT/epoxy blend	180	Lamination and drilling	1.6
BT	225	Lamination and drilling	1.8
Sycar	180	Lamination	3.2
Polyimide/epoxy blend	240	Lamination and drilling	2.5
Cyanate ester	250	Lamination, drilling, electroless, and copper oxide reduction	2.7

Note: Process deviation and material cost are compared with the standard FR-4 PCB.

the heat away. The junction temperatures between the chips and board could limit the COB and MCM-L applications if heat is not properly managed. Thus, the use of high thermal conductivity laminate PCBs can increase thermal performance. Also, thermal vias in the PCBs for local heat dissipation and mechanical designs such as thermal cutouts for direct heat sink attachment can expand the application limits of COB and MCM-L.

Low Coefficient of Thermal Expansion Laminate Materials

The coefficient of linear thermal expansion (CTE) is a material constant (possibly temperature dependent) defined as the change in length that a bar of unit length undergoes when its temperature is changed by one degree.[13] It plays a very important role in electronic packaging.

The CTE of PCBs can be obtained from the TMA. Typical results are shown in Figs. 1-46 and 1-47. It can be seen that for temperatures less than the T_g, the CTE in the Y-direction is 18.97 p.p.m./°C and 69.09 p.p.m./°C in the Z-direction. However, for temperatures greater than the T_g, the CTE in the Y-direction is 10.63 p.p.m./°C and 265.6 p.p.m./°C in the Z-direction. The reason for such large values of the CTE in the Z-direction is that there are no glass fabric constraints in the Z-direction. This large CTE in the Z-direction could introduce PCB reliability problems (e.g., copper barrel cracking). Table 1-7 lists some low CTE materials.

Under normal operations the CTE of silicon chips is between 2.5 and 3 p.p.m./°C[170] and the in-plane CTE of the standard FR-4 PCBs can be as high as 20 p.p.m./°C. Thus, for COB and MCM-L applications, there is a

Table 1-7   Low TCE Laminate Materials

Materials	Type of Reinforcement	TCE (1/°C) XY/Z	Process Deviations	Material Cost (Ratio to FR-4)
FR-4/E-glass	Fabrics	16/74	None	1
Epoxy/Aramid fiber	Paper	7/90	Lamination, drilling, and routing	7
Epoxy/Aramid fiber	Fabrics	5/95	Lamination, drilling, and routing	10
Cyanate ester/ S-glass	Fabrics	9/47	Lamination, drilling, electroless, and copper oxide reduction	7

Note: Process deviation and material cost are compared with the standard FR-4 PCB.

very large thermal expansion mismatch (TEM) between the chips and the PCB. This TEM creates a high magnitude of stresses and strains at the chips, interconnects (e.g., wire bonds, solder joints, gold bumps, and chip attachment), and board. These stresses and strains produce the driving force for COB and MCM-L interconnection failures.[13] Thus, the use of low CTE PCBs (e.g., epoxy/Aramid paper and epoxy/Aramid fabrics) can reduce the TEM and enhance the COB and MCM-L reliability. Also, encapsulation of the chips (Fig. 1-2) can lift the reliability levels.

High Copper Ductility Laminate Materials

Both the modeling and experimental results indicate the need for high copper ductility in the PCB to avoid significant cracking problems.[1,2,13] This is because the low cycle thermal fatigue life of copper used in PCBs is nonlinearly related to the ductility of the copper.

The ductility of a material can be described by the reduction of area (RA) which is defined as the ratio of the decrease in area to the initial area (A_0), i.e., $RA = (A_0 - A_f)/A_0$, where A_f is the area at fracture.[171-174] For a clamped circular copper film (radius $= a$ and initial thickness $= t_0$) subjected to a uniform pressure P (Fig. 1-48), the fracture should occur at the center (with a maximum displacement equal to h_f and the fracture thickness equal to t_f). The RA (in this case, the thickness) or ductility of the

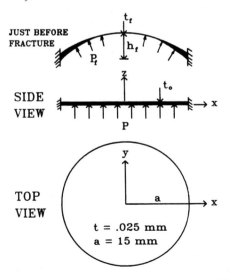

Figure 1-48   Ductility of a copper film.[176]

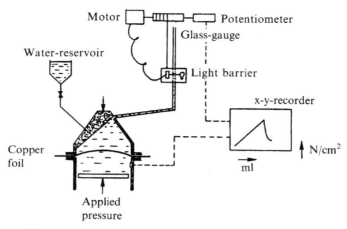

Figure 1-49 DUKTILOMAT hydraulic deep drawing tester.[175]

copper can be determined by combining the Duktilomat hydraulic deep-drawing test[175] and the finite element (FE) results.[176]

Figure 1-49 shows a schematic drawing of the Duktilomat hydraulic deep-drawing tester designed by Schering, Germany.[175] It can be seen that the film to be tested is clamped on to the top of a hydraulically pressurized cylinder and pressed into a spherical bulge (it has been shown[176] that this was a reasonable assumption). This test provides the applied pressure and volume of the bulge of the copper film prior to fracture.

With the measured volume of bulge (V) of the copper film and the assumption of spherical bulge, the center deflection of the bulge can be calculated by the following equation:

$$h = \left(\frac{3V}{\pi} + \sqrt{a^6 + \frac{9V^2}{\pi^2}} \right)^{1/3} + \left(\frac{3V}{\pi} - \sqrt{a^6 + \frac{9V^2}{\pi^2}} \right)^{1/3} \qquad (1\text{-}39)$$

Recording the applied pressure and the volume of the bulge (consequently, the center deflection) at time of rupture, the ductility (reduction of thickness) of the copper film can be determined by Fig. 1-50. Examples can be found in ref. 176.

Good Dimensional Stability and Low Moisture Absorption Laminate Materials

Dimensional stability of PCBs is very important for COB and MCM-L applications, since the pads and pitches on the PCB for the corresponding bare chips are very small. However, due to etching and heat treatment

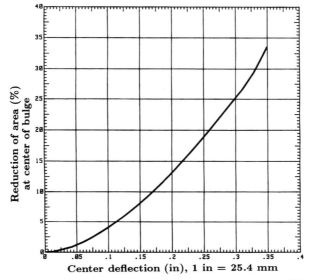

Figure 1-50 Ductility vs deflection at center of bulge.[176]

during PCB fabrication, the X- and Y-dimensions of the PCB may change. Depending on the type of laminate and prepreg, multilayer construction and processing, the dimensions of the PCB may shrink by 0.2%. In order to have good dimensional stability, improvements should be made in the resin systems (e.g., by fillers and low-shrink additives), the reinforcement systems (e.g., by use of different glass fabrics), and in process engineering.[169]

Laminate PCBs are not hermetic seals but permeable ones. For this reason, moisture absorption of PCBs is a long-standing reliability and process issue due to the possibility of metal conductor electromigration, prepreg delamination, T_g declination, Tan δ incrementation, dimensional instability, and misregistration. Thus, by using the low moisture absorption laminate PCBs (e.g., Sycar resin and liquid crystal polymer), the baking process could be reduced or even eliminated.

1.5.6 Flexural Rigidity and Young's Modulus Measurement of Board

During manufacturing, handling, shipping, rework, etc., the PCB is subjected to bending. Excess bending of the PCB may crack the chips, solder joints, chip attachments, wires, beam leads, etc. Thus, the bending stiffness (also called flexural rigidity) of the PCB is one of the important parameters for COB and MCM-L manufacturing and reliability. There are many methods proposed for measuring the flexural rigidity of structures.[171-174] The two following methods have been successfully applied to PCBs.[177]

Cantilever Beam Method

The set-up of the cantilever beam method is shown in Fig. 1-51. The specimen is clamped at one end and is subjected to a load (P) at the free end. During loading, the deflections (Δ) at the free end are recorded. The flexural rigidity (EI) is given by

$$EI = \frac{L^3}{3}\left(\frac{P}{\Delta}\right) \tag{1-40}$$

In Eq. (1-40), L is the length of the specimen, E is the Young's modulus, and I is the moment of inertia of the material. For example, Fig.   1-52 shows the $P - \Delta$ curve of a PCB specimen with length $L = 6$ in. (15.24 cm), width $w = 1$ in. (2.54 cm), and thickness $t = 0.058$ in. (1.47 mm). Then, $EI = 47.2$ in.^{2}/lb (1355 cm^2/N). Since $I = (1/12)wt^3 = 1.626 \times 10^{-5}$ in.^{4} (67.68 $\times$ 10^{-5} cm^4), then the Young's modulus, E, can be determined as $E = 2.9 \times 10^6$ lb/in.$^2 = 20$ GPa. It should be noted that a more accurate value of the Young's modulus can be obtained by $(1 - v_1 v_2)E$, where v_1 and v_2 are the Poisson's ratios of the PCB in the material principal directions 1 and 2, respectively.

Simply Supported Beam Method

Figure 1-53 shows the set-up of the simply supported beam method. In this case, the specimen is simply supported at both ends and is subjected to a load (P) at its mid span. During loading, the deflection at the center of the specimen (Δ) is recorded. The flexural rigidity (EI) is given by

$$EI = \frac{L^3}{48}\left(\frac{P}{\Delta}\right) \tag{1-41}$$

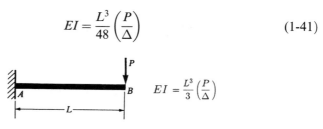

Figure 1-51   Cantilever beam specimen.

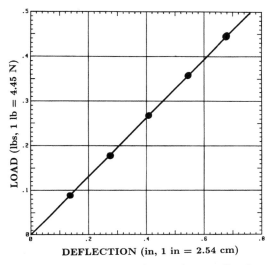

Figure 1-52 Load-deflection curve of a PCB from cantilver beam specimen.[177]

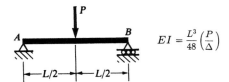

Figure 1-53 Simply supported beam specimen.

For example, Fig. 1-54 shows the load-deflection curve of a PCB specimen with length $L = 9$ in. (22.86 cm), width $w = 1$ in. (2.54 cm), and thickness $t = 0.058$ in. (1.47 mm). Then, $EI = 52.03$ in.^{2}/lb (1494 cm^2/N). Since $I = 67.68 \times 10^{-5}$ cm^4, then the Young's modulus is $E = 3.2 \times 10^6$ lb/in.$^2 = 22$ GPa. The reason why these values of the flexural rigidity and Young's modulus are so close to those from the cantilever beam method is that they used the same PCB. Usually, the average value of these two tests yields the flexural rigidity and Young's modulus of the PCB. Again, a more accurate value of the Young's modulus can be obtained by $(1 - v_1 v_2)E$.

1.6 CHIP ON BOARD

As mentioned earlier, COB is defined as the direct attachment of bare chips on the second level PCBs by either wire bonding, or TAB, or flip chip methods.

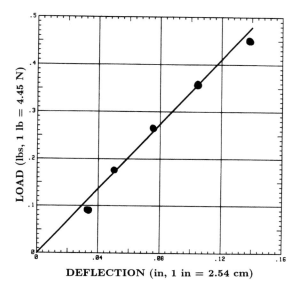

Figure 1-54   Load-deflection curve of a PCB from simply supported beam specimen.[177]

1.6.1 Wire Bonding Chips on Board

Figure 1-2 shows schematically a wire bonding chip on board. The process starts with attaching the back of a bare chip directly to the PCB with conductive or non-conductive epoxy chip attach materials. (For more information about chip attachment, see Chapter 6). This is followed by either thermocompression of gold or ultrasonic of aluminum wire bonding each chip bond pad to the corresponding bond pad on the PCB. Test and burn-in (if necessary) can be done after wire bonding, and one or two reworks are possible, though difficult.

In order to protect the chip, wire, and wire bonds from moisture, ionic contaminants, radiation, and hostile operating environments such as mechanical shock and vibration, the entire chip, bond wires, and pads on the PCB are covered with an epoxy resin. This is followed by curing the epoxy resin to form a protective encapsulant which is usually not to extend more than 3 mm from each edge of the bare chip and 2 mm from the PCB surface. After this stage rework is almost impossible.

During curing and cooling of the epoxy resin, thermal stresses will occur in the encapsulant. Excessive stresses may affect the reliability of the chip, chip bond pads, bond wires, bond pads on PCB, and epoxy. By integrating 11 piezoresistive silicon strain gauges in a test circuit on a 9 mm square silicon chip surface,[111] the polymerization stress during curing of the resin

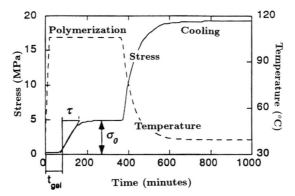

Figure 1-55   Polymerization stress during curing and cooling of the resin.[111]

and the thermomechanical stress during cooling after polymerization have been measured.

Figure 1-55 shows the measurement of the normal polymerization stress during curing of the resin and the thermomechanical stress during cooling down to room temperature.[111] It can be seen that the polymerization stress (σ_0) is more than 25% of the thermomechanical stress afer cooling. The gel time (t_{gel}) necessary to start the reaction and the time constant (τ) describing the evolution of the polymerization stress are given by, respectively,

$$t_{gel} = t_0 \exp\left(\frac{E_a}{kT}\right) \tag{1-42}$$

and

$$\tau_{gel} = \tau_0 \exp\left(\frac{E_a}{kT}\right) \tag{1-43}$$

where t_0 and τ_0 are constants, $k = 8.617 \times 10^{-5}\,\text{eV/K}$ is the Boltzmann's constant, T is the temperature measured in Kelvin (K), and E_a is the activation energy for both t_{gel} and τ_{gel} and is found to be 13.3 kcal/mol from Fig. 1-56.[111]

The effect of resin layer thickness on the polymerization stress is shown in Fig. 1-57. It can be seen that the polymerization stress is 1 MPa with 1 mm thick resin layer, and about 4 MPa with 2 mm thick. Thus, the polymerization stress in the resin is approximated to the square of the resin thickness. Figure 1-58 shows the effects of temperature on the curing stress and time. It can be seen that the curing time needed to obtain complete polymerization is about 80 min. at 125°C and about 4 h at 100°C. The final stress after cooling does not depend on curing temperature.

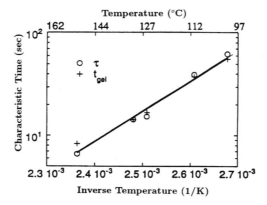

Figure 1-56   Characteristic time vs temperature.[111]

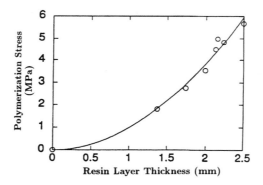

Figure 1-57   Effect of resin layer thickness on polymerization stress.[111]

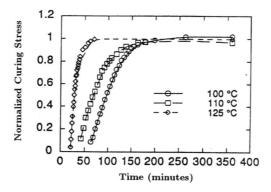

Figure 1-58   Effect of temperature and time on curing stress.[111]

Table 1-8 Wire Bonding Chips on Board: Typical Reliability Results

Tests	Test Conditions	Results
Humidity	85°C, 85% RH, 5 V for 1000 h	0/20
High temperature	125°, 5 V for 1000 h	0/20
Temperature cycling	−55°C to 125°C for 300 cycles	0/20
Pressure cooker	121°C, 2 atm for 200 h	0/20

Thermal management is usually a critical issue in the applications of wire bonding chip on board. This is because the chip surface is encapsulated by epoxy resin and the FR-4 PCB is not a good thermal conductive material. By using a high thermal conductivity laminate, PCB with thermal vias and thermal cutouts for direct heat sink attachment can enhance wire bonding chip on board thermal performance. Figures 1-59a and 1-59b show a couple of IBM Power Systems' thermal designs (KELLS technology). The difference between these two designs is the thermal path. In Fig. 1-59b, there is no dielectric material below the back of the chip, thus giving better thermal performance (thermal resistance can be as low as 0.5°C/W) than the design in Fig. 1-59a. It should be noted that even with the design in Fig. 1-59a, the thermal resistance is excellent and can be as low as 4.5°C/W.

In the past few years the reliability of wire bonding chips on board has been enhanced through the use of gold ball wire bonding and encapsulants. Table 1-8 shows a typical reliability test result. It can be seen that there are no failures even with very tough test conditions. For more information about wire bonding encapsulants, see Chapter 11. For more information about wire bonding bare chips on PCBs and other substrates, see Chapters 3 and 7.

1.6.2 Tape Automated Bonding Chips on Board

TAB is a technique allowing full automation of the bonding of one end of the copper beam lead (with either electroplated gold or tin, or electroless tin) to a semiconductor chip and the other end of the lead to a PCB (Fig. 1-2). The primary characteristic of TAB technology is a bonding projection or "bump" between the chip and the beam lead. The bump provides both the necessary bonding metallurgy for inner lead bonding and a physical stand-off to prevent lead-chip shorting.[14]

TAB offers many advantages over conventional wire bonding. The more prominent benefits of TAB are:[178-184] smaller bonding pad and pitch on chip, elimination of wire loop, more precise geometry, hermetically sealed passivation opening by the bump metallurgy so that no portions of the

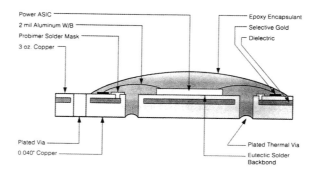

Figure 1-59a IBM Power Systems' thermal carrier (not to scale).

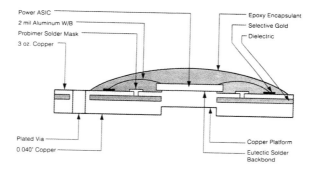

Figure 1-59b IBM Power Systems' thermal carrier (not to scale).

monolithic circuit are exposed, opportunity for gang bonding, opportunity for test and burn-in prior to board assembly, stronger and more uniform inner lead bond-pull strength, improved conduction heat transfer, better electrical and high frequency performance, attachment of chip in either a face-up or a face-down configuration, closer chip-to-chip distance in a multichip module, higher I/Os, less gold, lighter weight, and less required PCB surface area.

The reliability of TAB assembly has been reported by many authors, e.g., refs 178-184. In general, TAB assembly is reliable to use for many of the packaging applications, even though some of the fundamental problems are not very well understood. One of these problems is barrier layer metallization (BLM) or UBM.

One of the most common UBMs for TAB is Ti-W-Au. Figure 1-60a shows the test results of a Ti-W barrier alloy between 300 and 450°C.[181] It can be seen that a 0.4 μm Ti-W barrier can withstand annealing at 350°C for

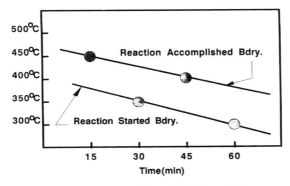

Figure 1-60a   Reaction kinetics of the Al/TiW/Au multilayer structure.[181]

about 30 min. Aluminum and gold were found to interdiffuse inhomogeneously and the reaction could be characterized phenomenologically to have nucleation, growth, and coalescence stages.The reaction tends to nucleate at aluminum hillocks. The resistance of Ti-W alloy to aluminum penetration is much greater than to gold. The diffusional difference caused large volume increase on the aluminum side and Kirkendall voids formation on the gold side (Fig. 1-60b). The reaction sites had volcanic configuration and expanded laterally while the reaction continued. The detailed reaction mechanism is not well understood and requires further investigation. For more information about TAB, see Chapters 4 and 8. Also, for more information about TAB thermal management, read Chapter 13 of ref. 14 and Chapter 19 of ref. 12.

1.6.3 Cost Comparison Between Wire Bonding and TAB

In the past few years cost analysis of packaging alternatives and process selections has become very popular. Based on science and engineering many cost models have been proposed.[77-88] A cost analysis between the wire

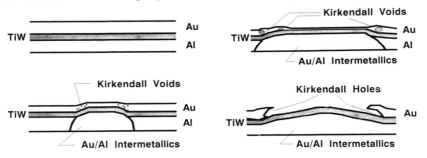

Figure 1-60b   Microstructural evolution of the Al/TiW/Au reaction.[181]

bonding and TAB of a few bare chips on PCB has been provided in ref. 80. The typical process model inputs and typical cost model outputs are shown in Tables 1-9 and 1-10, respectively. Figures 1-61 and 1-62 show the typical process flowchart of wire bonding and TAB, respectively. In Fig. 1-61, the wafer test block is highlighted because the cost is highly dependent upon test coverage during wafer test and burn-in. Test coverage $(1 - \%$ escapes) is defined as the percentage of defects that are detected at test.

Based on a seven chips wire bonding COB model,[80] Mavroides shows that relatively small changes in chip yield (wafer test yield) will lead to a visible steepening of the cost curve (Fig. 1-63). This follows the logic that the more bad chips (lower chip yield) there are, the higher the chance that they will escape into COB assembly.

Most cost penalties of TAB are the wafer bumping costs and additional testing costs. Based on a seven chips TAB COB model,[80] Mavroides shows the cost comparison between the wire bonding assembly (with a 90% chip yield) and the TAB assembly (with a 98% chip level test coverage, Fig. 1-64). PCB level test coverage for both wire bonding and TAB is assumed to be 95%. The upper TAB line shows the price for on-chip testing and burn-in at high yield. The lower TAB line shows the window for cost saving if the test and burn-in are performed at the PCB level instead of at the chip level. The crossover point is dependent on the chip yield, number of chips, number of I/Os per chip, KGD/UBD/KGS/UBS strategies, rework and repair methods and costs, and business needs.

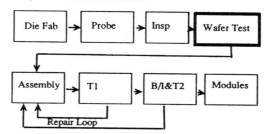

Figure 1-61   COB process flowchart using wire bonding.[80]

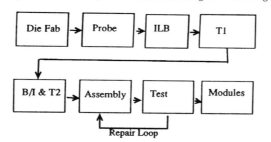

Figure 1-62   COB process flowchart using TAB.[80]

Table 1-9 Typical Process Model Inputs[80]

Total number of MCMs (volume)

Number of chips/MCM

I/O per chip

Wafer start fab cost

Process time and equipment cost of:

probe, at-speed test, assembly, test, rework, diagnostics, mechanical assembly

Substrate cost

Heat sink cost

Latent cost/module

Die/wafer

Process yields

Test yields on wafer and MCM

Test coverage

Rework yields

Operator cost

Overhead cost

Space cost

Machine utilization

Table 1-10 Typical Cost Model Outputs[80]

Total MCM cost

Dies (with probe)

At speed wafer test cost

MCM assembly and test cost

Rework cost

Module scrap cost

Component scrap cost

Latent cost

Mechanical assembly cost

Good modules shipped

DOA modules shipped

1.6.4 Solder Bumped Flip Chip on Board

Figure 1-2 shows schematically a solder bumped flip chip on board. It consists of the bare chip, solder bumps, underfill epoxy, and PCB. Basically, there are two groups of solder bumped flip chip applications, one with high temperature solders and the other with low temperature solders.

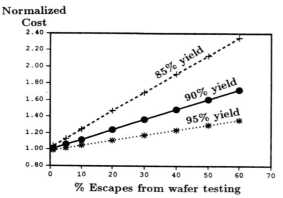

Figure 1-63   Effect of chip yield and wafer testing on costs.[80]

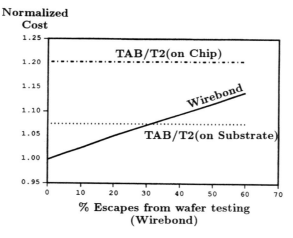

Figure 1-64   Cost comparison between wire bonding and TAB.[80]

High and Low Temperature Solders

There are at least three popular kinds of solders used for the flip chip bumps, the 42wt%Sn/58wt%Bi, 63wt%Sn/37wt%Pb, and 5wt%Sn/95wt%Pb. The solidus and liquidus temperatures (°C) for 42wt%Sn/58wt%Bi solder are 138 and 138, for 63wt%Sn/37wt%Pb solder are 183 and 183, and for 5wt%Sn/95wt%Pb solder are 308 and 312.[15] Their creep curves, i.e., strain vs time, at room temperature are shown in Figs. 1-65 and 1-66. It can be seen that the 5wt%Sn/95wt%Pb solder has the best creep resistance while the 42wt%Sn/58wt%Bi has the worst.

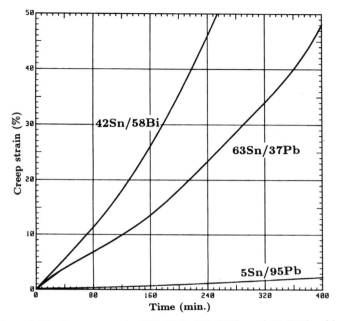

Figure 1-65 Creep curves for 42Sn/58Bi, 63Sn/37Pb, and 5Sn/95Pb solders.

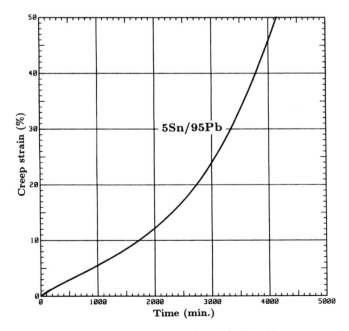

Figure 1-66 Creep curve for 5Sn/95Pb solder.

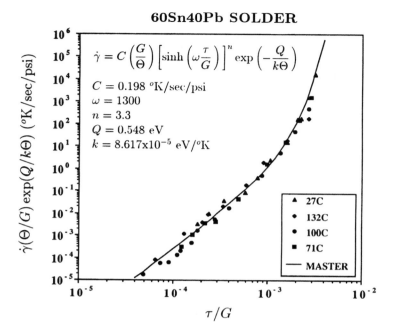

Figure 1-67   Steady state creep curve for 60Sn/40Pb solder.[185]

Figures 1-67 and 1-68, respectively, show the steady state creep curves for the low temperatue 60wt%Sn/40wt%Pb solder and the high temperature 2.5wt%Sn/97.5wt%Pb solder at various temperatures.[185] The Garofalo-Arrhenius steady-state creep[13] is generally expressed by

$$\dot{\gamma} = \frac{d\gamma}{dt} = C\left(\frac{G}{\Theta}\right)\left[\sinh\left(\omega\ \frac{\tau}{G}\right)\right]^{n} \exp\left(\frac{-Q}{k\Theta}\right) \qquad (1\text{-}44)$$

where γ is the steady-state creep shear strain, $\dot{\gamma}$ is the steady-state creep shear strain rate, t is the time, C is a material constant, G is the temperature dependent shear modulus, Θ is the absolute temperature (K), ω defines the stress level at which the power law stress dependence breaks down, τ is the shear stress, n is the stress exponent, Q is the activation energy for a specific diffusion mechanism, e.g., dislocation diffusion, solute diffusion, lattic self-diffusion, and grain boundary diffusion, and k is the Boltzmann's constant. For 60wt%Sn/40wt%Pb solder, the material constants of Eq. (1-44) have been experimentally determined by Darveaux and Banerji[185] with a single hyperbolic sine function. By using the test data of ref. 185 on the 60wt%Sn/40wt%Pb solder, Eq. (1-44) can be written as

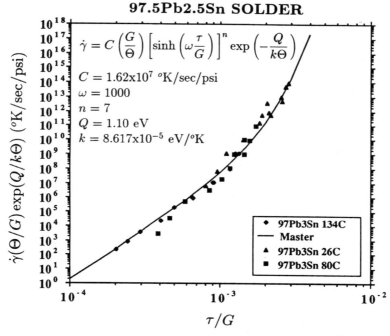

Figure 1-68 Steady state creep curve for 97.5Pb/2.5Sn solder.[185]

$$\dot{\gamma} = \gamma_0 \left[\sinh\left(\frac{\tau}{\tau_0}\right) \right]^{3.3} \tag{1-45}$$

where

$$\gamma_0 = \frac{1604(508 - \Theta)}{\Theta} \exp\left(\frac{-6360}{\Theta}\right) \tag{1-46}$$

and

$$\tau_0 = 3163 - 6.23\Theta \tag{1-47}$$

If the 60wt%Sn/40wt%Pb solder obeys the von Mises criterion,[13] then Eq. (1-45) can be written as

$$\dot{\varepsilon} = \varepsilon_0 \left[\sinh\left(\frac{\sigma}{\sigma_0}\right) \right]^{3.3} \tag{1-48}$$

where

$$\varepsilon_0 = \frac{926(508 - \Theta)}{\Theta} \exp\left(\frac{-6360}{\Theta}\right) \tag{1-49}$$

and

$$\sigma_0 = 5478 - 10.79\Theta \tag{1-50}$$

In Eq. (1-48), σ is the uniaxial stress, $\dot{\varepsilon}$ is the uniaxial steady-state creep strain rate. The unit for σ, τ, σ_0, and τ_0 is in lb/in.^{2} (psi), the unit for γ_0 and ε_0 is in 1/sec, and the unit for the temperature (Θ) is in Kelvin (K) which is obtained by adding 273.16 to temperature in degrees Celsius (°C).

Equations (1-45) and (1-48) can only be applied to, respectively, pure shear and uniaxial tension conditions. For combined stresses state, it is necessary to define an effective steady-state creep strain rate ($\dot{\varepsilon}_e$) and an effective stress (σ_e) as follows:[13]

$$\dot{\varepsilon}_e = \sqrt{\frac{2}{3} \, \dot{\varepsilon}_{ij}\dot{\varepsilon}_{ij}} \tag{1-51}$$

$$\sigma_e = \sqrt{\frac{3}{2} \, S_{ij}S_{ij}} \tag{1-52}$$

where

$$S_{ij} = \sigma_{ij} - \frac{1}{3} \, \sigma_{\beta\beta}\delta_{ij} \tag{1-53}$$

In Eq. (1-51), $\dot{\varepsilon}_{ij}$ is the steady-state creep strain rate tensor. In Eqs. (1-52) and (1-53), S_{ij} is the deviatoric stress tensor, σ_{ij} is the stress tensor and δ_{ij} is the Kronecker delta. Assuming that there exists a universal stress-strain rate curve and it coincides with the uniaxial curve, Eq. (1-48), then, we have

$$\dot{\varepsilon}_e = \varepsilon_0 \left[\sinh\left(\frac{\sigma_e}{\sigma_0}\right) \right]^{3.3} \tag{1-54}$$

Equation (1-54) has been used for studying the creep responses of solder interconnects, e.g., refs. 186-189.

For the 97.5wt%Pb/2.5wt%Sn solder, the material constants of Eq. (1-44) have also been experimentally determined by Darveaux and

Banerji[185] with a single hyperbolic sine function. By using the test data of ref. 185 on the 97.5wt%Pb/2.5wt%Sn solder, Eq. (1-44) can be written as

$$\dot{\gamma} = \gamma_0 \left[\sinh\left(\frac{\tau}{\tau_0}\right)\right]^7 \qquad (1\text{-}55)$$

where

$$\gamma_0 = \frac{1.62 \times 10^7 (1140 - \Theta)}{\Theta} \exp\left(\frac{-12\,765}{\Theta}\right) \qquad (1\text{-}56)$$

and

$$\tau_0 = 1710 - 1.5\Theta \qquad (1\text{-}57)$$

If the 97.5wt%Pb/2.5wt%Sn solder obeys the von Mises criterion,[13] then Eq. (1-55) can be written as

$$\dot{\varepsilon} = \varepsilon_0 \left[\sinh\left(\frac{\sigma}{\sigma_0}\right)\right]^7 \qquad (1\text{-}58)$$

where

$$\varepsilon_0 = \frac{1.4 \times 10^7 (1140 - \Theta)}{\Theta} \exp\left(\frac{-12\,765}{\Theta}\right) \qquad (1\text{-}59)$$

and

$$\sigma_0 = 2962 - 2.6\Theta \qquad (1\text{-}60)$$

Assuming that there exists a universal stress-strain rate curve and it coincides with the uniaxial curve, Eq. (1-58), then, we have

$$\dot{\varepsilon} = \varepsilon_0 \left[\sinh\left(\frac{\sigma_e}{\sigma_0}\right)\right]^7 \qquad (1\text{-}61)$$

Equation (1-61) has been used for studying the creep responses of solder interconnects, e.g., refs. 186-189.

Underfill Epoxy Encapsulants

One of the most important reasons why solder bumped flip chip on low cost PCB works is the underfill epoxy encapsulant. It reduces the effect of the global thermal expansion mismatch between the silicon chip and the organic substrate, i.e., it reduces the stresses and strains in the flip chip solder joints and redistributes the stresses and strains over the entire chip area that would otherwise be increasingly concentrated near the corner solder joints of the chip.

Other advantages of the encapsulant are to protect the chip from moisture, ionic contaminants, radiation, and hostile operating environments such as mechanical shock and vibration. The most desirable epoxy encapsulants should have a high glass transition temperature ($> 150°C$) and low thermal coefficient of linear expansion ($< 27 \times 10^{-6}/°C$). For more information about underfill encapsulant, see Chapter 12.

High Temperature Solder Bumped Flip Chip on Board

Figure 1-69 shows schematically a cross-section of a 5wt%Sn/95wt% Pb solder bumped chip on board,[141] which was developed by IBM in

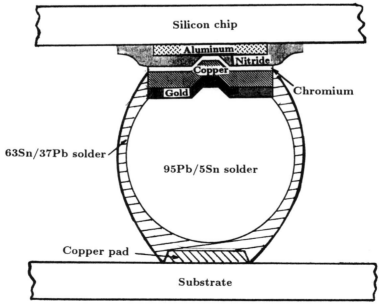

Figure 1-69   Cross section of a high-temperature solder bumped flip chip on eutectic solder coated board.[141]

Japan.[138-140] They called it flip chip attach (FCA) on surface laminar circuits (SLC) technology. SLC extends the classical PCB to very high density substrates for either single-chip or multichip applications. An SLC board consists of two major parts, namely, the ordinary FR-4 epoxy/glass substrate and the surface laminar layers for signal wiring. The dielectric layers are made of photosensitive epoxy and the wiring planes are made by copper plating. Small via holes (ranging from 100 to 127 μm) are made by a photo process on each dielectric layer and are used for the signal interconnection between wiring planes. Because of its high wireability, the SLC board can fan-out many very fine pitch I/O connection lines for a solder bumped flip chip.

The process of attaching a bare chip on SLC board technology is very similar to the IBM C4 technology[134-137] except that the SLC technology reflows at the lower melting point (183°C) of eutectic 63wt%Sn/37wt%Pb solder on the SLC board, and thus provides the opportunity to use low-cost substrate materials such as epoxy PCB. (At this stage, if it is necessary, test, burn-in, and rework should be done). The problem of a large thermal expansion mismatch between the silicon chip and the epoxy PCB is reduced by filling the gap between the chip and PCB with an epoxy resin. This resin completely fills the space underneath the chip (because of capillary action), and the chip and PCB are firmly bonded by curing the resin. At this stage, rework is possible but very difficult.[138] For more information about FCA, SLC and MCM-L, see Chapter 9.

Figure 1-70 shows schematically a 5wt%Sn/95wt%Pb solder bumped flip chip on a 63wt%Sn/37wt%Pb solder coated PCB with an underfill epoxy. For a 60°C temperature change, the U (displacement in the X-direction) and V (displacement in the Y-direction) fringe patterns of the whole cross-section have been determined by a Moiré interferometry method[190] and are shown in Fig. 1-71. From these fringe patterns, Guo, Chen, and Lim determined the relative displacements, U and V, and shear strain. It can be

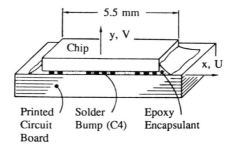

Figure 1-70 High-temperature solder bumped flip chip on eutectic solder coated board.[190]

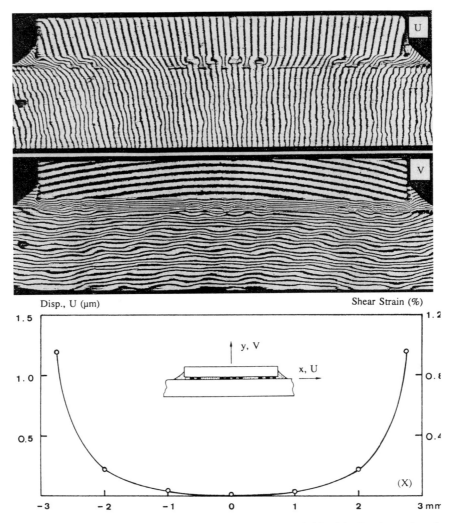

Figure 1-71   The fringe patterns of the U and V displacement fields of the flip chip on board. The plot shows the relative displacement in the X-direction between the chip and the PCB, and the distribution of average shear strains in the interconnection.[190]

seen that the corner of the chip moved relatively to the PCB in the X-direction about 1.2 µm and the corresponding shear strain is about 1%. It should be noted that, due to the edge effects,[13] the shear strain distribution in the X-direction is not linear and rises even faster when approaching the edge of the chip.

There are at least three major systems of thermal stresses and strains acting at the solder joints and encapsulant. The first is due to the local TEM between the silicon chip, solder joint, epoxy encapsulant, and PCB. The second is due to the global TEM between the silicon chip and the PCB. The last one is due to the overall bending of the PCB assembly. In ref. 141 the total (local, global, and bending) thermal response of the PCB assembly was analyzed using a local finite element analysis model, Fig. 1-72, with the global TEM and bending results[190] as the imposed displacement boundary conditions. The material properties for the analysis are shown in Table 1-11. The 63wt%Sn/37wt%Pb and 5wt%Sn/95wt%Pb solders are assumed to be elastoplastic materials. The yield stress and strain-hardening parameters for the 63wt%Sn/37wt%Pb solder are $3.74\,MN/m^2$ and 0.1, and for the 5wt%Sn/95wt%Pb solder are $8.3\,MN/m^2$ and 0.

The finite element model for the analysis of the corner portion (which includes the chip, solder joint, encapsulant, and PCB) of the encapsulated flip chip assembly is shown in Fig. 1-72. High order, 2-D plane stress elements were used for the model. Each element had eight nodes. Each node had two degrees of freedom. The present boundary-value problem was to determine the elastoplastic stresses and strains of the corner solder joint when it was subjected to an incrementally increased temperature from 22°C to 82°C ($\Delta T = 60$°C). The displacement boundary conditions were: (1) at the left-hand side of the silicon chip all points were fixed in the X-direction; (2) all points of the PCB moved 0.0015 mm in the positive X-direction; and (3) at the bottom of the PCB all points were fixed in the Y-direction. The temperature and displacements were applied to the encapsulated flip chip assembly incrementally. A total of eight increments were applied with approximately 11 interations per increment.

Table 1-11 Material Properties of an SLC Assembly[141]

	Young's Modulus (GPa)	Poisson's Ratio	Thermal Coefficient of Linear Expansion 10^{-6} (m/m–°C)
95Pb/5Sn	8.13	0.40	30
63Sn/37Pb	10	0.40	21
Encapsul.	6	0.35	30
Silicon	131	0.30	2.8
SLC XY	22	0.28	18
Board Z	22	0.28	70
Copper	121	0.35	17

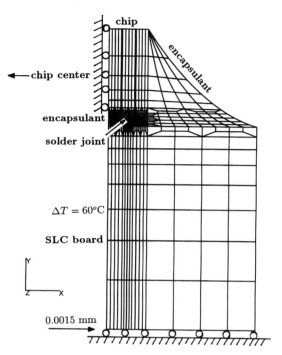

Figure 1-72 A local (corner) finite element model for the chip, solder joint, underfill encapsulant, and PCB.[141]

The accumulated effective plastic strain $(\overline{\varepsilon_p})$ distribution in the 95wt%Pb/5wt%Sn and 63wt%Sn/37wt%Pb solders are shown in Figs. 1-73 and 1-74, respectively. It is defined as follows

$$\overline{\varepsilon_p} = \frac{\sqrt{2}}{3} \int \sqrt{(d\varepsilon_x^p - d\varepsilon_y^p)^2 + (d\varepsilon_x^p)^2 + (d\varepsilon_y^p)^2 + \frac{3}{2}(d\gamma_{xy}^p)^2} \qquad (1\text{-}62)$$

where $d\varepsilon_x^p$ is the incremental plastic normal strain acting in the X-direction, $d\varepsilon_y^p$ is the incremental plastic normal strain acting in the Y-direction, and $d\gamma_{xy}^p$ is the incremental plastic shear strain acting in the Y-direction of the plane normal to the X-axis, which are not shown here. It can be seen from Fig. 1-73 that the maximum $\overline{\varepsilon_p}$ occurred near the corners of the 95wt%Pb/5wt%Sn solder joint with a value of 0.00932. The maximum $\overline{\varepsilon_p}$ in the 63wt%Sn/37wt%Pb solder joint occurred near the upper end (i.e., near the chip) with a value of 0.0129 (Fig. 1-74). Thus, if any solder joint cracking would occur, it should start at the 63wt%Sn/37wt%Pb solder near the chip.

In Coffin-Manson's relation[13,191]

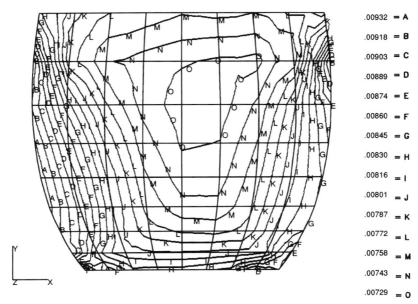

.00932 = A
.00918 = B
.00903 = C
.00889 = D
.00874 = E
.00860 = F
.00845 = G
.00830 = H
.00816 = I
.00801 = J
.00787 = K
.00772 = L
.00758 = M
.00743 = N
.00729 = O

Figure 1-73 Accumulated effective plastic strain in the 5Sn/95Pb solder.[141]

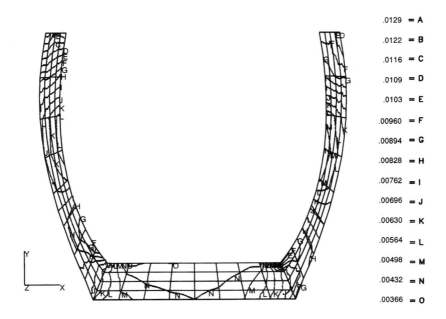

.0129 = A
.0122 = B
.0116 = C
.0109 = D
.0103 = E
.00960 = F
.00894 = G
.00828 = H
.00762 = I
.00696 = J
.00630 = K
.00564 = L
.00498 = M
.00432 = N
.00366 = O

Figure 1-74 Accumulated effective plastic strain in the 63Sn/37Pb solder.[141]

$$N_f = \theta(\Delta\gamma_p)^\xi \tag{1-63}$$

where N_f is the average number of cycles to failure, $\Delta\gamma_p$ is the plastic strain range, and θ and ξ are material constants. For 60wt%Sn/40wt%Pb or 63wt%Sn/37wt%Pb solders, θ and ξ have been determined by Solomon[192] at $-50°C$, $35°C$, and $125°C$. The average values are: $\theta = 1.2928$ and $\xi = -1.96$. Thus, in our case, $\Delta\gamma_p = 0.0129$ and $N_f = 6500$ cycles for the 63wt%Sn/37wt%Pb solder.

The von Mises (effective) stress (σ_e) (MPa) distribution in the encapsulant is shown in Fig. 1-75. It is defined as follows:

$$\sigma_e = \frac{\sqrt{2}}{2}\sqrt{(\sigma_x - \sigma_y)^2 + (\sigma_x)^2 + (\sigma_y)^2 + 6(\tau_{xy})^2} \tag{1-64}$$

where σ_x is the normal stress in the X-direction, σ_y is the normal stress in the Y-direction, and τ_{xy} is the shear stress acting in the Y-direction of the plane normal to the X-axis, which are not shown. It can be seen from Fig. 1-75 that the maximum stress (104 MPa) occurred at the left-hand free edge corners of the local finite element model. However, it should be noted that this large stress was due to the limitation of the present boundary value problem (there are very large thermal stresses at the free edge of the interface of any composite structures[193,194]) and this would not happen under real conditions. Thus, the maximum stress (61.6 MPa) occurred at Mark G, the

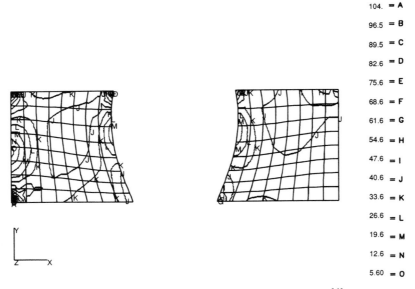

104.	= A
96.5	= B
89.5	= C
82.6	= D
75.6	= E
68.6	= F
61.6	= G
54.6	= H
47.6	= I
40.6	= J
33.6	= K
26.6	= L
19.6	= M
12.6	= N
5.60	= O

Figure 1-75 Effective stress in the underfill encapsulant.[141]

interface between the solder joint and the encapsulant. This stress acted at a very small area (Fig. 1-75) due to stress concentration and decreased rapidly to a much smaller value (40.6 MPa at Mark J). This stress was lower than the yield stress (45 MPa) of the encapsulant suggested by the manufacturer's data. It should be noted that the epoxy encapsulant is highly temperature- and rate-dependent.[195,196] These variables should be included in the analysis for more accurate results. For more information about epoxy encapsulants, see Chapters 11 and 12.

Figures 1-76 and 1-77 show the average (in-plane) load-displacement curve of the solder bumped flip chip assembly without and with the encapsulant. It can be seen that, (1) the failure load (466 lb or 2074 N) of the encapsulated interconnect was more than 10 times the one without encapsulant (33 lb or 147 N), and (2) the stiffness (slope of the load-displacement curve) of the encapsulated interconnect was more than three times the one without encapsulant. Cross-sections of the failed specimen with encapsulant showed that [141] the solder joint cracked near the chip but not on the chip pad. However, cross sections of the failed encapsulated specimen showed that the solder joint cracked near the chip and on the chip pad. The encapsulant was broken at the chip interface and the chip surface was free of encapsulant which indicates excellent encapsulant adhesion results.

Low Temperature Solder Bumped Flip Chip on Board

A large second level PCB consists of many different SMT packages. The standard SMT assembly process[12] starts with printing the 63wt%Sn/37wt%Pb or 60wt%Sn/40wt%Pb solder pastes on the PCB and then picking and placing the SMT packages on the pasted PCB. This is followed by reflowing the whole board. Thus, it would be a welcome feature for the board level manufacturers if bare chips were bumped with the 63wt%Sn/37wt%Pb or 60wt%Sn/40wt%Pb solders. In that case, the solder bumped bare chip is just like another "SMT package" for picking and placing on the PCB and mass reflowing with all other packages, power suppliers, connectors, etc. Higher cost eutectic solder coated PCBs for just a few bare chips are not necessary.

Figure 1-78 shows a regular low-cost PCB[97,98] for the 60wt%Sn/40wt%Pb solder bumped flip chip shown in Fig. 1-29. Thermal cycling (−45°C to 100°C, 1 cycle/h) results of five different square flip chips (3 mm, 6 mm, 9 mm, 12 mm, and 15 mm) assemblies with and without underfill resin have been reported by Rai et al.[97,98] and are shown in Fig. 1-79. It can be seen that the solder joints without underfill resin failed very early and those with larger chip size failed even earlier. For all size of chips with underfill

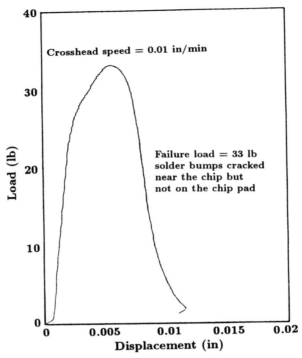

Figure 1-76 In-plane shear load deflection curve (without underfill encapsulant).[141]

resin, no failures have been observed at 1000 cycles. For the smaller chip sizes (3 mm, 6 mm, and 9 mm) assemblies with underfill resin, no failures have been observed even after 9000 cycles. Thus, the solder joints of flip chip on PCB with encapsulant are reliable to use for most of the packaging applications, because the underfill resin "cements" the interconnections.

1.7 PACKAGING ALTERNATIVES FOR MINIATURE PRODUCTS

As mentioned earlier, the trend in electronic products is low cost and miniaturization and advanced packaging technology will help to achieve these product design goals. But which packaging technology will make both business and technical sense for a particular miniature product? The answer is not simple and depends on applications. Table 1-12 shows the advantages and disadvantages of various packaging technologies for miniature products, for which cost, size, weight, performance, etc., are the most important requirements. In this section, we will discuss two particular

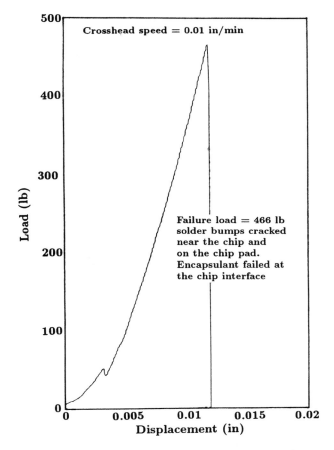

Figure 1-77   In-plane shear load deflection curve (with underfill encapsulant).[141]

applications, the PCMCIA PC Cards[197-201] by COB technology and the liquid crystal display (LCD) modules by chip on glass (COG) technology.[202-208]

1.7.1 PCMCIA PC Cards by COB Technology

In the past few years the market growth of portable electronic products has greatly increased. One of the key elements that is helping to make the growth possible is the PCMCIA PC cards. The physical specification, Table 1-13, defines the mechanical package of a PCMCIA PC card. It can be seen that there are three package types: Type I is a 3.3 mm thick package, which

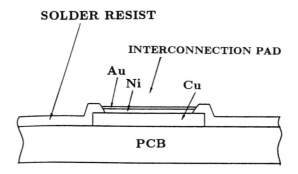

SOLDER RESIST

INTERCONNECTION PAD

Figure 1-78 SHARP's PCB for low temperature solder bumped flip chip applications.

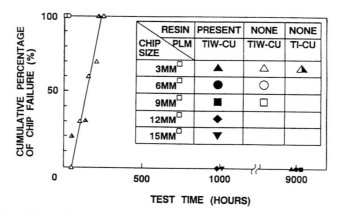

Figure 1-79 SHARP's flip chip on board thermal cycling results (−45°C to 100°C, 1 cycle/h).

can be used for memory applications, Type II is a 5 mm thick package which can be used for LANs and MODEMs applications, and Type III is a 10.5 mm thick package which can be used for rotating disk and complex structure applications.

Due to the physical constraints of PCMCIA PC cards the logical packaging choice is COB, but TSOP, TQFP, and TCP packages are in use today. One of the reasons is that these first level packages are available now and not expensive (Table 1-12). However, these packages have I/O limitations which may limit the functions that can be added to a card. Thus, in order to design new products with more functionality, personality, power, reliability, and cost saving, directly attaching bare chips to the cards (boards) is necessary. (Some PCMCIA PC cards produced in Japan are already using the wire bonding bare chips on board technology.) COB

Table 1-12   Packaging Alternatives for Miniature Products

Packaging Alternatives	Advantages	Disadvantages
TSOP, TQFP, and TCP on Board	• SMT compatible • Availability of packages • Module tests • Module burn-in • Reworkable • Low assembly cost	• Lead coplanarity • Package cracking • I/O limitations • Fine pitch limitations • Poorer electrical performance • Most board "real estate"
Wire Bonding Chip on Board	• Mature technology • Infrastructure exists • Flexible for new devices • Flexible for new bonding patterns	• Availability of wafers • Availability of dies • Availability of KGD • Tests and burn-in • I/O limitation • Peripheral technology • Sequential process • Additional equipment • Additional processes • Rework is difficult
Tape Automated Bonding Chip on Board	• Higher I/Os • Thinner profile • Better thermal performance • Better electrical performance • Lead compliance • Module tests • Module burn-in	• Availability of wafers • Availability of dies • Availability of KGD • Wafer bumping • Additional equipment • Additional processes • More board "real estate" compared to flip chip
Solder Bumped Flip Chip on Board	• High density • High I/Os • High performance • Noise control • Thinnest profile • SMT compatible • Area array technology • Smallest device foot prints • Self alignment	• Availability of wafers • Availability of dies • Availability of KGD • Wafer bumping • Test and burn-in • Underfill encapsulant • Additional equipment • Additional processes • Rework after encapsulant is difficult • Single joint touch-up is impossible

Table 1-13 PCMCIA Card Physical Specification

Type	Thickness (mm)	Applications
Type I	3.3	Memory
Type II	5.0	Simple I/O, e.g., LAN, MODEM
Type III	10.5	Disk, complex structures

technology will multiply the growth of PCMCIA PC cards, but as discussed in the previous Sections, cost, infrastructure, and assembly capability become a trade-off.

1.7.2 LCD Modules by COG Technology

LCDs are now widely used in various display applications, such as flat panel display, because LCDs provide excellent display quality, high resolution, and large capacity. Unlike PCMCIA PC cards which are dependent on the memory/ASIC/microprocessor modules, the primary focus of LCD modules is the functional substrate (mostly glass) and the construction of the pixels. Today, more than half of the LCD modules are assembled by TAB bare chips on glass. More than 40% of the LCDs use standard PQFPs/TQFPs and less than 5% use wire bonding and flip chips on glass. Table 1-14 summarizes the COG activities in Japan. It can be seen that more and more companies are interested in using flip chip on glass with both isotropic and anisotropic conductive adhesives. Chapter 10 presents a very high density micro bump bonding using gold bumps and a UV light curing resin.

Figure 1-80 shows schematically the cross-section of an LCD by NEC's COG technology.[64] Because of the photo-thermosetting resin to prevent circuit shorts between the adjacent bumps, 50 μm bump pitch is possible and has been verified by Matsui et al.[64] The bump material is indium, the bump heights are between 8 and 15 μm, and the bumping process is electroplated. The UBM are Au/Ti-W and are sputter-deposited. The bumping process with photo-thermosetting resin is shown in Fig. 1-81.

NEC's COG is bonded face down. By using vacuum, the indium bumped chip with photo-thermosetting resin is placed to a pressing head with heating capability, and the glass substrate is placed on a supporting plate. With a vision system, the chip (on the pressing head) is aligned on top of the substrate. With pressure (2-10 kg/chip), temperature (80-150°C), and less than 15 s, the chip is bonded to the glass.

Table 1-14 Chip on Glass (COG) Methods by Japanese Companies, after Matsushita

COG Technology	Bump Shape	Pressure Heat Processing	Current Bond Pitch (µm)	Future Bond Pitch (µm)	Bond Method	Chip Process	Repair Method
Citizen Watch	Au (Cu) bump on chip side	80–120°C 1.4 g/bump	216	100	Ag/Pd paste	Bump formation	Chip heating
Matsushita's MBB	Au bump on chip side	Room temperature 10–20 g/bump	50	1	Resin contraction force	Bump formation	Chip heating
Seiko Epson	Spherical bump printed on substrate	150–200°C 10–25 g/bump	130	60	Gold-plated resin ball	None	Chip heating
Toshiba	Au/In bump on chip side	Substrate: 55–60°C Chip: 120–150°C 20 g/bump	130	50	In alloy	Bump formation	Chip heating
Matsushita's SBB	Ball bump on chip side	100–120°C 50–70 g/bump	100–110		Ag paste	Ball bump formation	Chip heating
Sharp	Au film on chip side	Room temperature 20–30 g/bump	300	50	Au-plated resin ball	Resin ball formation Adhesive coating	Chip heating
Oki Electric	Au bump on chip side	170–200°C 20–50 g/bump	200	150	Anisotropic conductive sheet	Bump formation	Chip heating
Casio	Rectangular Au bump	180°C	80	40	Au-plated resin ball	Bump formation	Chip heating

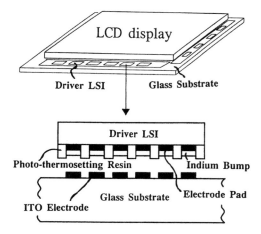

Figure 1-80 NEC's LCD with indium solder bumped flip chip on glass.

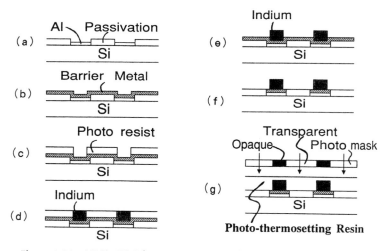

Figure 1-81 NEC's COG bumping process with photo-thermosetting resin.

1.8 SUMMARY

A brief introduction to the design, material, process, and reliability of wire bonding, TAB, and flip bare chips on board has been presented. Some important COB parameters such as chip (e.g., performance, possible and required I/Os, KGD, UBD, and fracture toughness measurement), chip level interconnects (e.g., wire bonding, TAB, and flip chip), first level packages (e.g., single-chip modules, MCM, MCCM, and the distinction between COB and MCM-L), and the board (e.g., wireability, propagation delay, characteristic impedance, KGS, UBS, material requirements, and flexural rigidity and Young's modulus measurements) have also been briefly mentioned. Furthermore, the applications of COB to PCMCIA PC cards and COG to LCDs have been discussed. For more useful and important COB, COG, and MCM information, see Chapters 2 through 12.

As the trend toward miniature and compact products continues, it seems clear that COB, MCCM, MCM-L, and COG are excellent electronic packaging and assembly technologies. It is our hope that the information presented in this book may assist in removing road blocks, avoiding unnecessary false starts, and accelerate design, material, and process development of these technologies. The COB, MCCM, MCM-L, and COG technologies are limited only by the business needs, ingenuity and imagination of engineers, visionless managements, and infrastructures.

ACKNOWLEDGMENTS

The author would like to thank Yutaka Tsukada of IBM in Japan and Tom Krulevitch, Wayne Schar, Matt Heydinger, Steve Erasmus, Jerry Gleason, July Glazer, and Anita Danford of Hewlett-Packard for the contribution to ref. 141. It has been a great pleasure and fruitful experience to work with them. The author also would like to thank Larry Moresco of Fujitsu Computer Packaging Technologies, Inc. for his review and useful comments.

REFERENCES

1. Seraphim, D. P., R. Lasky, and C. Y. Li, *Principles of Electronic Packaging*, McGraw-Hill Book Company, New York, 1989.
2. Tummala, R. R., and E. Rymaszewski, *Microelectronics Packaging Handbook*, Van Nostrand Reinhold, New York, 1989.
3. Vardaman, J., *Surface Mount Technology, Recent Japanese Developments*, IEEE Press, New York, 1992.
4. Johnson, R. W., R. K. Teng, and J. W. Balde, *Multichip Modules: System Advantages, Major Construction, and Materials Technologies*, IEEE Press, New York, 1991.
5. Senthinathan, R., and J. L. Prince, *Simultaneous Switching Noise of CMOS Devices and Systems*, Kluwer Academic Publishers, New York, 1994.
6. Sandborn, P. A., and H. Moreno, *Conceptual Design of Multichip Modules and Systems*, Kluwer Academic Publishers, New York, 1994.

7. Nash, F. R., *Estimating Device Reliability: Assessment of Credibility*, Kluwer Academic Publishers, New York, 1993.

8. Gyvez, J. P., *Integrated Circuit Defect-Sensitivity: Theory and Computational Models*, Kluwer Academic Publishers, New York, 1993.

9. Doane, D. A., and P. D. Franzon, *Multichip Module Technologies and Alternatives*, Van Nostrand Reinhold, New York, 1992.

10. Messuer, G., I. Turlik, J. Balde, and P. Garrou, *Thin Film Multichip Modules*, International Society of Hybrid Microelectronics, Silver Spring, MD, 1992.

11. Wong, C. P., *Polymers for Electronic and Photonic Applications*, Academic Press, San Diego, CA, 1993.

12. Lau, J. H., *Handbook of Fine Pitch Surface Mount Technology*, Van Nostrand Reinhold, New York, 1993.

13. Lau, J. H., *Thermal Stress and Strain in Microelectronics Packaging*, Van Nostrand Reinhold, New York, 1993.

14. Lau, J. H., *Handbook of Tape Automated Bonding*, Van Nostrand Reinhold, New York, 1992.

15. Lau, J. H., *Solder Joint Reliability: Theory and Applications*, Van Nostrand Reinhold, New York, 1991.

16. Frear, D., H. Morgan, S. Burchett, and J. Lau, *The Mechanics of Solder Alloy Interconnects*, Van Nostrand Reinhold, New York, 1993.

17. Manzione, L. T., *Plastic Packaging of Microelectronic Devices*, Van Nostrand Reinhold, New York, 1990.

18. Hwang, J. S., *Solder Paste in Electronics Packaging*, Van Nostrand Reinhold, New York, 1989.

19. Gilleo, K., *Handbook of Flexible Circuits*, Van Nostrand Reinhold, New York, 1991.

20. Engel, P. A., *Structural Analysis of Printed Circuit Board Systems*, Springer-Verlag, New York, 1993.

21. Suhir, E., *Structural Analysis in Microelectronic and Fiber Optics Systems*, Van Nostrand Reinhold, New York, 1991.

22. Matisoff, B. S., *Handbook of Electronic Packaging Design and Engineering*, Van Nostrand Reinhold, New York, 1989.

23. Prasad, R. P., *Surface Mount Technology*, Van Nostrand Reinhold, New York, 1989.

24. Manko, H. H., *Soldering Handbook for Printed Circuits and Surface Mounting*, Van Nostrand Reinhold, New York, 1986.

25. Morris, J. E., *Electronics Packaging Forum*, Volume 1, Van Nostrand Reinhold, New York, 1990.

26. Morris, J. E., *Electronics Packaging Forum*, Volume 2, Van Nostrand Reinhold, New York, 1991.

27. ASM International, *Electronic Materials Handbook*, Vol. 1, *Packaging*, Materials Park, OH, 1989.

28. Hollomon, J. K., Jr., *Surface-Mount Technology*, Howard W. Sams & Company, Indianapolis, IN, 1989.

29. Solberg, V., *Design Guidelines for SMT*, TAB Professional and Reference Books, New York, 1990.

30. Hutchins, C., *SMT: How to Get Started*, Hutchins and Associates, Raleigh, NC, 1990.
31. Bar-Cohen, A., and A. D. Kraus, *Advances in Thermal Modeling of Electronic Components and Systems*, Volume 1, Hemisphere Publishing Corp., New York, 1988.
32. Bar-Cohen, A., and A. D. Kraus, *Advances in Thermal Modeling of Electronic Components and Systems*, Volume 2, ASME Press, New York, 1990.
33. Kraus, A. D., and A. Bar-Cohen, *Thermal Analysis and Control of Electronic Equipment*, Hemisphere Publishing Corp., New York, 1983.
34. Harper, C. A., *Handbook of Microelectronics Packaging*, McGraw-Hill Book Company, New York, 1991.
35. Pecht, M., *Handbook of Electronic Package Design*, Marcel Dekker, New York, 1991.
36. Harman, G., *Wire Bonding in Microelectronics*, International Society for Hybrid Microelectronics, Reston, VA, 1989.
37. Lea, C., *A Scientific Guide to Surface Mount Technology*, Electrochemical Publications, Ayr, Scotland, 1988.
38. Lea, C., *After CFCs? Options for Cleaning Electronics Assemblies*, Electrochemical Publications Ltd., Isle of Man, 1992.
39. Wassink, R. J. K., *Soldering in Electronics*, Electrochemical Publications, Ayr, Scotland, 1989.
40. Pawling, J. F., *Surface Mounted Assemblies*, Electrochemical Publications, Ayr, Scotland, 1987.
41. Ellis, B. N., *Cleaning and Contamination of Electronics Components and Assemblies*, Electrochemical Publications, Ayr, Scotland, 1986.
42. Sinnadurai, F. N., *Handbook of Microelectronics Packaging and Interconnection Technologies*, Electrochemical Publications, Ayr, Scotland, 1985.
43. Lau, J. H., and S. Erasmus, "Review of Packaging Methods to Complement IC Performance," *Electronic Packaging & Production*, pp. 51-56, June 1993.
44. Motorola, *C4 Product Design Manual*, Vol. 1: *Chip and Wafer Design*, 1993.
45. Moresco, L., "Electronic System Packaging: The Search for Manufacturing the Optimum in a Sea of Constraints," *IEEE Transactions on Components, Hybrids, and Manufacturing Technology*, **13**(3), pp. 494-508, September 1990.
46. Murphy, B. T., "Cost-Size Optima of Monolithic Integrated Circuits," *IEEE Proceedings*, **52**, pp. 1537-1545, 1964.
47. Seeds, R. B., "Yield, Economics & Logistical Models for Coupled Digital Arrays," *IEEE International Convention Records*, Part 6, pp. 60-61, 1967.
48. Wessely, H., O. Fritz, P. Klimke, W. Koschnick, and K. H. Schmidt, "Electronic Packaging in the 90s – A Perspective from Europe," *Proceedings of the 40th IEEE Electronics Components and Technology Conference*, pp. 16-33, May 1990.
49. Donath, W. E., "Placement and Average Interconnection Lengths of Computer Logic," *IEEE Transactions on Circuits and Systems*, **CAS-26**, pp. 272-277, April 1979.
50. Turlik, I., and A. Reisman, "FORWARD," *Journal of Electronic Materials*, **18**(2), pp. 215-216, March 1989.

51. Radke, C. E., L. S. Su, Y. M. Ting, and J. Vanhorn, "Known Good Die and its Evolution – Bipolar and CMOS," *Proceedings of the 2nd International Conference and Exhibition on Multichip Modules*, pp. 152-159, April 1993.
52. Bracken, R. C., B. P. Kraemer, R. Paradiso, and A. Jensen, "Multichip-Modules, Die and MCM Test Strategy: The Key to the MCM Manufacturability," *Proceedings of the 1st International Conference and Exhibition on Multichip Modules*, pp. 456-460, April 1992.
53. Trent, J. R., "Test Philosophy for Multichip Modules," *Proceedings of the 1st International Conference and Exhibition on Multichip Modules*, pp. 444-452, April 1992.
54. Smitherman, C. D., and J. Rates, "Methods for Processsing Known Good Die," *Proceedings of the 1st International Conference and Exhibition on Multichip Modules*, pp. 436-443, April 1992.
55. Martin, S., D. Gage, T. Powell, and B. Slay, "A Practical Approach to Producing Known-Good Die," *Proceedings of the 2nd International Conference and Exhibition on Multichip Modules*, pp. 139-151, April 1993.
56. Corbett, T., "A Process Qualification Plan for KGD," *Proceedings of the 2nd International Conference and Exhibition on Multichip Modules*, pp. 166-171, April 1993.
57. Yamada, K., A. Tanaka, H. Shinohara, M. Honda, T. Hatada, A. Yamagiwa, and Y. Shirai, "A CPU Chip-On-Board," *Proceedings of the 43rd IEEE Electronic Components and Technology Conference*, pp. 8-11, June 1993.
58. Narayanan, R., P. M. Hall, and R. Chanchani, "Thermal and Mechanical Analysis of Flip-Chips on a Liquid-Cooled Multichip Module," *Proceedings of the 43rd IEEE Electronic Components and Technology Conference*, pp. 22-31, June 1993.
59. Lin, A. W., A. M. Lyons, and P. G. Simpkins, "Reliability and Thermal Characterization of a 3-Dimensional Multichip Module," *Proceedings of the 43rd IEEE Electronic Components and Technology Conference*, pp. 71-79, June 1993.
60. Clementi, J., J. McCreary, T. M. Niu, J. Palomaki, J. Varcoe, and G. Hill, "Flip-Chip Encapsulation On Ceramic Substrates," *Proceedings of the 43rd IEEE Electronic Components and Technology Conference*, pp. 175-181, June 1993.
61. Powell, D. O., and A. K. Trivedi, "Flip-Chip on FR-4 Integrated Circuit Packaging," *Proceedings of the 43rd IEEE Electronic Components and Technology Conference*, pp. 182-186, June 1993.
62. Suryanarayana, D., T. Y. Wu, and J. A. Varcoe, "Encapsulants Used in Flip-Chip Packages," *Proceedings of the 43rd IEEE Electronic Components and Technology Conference*, pp. 193-198, June 1993.
63. Cappo, F., J. Milliken, and J. Mosley, "Highly Manufacturable Multi-Layered Ceramic Surface Mounted Package," *Proceedings of the 11th IEEE IEMTS*, pp. 424-428, September 1991.
64. Matsui, K., K. Utsumi, H. Ohkubo, and C. Sugitani, "Resin and Flexible Metal Bumps for Chip-on-Glass Technology," *Proceedings of the 43rd IEEE Electronic Components and Technology Conference*, pp. 205-210, June 1993.

65. Chang, D., J. Fulton, H. Ling, M. Schmidt, R. Sinitski, and C. Wong, "Accelerated Life Test of Z-axis Conductive Adhesives," *Proceedings of the 43rd IEEE Electronic Components and Technology Conference*, pp. 211–217, June 1993.

66. Okuno, A., N. Oyama, T. Hashimoto, H. Kinoshita, and H. Shida, "New Metal Board for COB, Multi-Chip Module, TAB, and Flip-Chip," *Proceedings of the 43rd IEEE Electronic Components and Technology Conference*, pp. 555–559, June 1993.

67. Tessier, T. G., and E. G. Myszka, "Approaches to Cost Reducing MCM-D Substrate Fabrication," *Proceedings of the 43rd IEEE Electronic Components and Technology Conference*, pp. 570-578, June 1993.

68. Chu, D., and D. S. Shen, "A Maskless Flip-Chip Solder Bumping Technique," *Proceedings of the 43rd IEEE Electronic Components and Technology Conference*, pp. 610-614, June 1993.

69. Ramakrishna, K., J. C. Lo, and Y. Guo, "An Assessment of Mechanical Reliability of a Die-Bonded Chip Package During Chip Encapsulation and Accelerated Thermal Cycling," *Proceedings of the 43rd IEEE Electronic Components and Technology Conference*, pp. 670-678, June 1993.

70. Hartman, D. H., "Optical Interconnects at the Multi-Chip Module Level: A View from the System Level," *Proceedings of the 43rd IEEE Electronic Components and Technology Conference*, pp. 701-704, June 1993.

71. Wong, C. P., and R. McBride, "Robust Titanate-Modified Encapsulants for High Voltage Potting Application of Multi-Chip Module," *Proceedings of the 43rd IEEE Electronic Components and Technology Conference*, pp. 748-754, June 1993.

72. Kimura, M., T. Shimoto, K. Matsui, K. Utsumi, T. Kusaka, and T. Koike, "High Density Multilayer Substrate for Si on Si High Speed Module," *Proceedings of the 13th IEEE International Manufacturing Technology Symposium*, pp. 353-357, September 1992.

73. Kusaka, T., N. Senba, A. Nishizawa, N. Takahashi, and T. Shimoto, "A Silicon-on-Silicon Packaging Technology for Advanced ULSI Chips," *Proceedings of the 43rd IEEE Electronic Components and Technology Conference*, pp. 941-947, June 1993.

74. Hashemi, H., M. Olla, D. Cobb, P. Sandborn, M. McShane, G. Hawkins, and P. Lin, "A Mixed Solder Grid Array and Peripheral Leaded MCM Package," *Proceedings of the 43rd IEEE Electronic Components and Technology Conference*, pp. 951-956, June 1993.

75. More, M., Y. Kizaki, M. Saito, and A. Hongu, "A Fine Pitch COG Technique for a TFT-LCD Panel Using an Indium Alloy," *Proceedings of the 43rd IEEE Electronic Components and Technology Conference*, pp. 992-996, June 1993.

76. Chen, Y., C. Wang, and C. Lee, "Directly Deposited Lead-Indium-Gold Composite Solder," *Proceedings of the 43rd IEEE Electronic Components and Technology Conference*, pp. 1004-1007, June 1993.

77. Messner, G., and W. Smith, "Equations for Selection of Cost-Efficient Interconnection Designs," *Proceedings of the 42nd IEEE Electronic Components and Technology Conference*, pp. 10-16, May 1992.

78. Thornberg, G., "Cost Analysis of a High-Performance CMOS Multichip Module," *Proceedings of the 42nd IEEE Electronic Components and Technology Conference*, pp. 17–21, May 1992.
79. Hawley, A., and R. Gayle, "TAB Multichip Modules: A Case Study," *Proceedings of the 11th IEEE International Manufacturing Technology Symposium*, pp. 27-30, September 1991.
80. Mavroides, J., "Cost Saving Opportunities with Multi-Chip Modules," *Proceedings of the 11th IEEE International Manufacturing Technology Syposium*, pp. 413-416, September 1991.
81. Ng, L. H., "What Drives the Cost of Thin Film Multi-Chip Modules?," *Proceedings of IEEE/Japan IEMTS*, pp. 381-385, June 1991.
82. Ng, L. H., "Economic Impact of Processing Technologies on Thin Film MCMs," *Proceedings of the 42nd IEEE Electronic Components and Technology Conference*, pp. 1042–1045, May 1992.
83. Vardaman, E., and L. Ng, "A Cost/Performance Analysis of Multichip Module Interconnects," *Proceedings of the ISHM Conference*, pp. 27–32, 1991.
84. Ho, C. W., and A. Deshpande, "A Low Cost Multichip Module-D Technology," *Proceedings of the 2nd International Conference and Exhibition on Multichip Modules*, pp. 483-488, April 1993.
85. Chong, F. C., B. Sadler, F. Swiatowiec, D. Fu, J. Zhang, J. Lam, M. Warder, and S. Westbrook, "The Evolution of MCM Test from High Performance Bipolar Mainframe Multichip Modules to Low Cost Work Station Multichip Modules," *Proceedings of the 2nd International Conference and Exhibition on Multichip Modules*, pp. 404-410, April 1993.
86. Frye, R. C., and A. Shah, "Targeting Low Cost, High Volume MCM Applications," *Proceedings of the 2nd International Conference and Exhibition on Multichip Modules*, pp. 12-17, April 1993.
87. Ho, C. W., and H. Green, "Impact of Low-Cost MCM-D on MCM Designers' Technology Choices," *Proceedings of the 2nd International Conference and Exhibition on Multichip Modules*, pp. 481-483, April 1993.
88. Tessier, T. G., and E. G. Myszka, "High Performance MCM-L/D Substrate Approaches for Cost-Sensitive Packaging Applications," *Proceedings of the 2nd International Conference and Exhibition on Multichip Modules*, pp. 200-207, April 1993.
89. Fleischner, P. S., "MCM-L Substrates: A Case Study," *Proceedings of the 2nd International Conference and Exhibition on Multichip Modules*, pp. 582-587, April 1993.
90. DiFrancesco, L., and C. Reynolds, "Socketing Chip-On-Board to a Multichip Module Using Particle Interconnect," *Proceedings of the 2nd International Conference and Exhibition on Multichip Modules*, pp. 335-340, April 1993.
91. Freda, M., and C. Reynolds,"MCM-L Enabling Technologies," *Proceedings of the 2nd International Conference and Exhibition on Multichip Modules*, pp. 214-219, April 1993.
92. Conte, A. S., "MCM-L, The Answer for Desktop Workstations," *Proceedings of the 2nd International Conference and Exhibition on Multichip Modules*, pp. 18-21, April 1993.

93. Ackaert, A., L. Van Dam, W. Delbare, and K. Allaert, "The Use of MCM-L Manufacturing and Assembly Technology for Broadband Application," *Proceedings of the 2nd International Conference and Exhibition on Multichip Modules*, pp. 134-138, April 1993.

94. Reynolds, C., M. Freda, and L. DiFrancesco, "MCM-L Structures, Base Technologies, and Design Guidelines," *Proceedings of the 2nd International Conference and Exhibition on Multichip Modules*, pp. 575-581, April 1993.

95. Brooks, R., D. Hendricks, R. George, and K. Wasko, "Direct Chip Attach – A Viable Chip Mounting Alternative," *Proceedings of the 2nd International Conference and Exhibition on Multichip Modules*, pp. 595-598, April 1993.

96. Messner, G., "Prognosis for MCM-L," *Proceedings of the 2nd International Conference and Exhibition on Multichip Modules*, pp. 599-604, April 1993.

97. Rai, A., Y. Dotta, H. Tsukamoto, T. Fujiwara, H. Ishii, T. Nukii, and H. Matsui, "COB (Chip On Board) Technology: Flip Chip Bonding Onto Ceramic Substrates and PWB (Printed Wiring Boards)," *ISHM Proceedings*, pp. 474-481, 1990.

98. Rai, A., Y. Dotta, T. Nukii, and T. Ohnishi, "Flip-Chip COB Technology on PWB," *Proceedings of IMC*, pp. 144–149, June 1992.

99. Berg, K., "The Sequential Process Advantage in MCM-L Construction," *Proceedings of the 2nd International Conference and Exhibition on Multichip Modules*, pp. 190-195, April 1993.

100. Kato, M., "Development and Application of MCM-L, Practical Solution for Commercial Type MCM," *Proceedings of the 1st International Conference and Exhibition on Multichip Modules*, pp. 485-489, April 1992.

101. Higgins, L. M., D. Robinson, E. Mammo, and J. Baum, "Glob-Top Encapsulant Suitability for Large Wire Bonded Die on MCM-L," *Proceedings of the 1st International Conference and Exhibition on Multichip Modules*, pp. 482-484, April 1992.

102. Higgins, L. M., and D. L. Robinson, "Development and Assembly of a 40 MHz MCM-L Based DSP-Fast Static RAM Multichip Module," *Proceedings of the 2nd International Conference and Exhibition on Multichip Modules*, pp. 563-574, April 1993.

103. Acocella, J., J. Benenati, T. Caulfield, and K. Puttlitz, "Mixed Wirebond, TAB, and Flipchip Interconnections on Multi-Layer Ceramic MCMs," *Proceedings of the 2nd International Conference and Exhibition on Multichip Modules*, pp. 358-365, April 1993.

104. Hirakawa, T., C. Kato, and K. Nishimura, "Optimizing The Mounting System for Flip-Chips by Means of Finite Element Analysis," *Proceedings of IEEE/ Japan IEMTS*, pp. 318–321, June 1991.

105. Lau, J. H., "Thermoelastic Solutions for a Semi-Infinite Substrate with a Powered Electronic Device," *J. of Electronic Packaging, Trans. of ASME*, **114**, pp. 353-358, September 1992.

106. Lau, J. H., "Thermoelastic Solutions for a Finite Substrate with an Electronic Device," *J. of Electronic Packaging, Trans. of ASME*, **113**, pp. 84-88, March 1991.

107. Lau, J. H., "Thermoelastic Problems for Electronic Packaging," *J. of the International Society for Hybrid Microelectronics*, No. 25, pp. 11–15, May 1991.
108. Marrs, R. C., B. Freyman, and J. Martin, "High Density BGA Technology," *Proceedings of the 2nd International Conference and Exhibition on Multichip Modules*, pp. 326-329, April 1993.
109. Caulfield, T., J. Benenati, and J. Acocella, "Surface Mount Array Interconnections for High I/O MCM-C to Card Assemblies," *Proceedings of the 2nd International Conference and Exhibition on Multichip Modules*, pp. 320-325, April 1993.
110. Marrs, R. C., "Recent Developments in Low Cost Plastic MCMs," *Proceedings of the 2nd International Conference and Exhibition on Multichip Modules*, pp. 220-229, April 1993.
111. Sarbach, P., L. Guerin, A.Weber, M. Dutoit, and P. Clot, "Stress Analysis and Reliability of Chip On Board Encapsulation Technology," *Proceedings of the 15th IEEE International Manufacturing Technology Symposium*, pp. 82-86, October 1993.
112. Scheifers, S. M., and C. Raleigh, "Flip Chip On Board (FCOB) Process Characterization," *Proceedings of the 15th IEEE International Manufacturing Technology Symposium*, pp. 143-156, October 1993.
113. Sandborn, P. A., M. Abadir, and C. Murphy, "A Partitioning Advisor for Studying the Tradeoff Between Peripheral and Area Array Bonding of Components in Multichip Modules," *Proceedings of the 15th IEEE International Manufacturing Technology Symposium*, pp. 271-276, October 1993.
114. Baba, S., W. Carlomagno, D. Cummings, and F. Guerrero, "Bonded Interconnect Pin (BIP) Technology, A New Bare Chip Assembly Technique for Multichip Modules," *Proceedings of IEEE/Japan IEMTS*, pp. 136-139, June 1991.
115. Chung, T., J. Chang, and A. Emamjomeh, "Tape Automated Bonded Chip on MCM-D," *Proceedings of the 15th IEEE International Manufacturing Technology Symposium*, pp. 282-293, October 1993.
116. Narayan, C., and S. Purushothaman, "Thin Film Transfer Process for Low Cost MCMs," *Proceedings of the 15th IEEE International Manufacturing Technology Symposium*, pp. 373-380, October 1993.
117. Roszel, L., "Laminate Substrates for High Speed Interconnect," *Proceedings of the 15th IEEE International Manufacturing Technology Symposium*, pp. 423-426, October 1993.
118. Clot, P., "Low Cost Multi-Chip Modules Based on Single and Double Sided Chip-On-Board Modules Following JEDEC Standard Specifications," *Proceedings of the 15th IEEE International Manufacturing Technology Symposium*, pp. 477-480, October 1993.
119. Ho, C. W., and H. Green, "Impact of Low-cost MCM-D on MCM Designers' Technology Choices," *Proceedings of the 15th IEEE International Manufacturing Technology Symposium*, pp. 481-483, October 1993.
120. Rogren, P. E., "The Current State of Laminate Based, Molded MCM Technology," *Proceedings of the 15th IEEE International Manufacturing Technology Symposium*, pp. 485–489, October 1993.

121. Machuga, S. C., S. E. Lindsey, K. Moore, and A. Skipor, "Encapsulation of Flip Chip Structures," *Proceedings of the 13th IEEE International Manufacturing Technology Symposium*, pp. 53-58, September 1992.

122. Charles, H. K., "Electronic Materials and Structures for Multichip Modules," *J. of Electronic Packaging*, pp. 226-233, June 1992.

123. Charles, H. K., "Packaging With Multichip Modules," *Proceedings of the 13th IEEE International Manufacturing Technology Symposium*, pp. 206-210, September 1992.

124. Chu, D., C. Reber, and D. Palmer, "Screening ICs on the Bare Chip Level: Temporary Packaging," *Proceedings of the 13th IEEE International Manufacturing Technology Symposium*, pp. 223-226, September 1992.

125. Newcombe, E. H., "Design Tradeoffs When Using SMT, COB, and/or TAB Packaging Technologies," *Proceedings of the 13th IEEE International Manufacturing Technology Symposium*, pp. 343-347, September 1992.

126. Knott, J. F., *Fundamentals of Fracture*, John Wiley and Sons, New York, 1973.

127. Logsdon, W. A., P. K. Liaw, and M. A. Burke, "Fracture Behavior of 63Sn-37Pb," *Engineering Fracture Mechanics*, **36**(2), pp. 183–218, 1990.

128. Lau, J. H., and D. Rice, "Thermal Fatigue Life Prediction of Flip Chip Solder Joints by Fracture Mechanics Method," *ASME/JSME Proceedings of Advances in Electronic Packaging*, pp. 385-392, April 1992.

129. Lau. J. H., "Thermomechanical Characterization of Flip Chip Solder Bumps for Multichip Module Applications," *Proceedings of the 13th IEEE International Electronics Manufacturing Technology Symposium*, pp. 293-299, September 1992.

130. Lau, J. H., "Thermal Fatigue Life Prediction of Flip Chip Solder Joints by Fracture Mechanics Method," *International Journal of Engineering Fracture Mechanics*, **45**(5), pp. 643-654, July 1993.

131. Dieter, G., *Mechanical Metallurgy*, McGraw-Hill Book Company, New York, 1986.

132. Meisser, C., "Modern Bonding Processed for Large-Scale Integrated Circuits, Memory Modules and Multichip Modules in Plastic Packages," *Proceedings of the IEEE International Electronic Manufacturing Technology Symposium*, pp. 80-84, September 1991.

133. Saboui, A., "Wire Bonding Limitations for High Density Fine Pitch Plastic Packages," *Proceedings of the IEEE International Electronic Manufacturing Technology Symposium*, pp. 75-79, September 1991.

134. Totta, P. A., and R. P. Sopher, "SLT Device Metallurgy and Its Monolithic Extension," *IBM Journal of Research and Development*, pp. 226-238, May 1969.

135. Goldmann, L. S., "Geometric Optimization of Controlled Collapse Interconnections," *IBM Journal of Research and Development*, pp. 251-265, May 1969.

136. Seraphim, D. P., and J. Feinberg, "Electronic Packaging Evolution," *IBM Journal of Research and Development*, pp. 617-629, May 1981.

137. Tummala, R., and B. Clark, "Multichip Packaging Technologies in IBM for Desktop to Mainframe Computers," *Proceedings of the 42nd IEEE Electronic Components and Technology Conference*, pp. 1-9, May 1992.

138. Tsukada, Y., Y. Mashimoto, and N. Watanuki, "A Novel Chip Replacement Method for Encapsulated Flip Chip Bonding," *Proceedings of the 43rd IEEE/ EIA Electronic Components & Technology Conference*, pp. 199-204, June 1993.

139. Tsukada, Y., Y. Maeda, and K. Yamanaka, "A Novel Solution for MCM-L Utilizing Surface Laminar Circuit and Flip Chip Attach Technology," *Proceedings of the 2nd International Conference and Exhibition on Multichip Modules*, pp. 252-259, April 1993.

140. Tsukada, Y., S. Tsuchida, and Y. Mashimoto, "Surface Laminar Circuit Packaging," *Proceedings of the 42nd IEEE Electronic Components and Technology Conference*, pp. 22-27, May 1992.

141. Lau, J. H., T. Krulevitch, W. Schar, M. Heydinger, S. Erasmus, and J. Gleason, "Experimental and Analytical Studies of Encapsulated Flip Chip Solder Bumps on Surface Laminar Circuit Boards," *Circuit World*, **19**(3), pp. 18-24, March 1993.

142. Yamada, H., Y. Kondoh, and M. Saito, "A Fine Pitch and High Aspect Ratio Bump Array for Flip-Chip Interconnection," *Proceedings of the IEEE International Electronic Manufacturing Technology Symposium*, pp. 288-292, September 1992.

143. Lau, J. H., S. Golwalkar, P. Boysan, R. Surratt, D. Rice, R. Forhringer, and S. Erasmus, "Solder Joint Reliability of a Thin Small Outline Package (TSOP)," *Circuit World*, **20**(1), pp. 12-19, November 1993.

144. Lau, J. H., S. Golwalkar, D. Rice, S. Erasmus, and R. Forhringer, "Experimental and Analytical Studies of 28-Pin Thin Small Outline Package Solder-Joint Reliability," *J. of Electronic Packaging, Trans. of ASME*, **114**, pp. 169-176, June 1992.

145. Lau, J. H., S. Golwalkar, and S. Erasmus, "Advantages and Disadvantages of Thin Small Outline Packages (TSOP) with Copper Gull-Wing Leads," *Proceedings of the ASME International Electronics Packaging Conference*, pp. 1119-1126, Binghamton, NY, September 1993.

146. Nakamura, Y., M. Ohta, T. Nishioka, A. Tanaka, and S. Ohizumi, "Effects of Mechanical and Flow Properties of Encapsulating Resin on the Performance of Ultra Thin Tape Carrier Package," *Proceedings of the 43rd IEEE Electronic Components and Technology Conference*, pp. 419-424, June 1993.

147. Lau, J. H., J. Miremadi, J. Gleason, R. Haven, and S. Ottoboni, "No Clean Mass Reflow of Large Over Molded Plastic Pad Array Carriers (OMPAC)," *Proceedings of the IEEE International Electronic Manufacturing Technology Symposium*, pp. 63-75, October 1993.

148. Lau, J. H., et al., "Reliability of Ball Grid Array Solder Joints Under Bending, Twisting, and Vibration Conditions," to be presented at the *ASME WAM*, December 1993.

149. Lau, J. H., Y. Pao, C. Larner, S. Twerefour, R. Govila, D. Gilbert, S. Erasmus, and S. Dolot, "Reliability of 0.4 mm Pitch, 256-Pin Plastic Quad Flat Pack No-Clean and Water-Clean Solder Joints," *Soldering & Surface Mount Technology*, No. 16, February 1994.

150. Lau, J. H., R. Govila, C. Larner, Y. Pao, S. Erasmus, S. Dolot, M. Jalilian, and M. Lancaster, "No-Clean and Solvent-Clean Mass Reflow Processes of 0.4 mm

Pitch, 256-Pin Fine Pitch Quad Flat Packs (QFP)," *Circuit World*, **19**(1), pp. 19-26, October 1992.

151. Lau, J. H., G. Dody, W. Chen, M. McShane, D. Rice, S. Erasmus, and W. Adamjee, "Experimental and Analytical Studies of 208-Pin Fine Pitch Quad Flat Pack Solder-Joint Reliability," *Circuit World*, **18**(2), pp. 13–19, January 1992.

152. Lau, J. H., "Thermal Stress Analysis of SMT PQFP Packages and Interconnections," *J. of Electronic Packaging, Trans. of ASME*, **111**, pp. 2-8, March 1989.

153. Lau, J. H., L. Powers, J. Baker, D. Rice, and W. Shaw, "Solder Joint Reliability of Fine Pitch Surface Mount Technology Assemblies," *IEEE Trans. of CHMT*, **13**, pp. 534-544, September 1990.

154. Lau, J. H., and C. A. Keely, "Dynamic Characterization of Surface Mount Component Leads for Solder Joint Inspection," *IEEE Trans. on CHMT*, **12**, pp. 594-602, December 1989.

155. Lau, J. H., and G. Harkins, "Stiffness of 'Gull-Wing' Leads and Solder Joints for a Plastic Quad Flat Pack," *IEEE Trans. on CHMT*, **13**(1), pp. 124-130, March 1990.

156. Lau, J. H., and G. Harkins, "Thermal Stress Analysis of SOIC Packages and Interconnections," *IEEE Trans. on CHMT*, **11**(4), pp. 380-389, December 1988.

157. Lau, J. H., R. Subrahmanyant, D. Rice, S. Erasmus, and C. Li, "Fatigue Analysis of a Ceramic Pin Grid Array Soldered to An Orthotropic Epoxy Substrate," *J. of Electronic Packaging, Trans. of ASME*, **113**, pp. 138-148, June 1991.

158. Lau, J. H., S. Leung, R. Subrahmanyant, D. Rice, S. Erasmus, and C. Li, "Effects of Rework on the Solder Joint Reliability of Pin Grid Array Interconnects," *J. of the Institute of Interconnection Technology (Circuit World)*, July 1991.

159. Lau, J. H., G. Harkins, D. Rice, J. Kral, and B. Wells, "Experimental and Statistical Analyses of Surface-Mount Technology PLCC Solder-Joint Reliability," *IEEE Trans. on Reliability*, **37**(5), pp. 524-530, December 1988.

160. Tummala, R. R., "Multichip Technologies From Personal Computers to Mainframes and Supercomputers," *Proceedings of NEPCON West*, pp. 637-643, 1993.

161. Kersten, P., and H. Reichl, "Integration of Thin Film Polyimide with PCB Technology," *Circuit World*, **19**, pp. 16–17, 20, 1993.

162. Chang, C. S., "Resistive Signal Line Wiring Net Designs in Multichip Modules," *Proceedings of the 43rd IEEE Electronic Components and Technology Conference*, pp. 516-522, June 1993.

163. Chang, C. S., "Electrical Design of Signal Lines for Multilayer Printed Circuit Boards," *IBM J. Res. Develop.*, **32**, pp. 647-657, 1988.

164. Chang, C. S., "Electrical Design Methodologies," in *Electronic Materials Handbook*, Vol. 1: *Packaging*, pp. 25-44, ASM International, Materials Park, Ohio, 1989.

165. Chang, C. S., "Printed Circuit Board Signal Line Electrical Design," in *Principles of Electronic Packaging*, ed. D. Seraphim, R. Lasky, and C. Li, McGraw-Hill, New York, 1989.

166. Chang, C. S., "Transmission Lines," *Circuit Analysis, Simulation and Design*, ed. A. E. Ruehli, Vol. 3, Part 2, pp. 292-332, North-Holland, 1987.

167. Dearholt, D. W., and W. R. McSpadden, *Electronic Wave Propagation*, McGraw-Hill, New York, 1973.

168. Ramo, S., J. Whinnery, and T. Duzer, *Fields and Waves in Communication Electronics*, John Wiley & Sons, New York, 1984.

169. Cygon, M., "High Performance Materials for PCBs," *Circuit World*, **19**, pp. 14–18, October 1992.

170. Lau, J. H., and L. L. Moresco, "Mechanical Behavior of Microstrip Structures Made of Y-Ba-Cu-O Superconducting Ceramics," *IEEE Trans. on CHMT*, **11**(4), pp. 419-426, December 1988.

171. Gere, J. M., and S. P. Timoshenko, *Mechanics of Materials*, 2nd ed., PWS Engineering, Boston, MA, 1984.

172. Tetelman, A. S., and A. J. McEvily, Jr., *Fracture of Structural Materials*, John Wiley & Sons, New York, 1967.

173. Fung, Y. C., *Foundations of Solid Mechanics*, Prentice-Hall, Englewood Cliffs, NJ, 1965.

174. Boresi, A. P., and O. M. Sidebottom, *Advanced Mechanics of Materials*, John Wiley & Sons, New York, 1984.

175. Rolff, R., *Duktilomat*, Schering, Galvanotechnik, Germany, 1985.

176. Lau, J. H., "Elasto-Plastic Large Deflection Analysis of a Copper Film," *J. of Electronic Packaging, Trans. of ASME*, **114**, pp. 221-225, June 1992.

177. Lau, J. H., and G. H. Barrett, "Stress and Deflection Analysis of Partially Routed Panels for Depanelization," *IEEE Trans. on CHMT*, **10**(3), pp. 411-419, September 1987.

178. Lau, J. H., S. J. Erasmus, and D. W. Rice, "An Introduction to Tape Automated Bonding Technology," in *Electronics Packaging Forum*, ed. by J. E. Morris, Van Nostrand Reinhold, pp. 1–83, New York, 1991.

179. Kang, S. K., "Gold-to-Aluminum Bonding for TAB Applications," *Proceedings of the 42nd IEEE Electronic Components and Technology Conference*, pp. 870-875, May 1992.

180. Lau, J. H., S. J. Erasmus, and D. W. Rice, "Overview of Tape Automated Bonding Technology," *Circuit World*, **16**(2), pp. 5-24, 1990.

181. Tung, C. H., and Y. S. Kuo, "Tape Automated Bonding (TAB) Failure Mechanism Studies – Lead Fatigue and Barrier Alloy Penetration," to be published at the *IEEE Transactions on CHMT*, December 1993.

182. Lau, J. H., S. J. Erasmus, and D. W. Rice, "Overview of Tape Automated Bonding Technology," *Electronic Materials Handbook*, Vol. 1: *Packaging*, ASM International, November 1989.

183. Chen, W. T., J. Raski, J. Young, and D. Jung. "A Fundamental Study of Tape Automated Bonding Process," *J. of Electronic Packaging, Trans. of ASME*, pp. 216-225, September 1991.

184. Lau, J. H., W. D. Rice, and G. Harkins, "Thermal Stress Analysis of TAB Packages and Interconnections," *IEEE Trans. on CHMT*, **13**(1), pp. 183-188, March 1990.

185. Darveaux, R., and K. Banerji, "Constitutive Relations for Tin-Based Solder Joints," *IEEE Transactions on Components, Hybrids, and Manufacturing Technology*, **15**(6), pp. 1013-1024, December 1992.

186. Lau, J. H., "Creep of 96.5Sn3.5Ag Solder Interconnects," *Soldering & Surface Mount Technology*, No. 15, pp. 45–49, September 1993.

187. Lau, J. H., "Creep of 62Sn36Pb2Ag Solder Interconnects," *Proceedings of the ASME International Electronics Packaging Conference*, pp. 995-1003, Binghamton, NY, September 1993.

188. Lau, J. H., "Creep of Solder Interconnections Under Combined Loads," *Proceedings of the IEEE Electronic Components & Technology Conference*, pp. 852-857, June 1993.

189. Lau, J. H., "Bending and Twisting of 63Sn37Pb Solder Interconnects with Creep," to be presented at the *ASME WAM*, December 1993.

190. Guo, Y., W. Chen, and K. Lim, "Experimental Determinations of Thermal Strain in Semiconductor Packaging Using Moiré Interferometry," *Proceedings of the 1st Joint ASME/JSME Conference on Electronic Packaging*, pp. 779-784, April 1992.

191. Manson, S. S., *Thermal Stress and Low Cycle Fatigue*, McGraw-Hill, New York, 1966.

192. Solomon, H. D., "Fatigue of 60/40 Solder," *IEEE Transactions on Components, Hybrids and Manufacturing Technology*, Vol. 9, pp. 423-432, December 1986.

193. Suhir, S., "Interfacial Stresses in Bimetal Thermostats," *ASME Journal of Applied Mechanics*, **55**, pp. 595-600, 1989.

194. Lau, J. H., "A Note on the Calculation of Thermal Stresses in Electronic Packaging by Finite Element Methods," *ASME Journal of Electronic Packaging*, **111**, pp. 313-320, December 1989.

195. Suhir, E., and J. M. Segelken, "Mechanical Behavior of Flip-Chip Encapsulants," *ASME Journal of Electronic Packaging*, **112**, pp. 327-332, December 1990.

196. Suryanarayana, D., D. Hsiao, R. Gall, and J. McCreary, "Flip Chip Solder Bump Fatigue Life Enhanced by Polymer Encapsulation," *Proceedings of the IEEE Electronic Components & Technology Conference*, pp. 338-344, May 1990.

197. Gecius, J., "PCMCIA – An Emerging Packaging Technology for Portable Computers," *Proceedings of NEPCON West*, pp. 733–736, February 1993.

198. Lowe, H., and W. Ford, "Overview of PCMCIA Packaging and Card Assembly Process," *Proceedings of NEPCON West*, pp. 737-742, February 1993.

199. Lyn, R. J., "Encapsulation for PCMCIA Assemblies – An Overview," *Proceedings of NEPCON West*, pp. 743-751, February 1993.

200. Roods, S., M. Okimura, and D. Urban, "Inspection of Very-Fine-Pitch Connections on PCMCIA Cards," *Proceedings of NEPCON West*, pp. 752-762, February 1993.

201. Derdall, B. C., J. McKaig, and P. Min, "IBM Developed Attach and Removal Tool," *Proceedings of NEPCON West*, pp. 763-766, February 1993.
202. Nukii, T., N. Kajimoto, H. Atarashi, H. Matsubura, K. Yamamura, and H. Matsui, "LSI Chip Mounting Technology For Liquid Crystal Displays," *Proceedings of International Symposium on Microelectronics*, pp. 257-262, 1990.
203. Hatada, K., and H. Fujimoto, "A New LSI Bonding Technology-Micron Bump Bonding Technology," *Proceedings of the IEEE Electronic Components & Technology Conference*, pp. 45-49, May 1989.
204. Saito, M., M. Mori, A. Hongu, and A. Niitsuma, "COG (Chip on Glass) Technique for LCD Using a Low Melting Point Metal," *Proceedings of International Symposium on Microelectronics*, pp. 263-271, 1990.
205. Masuda, M., K. Sakuma, E. Satoh, Y. Yamasaki, H. Miyasaka, and J. Takeuchi, "Chip on Glass Technology for Large Capacity and High Resolution LCD," *Proceedings of IEEE IEMTS*, pp. 57-60, September 1989.
206. Tsuboi, K., H. Kobayashi, S. Yagi, and S. Nishi, "Application of Chip on Board Assembly Technology Onto LED Lamp Unit for Copy Machines," *Proceedings of IEEE/Japan IEMTS*, pp. 152-155, June 1991.
207. Uchida, M., K. Nozawa, and Y. Karasawa, "Interconnective Technology with Metallization for LCD," *Proceedings of IEEE/Japan IEMTS*, pp. 97-100, June 1991.
208. Kubota, T., T. Kimura, and S. Ushiki, "COG (Chip-On-Glass) Mounting of Si and GaAs Devices," *Proceedings of IEEE/Japan IEMTS*, pp. 188-191, June 1991.

2

Making COB Testing Tractable: Chip Pretest and System Diagnostics

Charles F. Hawkins, David W. Palmer, and R. Keith Treece

2.1 INTRODUCTION

Most Chip on Board (COB) and Multichip Module (MCM) designs will be complex having many integrated circuits (ICs) and discretes with dense interconnections. Such systems' electrical testing and diagnostics have been characterized traditionally by an *ad hoc* dependence on the particulars of each module layout; not by an integrated, universal methodology. The COB product has an additional hurdle compared to traditional Printed Wiring Board (PWB) in that the individual ICs are not first placed in robust packages which can undergo extensive pretesting before commitment to the board (such as high frequency, high and low temperature, and extended burn-in testing).

For decades the ability to test extensively the individual component piece parts as well as the unpopulated interconnection board before assembly has been a major contributor to low defect levels and good field reliability. If all individual components are tested to zero defects then only errors in assembly and reliability defects should lead to failure. However, the drive for higher density, lower cost, and higher system performance has replaced individual robust packages with COB. To compensate for this lack of package convenience, it is necessary to modify traditional PWB testing methodology. A new testing perspective has roughly three separate procedures: bare chip pretesting, unpopulated substrate verification, and system diagnostics.

The absence of robust single chip packages requires alternative approaches for bare chip testing. Candidates include: (1) more intensive wafer level probe tests, (2) temporary packaging of the individual chips in a robust carrier or TAB frame (Tape Automated Bonding), (3) using easy rework capability of flip chip and HDI (General Electric Company–High Density Interconnect) for temporary packaging for test, and (4) refining wafer level sampling and accelerated aging wafer test structures.

COB also places new demands on present practices of electrical system diagnostics. For example, after each piece part and substrate has been individually pretested and assembled there are still system testing challenges to (1) assure functionality within requirements, and (2) localize the defects for rework if requirements are not immediately met. (We are ignoring the case where economics allow discarding units that do not pass functional tests, since these systems do not require diagnostics.) Many methods are being refined in industry to meet these system level testing challenges such as boundary scan on the ICs to validate the IC connection and the IC functionality after assembly, layout with extra pads for probe and rework, built-in self test to allow inexpensive board test equipment to perform system frequency testing, and I_{DDQ} testing, monitoring to detect defects buried deep in the multichip circuitry.

Many directions for such complex product testing are seen today, but no universal test and diagnostic standards exist. A strong trend that addresses electronic complexity is the industry acceptance of test and diagnostic standards. The IEEE 1149.1 boundary scan standard was written to solve board node density problems that prevented electromechanical probe access by bed-of-nails board testers.[1-3] Boundary scan is an electronic equivalent to mechanical probing. The 1149.1 standard is essential for most Multichip Module (MCM) products which usually have higher node density than boards. COB can be considered a printed wiring board approach to high density MCMs.

Other IEEE standards are proposed that address system hierarchical testing and diagnostics (P1149.5)[4] and analog and mixed signal testing (P1149.4).[5] These latter two standards are being designed to link with the 1149.1 boundary scan design to provide an electronic means to examine all system (and MCM) products down to the smallest component level. These standards and proposed standards should be carefully considered by MCM/COB design, test, and manufacturing teams. They are described in more detail later in this chapter.

2.2 KNOWN GOOD DIE

The longstanding preference for individually, robustly packaged ICs has been driven by the opportunity to test extensively the packaged part before commitment to the printed wiring board. The growth in COB has often

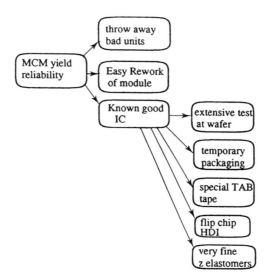

Figure 2-1   Three approaches to achieve reliability for bare die MCMs.

forced companies either to invent a system for electrically qualifying bare die or to ignore qualification and risk rework and testability problems. There are no standards yet for achieving Known Good Die (KGD), resulting in each company developing or buying their own *ad hoc* solution (Fig. 2-1). Several of the most pursued methods are described:

Testing at the Wafer Level   Most ICs receive only a room temperature partial functional test at the wafer level. Those that pass go through single chip assembly and receive more extensive testing when packaged. Testing can be enhanced at the wafer level in a number of ways. For example, more complete defect oriented testing can be performed at room temperature, but at the expense of tying up a million dollar wafer probe tester (Sentry, HP, Advantest, TI, LTX). Controlled impedance (wave guided) probes with closely mounted driver and measurement circuitry, can perform wafer level testing at the designed IC clock frequency, thus decreasing functional test time. Membrane probes have also been used to provide controlled impedance and high frequency testing; they minimize damage to bond pads allowing multiple probing that may be necessary to perform high temperature and low temperature testing after the room temperature pass. Normal rigid probes considerably damage standard aluminum alloy pads, and are usually limited to two or three contacts. On the other hand, non-destructive probes sometimes have trouble breaking through insulating oxide layers on bond pads. Because of the difference in IC operation when loaded by probes compared to when embedded in a circuit, a standard transformation between wafer results and anticipated COB performance must be established for each chip.

For CMOS ICs, quiescent current testing (I_{DDQ}) is the single most efficient test methodology for defect detection.[6,7] I_{DDQ} testing rejects ICs that have a power supply quiescent current above a certain level, such as 1 µA or 10 µA. thus at the cost of considerably more time on multi-million dollar wafer test machines, it is possible to achieve extensive pretesting with the exception of burn-in. A specification for such a wafer KGD approach is under development.[8]

Burn-in is a challenge for wafer level testing. At least two approaches have been discussed: (1) an application of a GE HDI type overlay interconnection on the whole wafer, (2) lot sample burn-in rather than 100% screening. The first approach would use the tape interconnection to supply DC power and digital inputs to all the ICs in the wafer during the burn-in. Local switching of DC power would be necessary to avoid shutting down the test if one device failed from high leakage. Individual chip I_{DDQ} could also be performed using the same individual wiring. As in most burn-ins the digital output would not be monitored during the testing, but the wafer would receive a room temperature I_{DDQ} and functional test after the full burn-in. However, the collective heat generated by a wafer of microprocessors might sometimes prohibit this interconnect overlay method, and robust power buses would be necessary to supply the DC voltage. (A cooling chamber may be needed to perform wafer burn-in instead of the usual oven.)

The second approach to accomplish burn-in uses extensive testing of packaged parts over a history of a long production run (memories) where it has been shown that less than 0.1% of the parts ever fail burn-in. This may, in turn, indicate that no burn-in need be performed given the small cost of the occasional rework. Another variation of this approach would perform burn-in on a small sample of packaged chips from each wafer once it has been established for a particular design and wafer fab that burn-in is consistent across each wafer. An extension of this approach is to design sensitive burn-in structures on test keys of each wafer (or on each chip) that can be quickly tested in a "wafer level reliability" sense. For example, if a particular design/fab had experienced burn-in failures due only to electromigration, test structures could be designed that were orders of magnitude more susceptible to electromigration than the product die so that a quick test could qualify all the chips on the wafer (baring a random gross defect). I_{DDQ} testing is highly recommended before and after burn-in. A reliability study by Ford Microelectronic Corp. showed that although I_{DDQ} screening did not eliminate the need for burn-in, it did reduce the number of die that failed burn-in by about 50%.[9] A post burn-in I_{DDQ} test is necessary to detect certain burn-in activated defects such as gate oxide shorts.[10]

Temporary Packaging   If manufacturers do not rapidly enhance their wafer testing, it will be necessary to perform the testing at the packaged chip level. One approach temporarily packages the chip for extensive testing, and then removes the package to leave the naked chip (or the chip on a carrier).

The temporary package method has depended on the testing infra-structure of each particular company. For example, facilities with hermetic military specified packages can perform temporary packaging by simply changing their die attach material and their wire bonder parameters.[11,12] No capital equipment or trained personnel need be changed to accomplish KGDs. The die attach material is changed to one that can be easily removed by solvent or elevated temperature. Wire bonding options include (1) aluminum ultrasonic wire bonding adjusted to slight deformation for easy mechanical removal, (2) thermosonic copper wire for easy chemical removal, or (3) electroless nickel plating on the bond pads so that chemical removal of the nickel will remove any temporary bond wire. Temporary lids are either affixed with Kapton tape or designed to snap in place depending on the package design or volume of parts. Such temporary packaging to date has shown only a 2% drop-out due to the extra handling of the ICs. Since the package and lids can be reused, the incremental cost for the temporary assembly and disassembly is typically a few dollars in a small volume semi-automatic line. This is acceptable for high cost chips and systems, but not for consumer goods.

Facilities with flip chip capability have used temporary packaging for some years to pretest their chips. Once a robust solder bumping process has been developed, the ability to rework or temporarily package an IC comes automatically. The IC can be removed with a hot forming gas blade and reapplication of any part can occur at least 10 times. Typically a ceramic carrier is designed for pretesting in which each IC bond pad pattern is compatible with the tester handlers. The yield on this extra handling is essentially 100% and the cost is less than one-third of that of the hermetic temporary packaging.

A variation or extension of this process has been demonstrated by Sandia National Laboratories in which standard ICs (intended for wire bonding) have their bond pads repositioned and converted to flip chip solder bumps (Fig. 2-2). This process is useful to system houses that use chips from many manufacturers for their modules since unifying the format to flip chip gains both easier pretesting and closer chip packing. In addition, some standardization in IC pad patterns and a decrease in needed drive current for MCM use are design advantages.

HDI Process   The GE HDI process for wafer level burn-in can provide temporary packaging for pretesting.[13] The metallization and polymer film can be easily removed after testing leaving only a small titanium square on

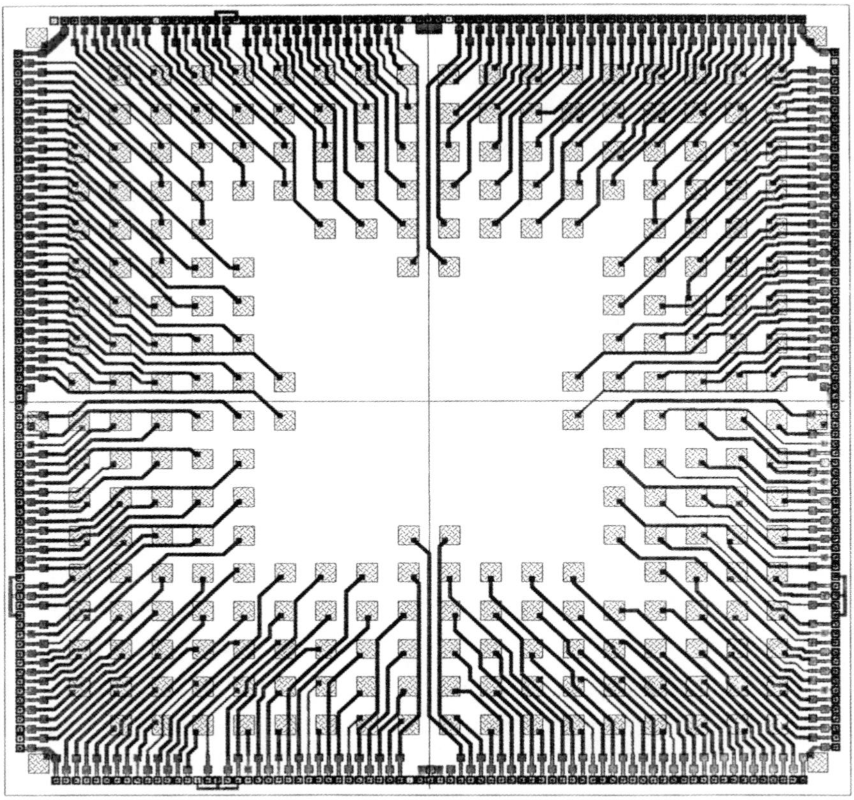

Figure 2-2   Repatterned Xilinx XC4010 Die. Additive polyimide/metal layers interconnect original bond pads to an area array of larger pads (.015″ pads on .030″ pitch) which are compatible with test/burn-in sockets.

each bond pad. If a company is already using the HDI process, pretesting is achieved with only a small incremental expense.

   Tape Automated Bonding (TAB)   For 20 years the advantage of pretest (including burn-in) has been touted for TAB parts. In fact only a small percentage of TAB parts see more testing than wafer probed parts given that they are used in low cost commercial product. However, MCMs have again raised this possible advantage.[14] To utilize this strength, the TAB part must be made of electrically separated frames so that individual DC power and digital signals can be monitored. In addition, alloy, plating (thick gold), and solder materials must be chosen so that the burn-in environment will not degrade them to the point that yields suffer during assembly. These custom frames can be expensive and a designer is limited to a subset of ICs that are available in TAB format. Today, most TAB technologies do not qualify for pretesting, but a company proficient in TAB could adapt its technology without undue hardship.

Elastomer   Most chips have bond pad pitches of 6-8 mils and pads at least 3 mil square. Thus a z-axis elastomer with a pitch of 1-2 mil could be used to make electrical contact temporarily with a chip carrier for testing. Such an elastomer is not readily available, but several companies are making product in the 4 mil range and a polyimide film with conductive through-buttons on 1-2 mil spacing has recently been announced.[15] A reusable carrier with controlled impedance leads and ability to survive burn-in must be designed to be compatible with the new compliant fine pitch film to provide pretesting.

Generalities   All of these pretest methods depend on a "hermetic" chip. That is, a very good ion and moisture barrier, such as defect-free silicon nitride, is needed over all the active silicon, and a barrier metal is needed on top of the bond pads. Thus any incidental moisture or ion contamination that occurs during temporary assembly and testing will remain on the surface and can be easily cleaned off just before chip commitment to the module.

Extensive visual inspection on die can often eliminate high risk ICs, particularly if previous failures have been traced to visually screenable defects. Although always important, minimization of electrostatic discharge has more reward in the COB eivironment than the individually packaged part since COB rework is more expensive than single package throw-away.[16,17] In addition, in order to double the speed of many ICs, their ESD protection may be removed and instead reside at the I/O for the PWB.

Not all components need pretest. Much depends on the number of ICs on the board design, the known reliability of the ICs, the ease of rework of the specific COB technology, and the size of the production run. Several COB/MCM vendors have pointed out that for modules below 10 IC count, exhaustive pretesting is cost effective only on chips with known large drop-outs (greater than 10%).[18] Memory chips from a frozen production line with known low burn-in drop-out are often used right out of the waffle pack.

2.3 COB DESIGN FOR TEST AND DIAGNOSTICS

The large numbers of potential hard-to-probe defects force COB design both to anticipate and to structure testing and diagnostic methodologies beyond that practiced by PWB or SMT designs. Several major issues must be integrated in the pre-design phase. These are

1. *Diagnostics*: Recognition that accurate diagnostics must be designed into the COB and this requirement overrides that of the design and manufacturing specifications.
2. *Design for Testability (DFT)*: DFT tools must be used strategically; the test and diagnostic requirements of each hierarchical level must be specified before any design begins.
3. *Defect Analysis*: An analysis must be made for all potential defects that might occur throughout the whole COB assembly and a test strategy designed for each defect type.
4. *Concurrent Engineering*: The COB will be developed by people organized in a tight concurrent engineering style; test and diagnostic considerations are concurrently addressed.

Enforcement of all these requirements will not ensure total success, but without them manufacturers will see serious delays if not prevention of the COB product reaching market. Each of the above topics is discussed below in more detail.

All test and diagnostic processes generate two types of costly errors: (1) Type I errors in which good product is misdiagnosed and then classified as defective, and (2) Type II errors in which defective product is classified as good.[19,20] Type I errors have existed for a long time without a systematic analysis of their effects on quality and cost. Both Type I and Type II errors are prevalent in COB test and diagnostics.

2.3.1 Diagnostics

All designs will have errors and all manufactured products will have defects; the common link is that both must be accurately diagnosed in a short time period. Accurate rework demands accurate diagnostics, which is something not reported for present conventional IC, board, and SMT products. The acronyms NTF, NFF, and RETOK are familiar diagnostic Type I error terms that manufacturing engineers use for no trouble found, no fault found or retest okay.[21] Data from test diagnostics at the system, board, and component levels indicate that faulty diagnosis (Type I error) can occur at levels from 15 to 70%.[22-24] A five-year study at one board and system manufacturer reported an average false diagnosis of 45%.[22] It is not unusual for manufacturing test engineers to assume that a failed test is really an error on the part of the test process; this is diagnostics out of control. These numbers, taken from ICs, boards, and systems, were from parts in which good electromechanical access was available.

COB Type I diagnosis error rates may be worse since electromechanical access for test probes is denied, product complexity is significant, and higher clock frequencies are considered. This poor level of diagnostic performance for present and past technology electronic products sends a serious warning to COB manufacturers since COBs have virtually no electromechanical access as do PWBs and even SMT products. The cost of rework is considerably higher with COBs so that diagnostics must be carefully considered at the beginning of the design. The tools for diagnosis are often those of testability that use design for test structures. In addition, diagnostics demands software that can integrate information from a variety of techniques.

2.3.2 Design for Testability Tools

Specifications for test and diagnostics begin at the highest hierarchical level. Several structural test and diagnostic techniques exist to aid electronic circuits and these should all be considered and coordinated with the various hierarchies in the COB/MCM design. Many of these structures have been used since the early 1970s and are usually lumped under the title design for testability (DFT) circuits.[25] DFT usually implies that special test and diagnostic circuits are added to the design to assist the controllability and observability of the circuit. These were developed initially for ICs, but DFT tools now exist (or are in a planning stage) for board and system level electronics. This section will only introduce these topics, put them in perspective for COB test and diagnostic issues, and provide references. DFT techniques at IC, board, and system levels are described with their relation to equivalent COB hierarchical levels. Typically, problems of test and diagnosis at higher technical levels of the COB are solved by DFT structures designed in at lower hierarchical levels. There is a strong need for communication over the technical spectrum of the product.

The two major IC DFT tools are circuit scan and built-in self-test (BIST). Power supply quiescent current monitoring (I_{DDQ}) is not a DFT structure, but is the most sensitive technique for IC defect detection and diagnosis.[6,7] I_{DDQ} testing implemented as an on-circuit feature is very relevant to COB/MCM test and diagnosis and is discussed later. IC circuit scan and BIST tools are relatively mature with many applications reported on their use.[26-28] Importantly, most ASIC suppliers now provide design software to implement circuit scan tools. There are DFT structures implemented on ICs that have little to do with IC test, but serve test and diagnostics at higher levels of assembly. An example is the boundary scan (BSCAN) methodology that places test storage cells at the I/O nodes of the IC.[1-3] These BSCAN cells allow electronic testing of the interconnects between components replacing the traditional electromechanical bed-of-nails probe testing. ASIC suppliers typically do not provide software design tools to assist with BIST implementation.

Each DFT tool is described with emphasis on its use in COB/MCMs. We stress that all DFT tools should be evaluated for applicability in the total MCM design. Each test tool provides a unique opportunity to assist the difficult diagnostic process once the COB/MCM is assembled. All DFT test tools add circuit overhead and may degrade MCM performance specifications so that trade-off decisions are necessary. However, it should be emphasized that without DFT assistance, diagnosis and test may be virtually impossible. If a COB cannot be tested or diagnosed quickly and accurately, then it should not be manufactured.

Circuit scan is an IC test structure that allows all sequential elements (flip-flops) in the design to be linked in a serial chain during test, but to assume a normal configuration during system operation.[26] It is one of the oldest IC DFT techniques. Figure 2-3 shows a general circuitry in both (a) the normal

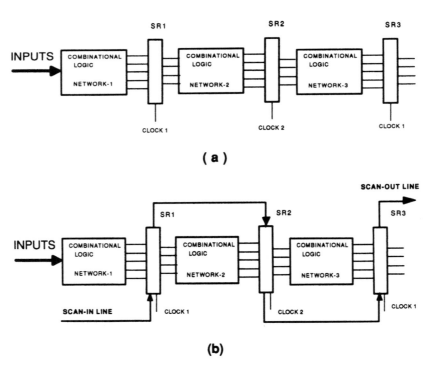

Figure 2-3 Circuit scan (a) normal mode, (b) scan test mode.

mode and (b) the test mode. Four control pins are required to control these test circuits. In test mode, test vectors are shifted serially into the scan chain where virtually all nodes in the circuit can be electrically controlled. The signal values on output nodes of the combinational subcircuits can then be observed by using the same scan chain to capture and shift out the results. Circuit scan allows the normally difficult to test sequential circuit elements to be tested by shifting of logic ones and zeros through the scan chain. The combinational logic blocks can have test vectors generated independent of the sequential cells, allowing improved test pattern generation.

Circuit scan has been popular since the late 1970s and it is now available as a design tool through most ASIC suppliers. The scan cell subcircuits are predesigned as standard cells, routing is automatic, and test vectors are automatically generated. The test vectors typically provide logic zero and one control of all logic gate input-output nodes and provide observability at the designated subcircuit output nodes. Circuit scan silicon overhead on the die is reported to range from 3% to as high as 30%. The variability is a function of the use of existing normal flip-flops in the design and how they can be coupled as scan flip-flops. Designer experience is often cited as a major factor in efficient scan circuit design.

Experience with scan designed ICs has consistently shown that the design verification phase is reduced by factors of two to four which is a consideration in the economics of getting market share with shorter time to market. The reduction in time of the design verification phase is attributed to the ability of scan to debug the circuit much more efficiently. Circuit scan can be used at the MCM stage to interrogate electronically the internal circuitry of individual ICs, thus allowing more accurate diagnostics. These strong reasons for including scan circuitry put the DFT decision process in a position to ask, "Why shouldn't circuit scan be used on all ICs used in the COB/MCM?" A few designs do not have circuit scan, such as memories and certain commercial ICs. However, all ASIC designs present an opportunity to incorporate test and diagnostic intelligence that will be used at a higher level of assembly.

Built-In Self-Test (BIST) is a DFT technique that designs intelligence into the IC so that, under external control, the IC can generate test stimuli and evaluate the responses for verification of the circuit functionality.[27-29] Test vectors can be generated from ROM stored control signals, but typically it is more convenient and efficient to use linear feedback shift registers (LFSRs) to generate pseudo-random test vectors. An LFSR circuit typically generates a set of output signals encompassing the truth table values for a set of input nodes. The digital responses resulting from the LFSR pseudo-random vectors are compacted in a single register using a check sum technique. This single register result provides a signature that is unique for good ICs. The LFSR circuits can be clocked at the maximum operating

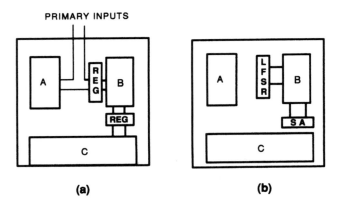

Figure 2-4 BIST using an LFSR circuit (a) normal mode, (b) circuit in test mode.[29]

frequency of the circuit so that BIST is a unique tool to test for timing defects. Figure 2-4*a* shows a generalized IC partition that has one of its storage registers reconfigured in test mode in Fig. 2-4*b* to function as an LFSR.[29] Circuit block B is stimulated by the LFSR patterns while the digital responses are evaluated by a signature register (SA). While BIST consumes overhead just as circuit scan, designers often use normal registers in a dual purpose design. Control circuitry is added around the registers to convert them to LFSRs during test mode.

When BIST exists on an IC, then a goal is to design test control circuitry at the COB/MCM level so that the BIST can be activated and the responses read out at MCM I/O signal nodes. BIST is an *ad hoc* technique in that its application must fit into existing portions of the IC subcircuits. Total BIST evaluation of an IC is not typically achieved so that the MCM product team must realize that BIST analysis of an IC usually refers only to portions of the IC. It may be desirable for the MCM team to consider a higher level BIST mode that drives and evaluates clusters of ICs on the MCM.

BIST is a unique at-speed DFT tool that allows access to certain portions of the MCM circuitry, however BIST limitations should be understood. IC defects such as certain bridges, memory faults, and delay faults may escape an at-speed BIST test even though the circuit experiences the truth table vectors. BIST will detect many of these defect types by chance, but will also by chance miss a significant number. A discussion of CMOS defects and test escapes can be found in references 7 and 30.

I_{DDQ} monitoring is a sensitive test and diagnosis technique using the property that the quiescent power supply current (I_{DDQ}) of a CMOS IC is typically in the low nA range.[6,7] Most CMOS defects cause I_{DDQ} elevations of several orders of magnitude that are easily detectable by off-chip monitoring. Figure 2-5 illustrates how I_{DDQ} is elevated for a defect in a

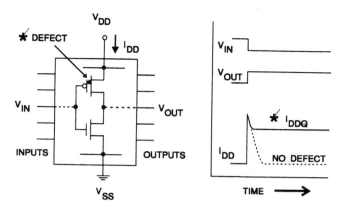

Figure 2-5   CMOS defects and I_{DDQ} elevation.[7]

CMOS inverter. If a source to gate bridging defect exists in the p-channel transistor, then elevated I_{DDQ} exists when the logic input signal is pulled low. The increase in I_{DDQ} over its normal value is typically three to seven orders of magnitude representing a clear, detectable signal. I_{DDQ} monitoring has been practiced by a few companies since the mid-1970s for sensitive detection of CMOS defects in order to achieve higher quality ICs than by conventional voltage sensing methods.

Defect detection with voltage sensing requires signal control of a defective node as well as subsequent observation of the signal at a primary output. However, I_{DDQ} testing requires only control of circuit nodes without the subsequent observation at a primary output node, therefore test vector requirements are greatly reduced. The observation is done through the current in the V_{DD} pin of the IC. Comparisons of defect detection with current versus voltage monitoring report vector count reductions for I_{DDQ} monitoring on the order of 1% or less.[31-33] Figure 2-6 shows a general IC schematic illustrating how defects can be detected without need to observe signals at the primary outputs (POs) of the IC. If the 2-NAND gate output was bridged to V_{DD}, then the test vector would only be required to set the 11 logic condition at the two inputs of the gate. The defect would be detected by the large increase in I_{DDQ} and no requirement exists for sensitizing a logic path between the defect and a primary output node. This feature simplifies the generation and reduces the number of test patterns.

Recent data on I_{DDQ} monitoring have shown its superior ability over conventional techniques used to test and diagnose ICs.[30-33] Aitken reported the efficient combined use of voltage and I_{DDQ} monitoring for defect diagnostics.[34] I_{DDQ} is particularly sensitive to many forms of electrical overstress or electrostatic damage (EOS/ESD) that are prevalent in

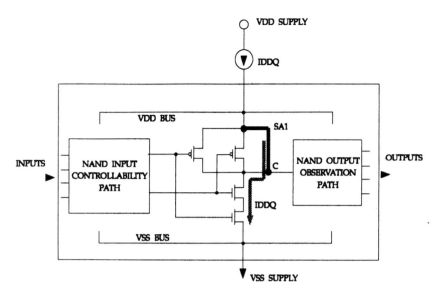

Figure 2-6 Controllability and observability with I_{DDQ} testing

assembly operations;[10] these EOS/ESD defects are often reliability hazards and can initially escape Boolean oriented testing.

I_{DDQ} monitoring has not yet been used for test and diagnostic purposes above the IC level, but is being considered as a tool for COBs because of its sensitive ability to detect and locate defects with small numbers of test vectors. COBs present problems for I_{DDQ} monitoring since DC background currents may be present due to pull up resistors (boundary scan) or because the IC designers paid insufficient attention to the advantages of designing CMOS ICs in a low current, fully static, fully complementary mode. The use of I_{DDQ} at higher levels of assembly therefore require special I_{DDQ} monitoring circuits that are placed in the MCM as individual, dedicated ICs or partitioned on the ASICs themselves. These built-in current (BIC) monitors must be activated by test control circuits to perform I_{DDQ} monitoring for designated vectors. Defect detection occurs from a signal generated by the BICs. Research on BICs has been reported since 1988 with emphasis on IC testing.[35,36] This novel test circuit is seriously considered despite its apparent complexity because of the lack of good alternatives.

Two final important properties of I_{DDQ} must be noted for their relevance to MCMs. The first is that a study of the effect of I_{DDQ} screening for IC burn-in was done and showed that while I_{DDQ} testing did not eliminate the requirement for burn-in, it did reduce the number of faulty ICs sent to burn-in by about 50%.[9] The second effect relates I_{DDQ} leakages and delay faults. Generically, any defect leakage into or out of a logic signal node will degrade either the rise or fall time of an IC. Elevated leakage into a node increases the signal fall time (increase signal rise time) and any leakage out of a node increases signal rise time (decrease signal fall time).

All of the observations discussed here emphasize that I_{DDQ} is probably the single most powerful test technique to employ on ICs prior to die mounting on substrates. I_{DDQ} testing alone is not sufficient for zero defect screening and the application of voltage sensing test techniques should be carefully targeted for certain open circuit and delay fault defects.[31] The careless design of CMOS ICs having unnecessary designed-in high current states must be avoided. High performance IC processor design for I_{DDQ} testability has been shown for a 100 MHz clock rate for a circuit with one million transistors.[37]

Board Level (MCM substrate) DFT tools exist that dominantly emphasize boundary scan (BSCAN, also known as the IEEE 1149.1 standard). BSCAN places boundary scan register cells at the signal and clock I/O pads on components placed on a board or MCM substrate.[1-3] These cells designed into the ICs are linked in a serial network during test mode so that test vector stimuli and results can be transported from off-board into difficult to access electrical nodes. Since virtually all MCM nodes are mechanically inaccessible, this test communication BSCAN bus can potentially reach all of the MCM nodes. The IEEE 1149.1 standard allows a test mode for : (1) testing chip-to-chip interconnections, (2) individual chip testing, (3) initiation and evaluation of chip BIST, (4) a bypass register mode that shortens the serial boundary scan path, (5) identification of individual chip JEDEC numbers upon interrogation, and (6) general user-defined other test instructions. Boundary scan requires four test pins that are designated collectively as the test access port (TAP); these are: the test data input (TDI), test data output (TDO), test mode select (TMS) and test clock (TCK) with an optional fifth pin for test reset (TRST). BSCAN requires a control circuit called the TAP controller that is part of the IC. BSCAN is implemented at the IC level to support higher test and diagnostics at the next level of assembly. BSCAN is seldom used to test the IC itself during IC production, but is directed toward accurate diagnostics of ICs when they are mounted at the board or MCM substrate level.

The boundary scan tool is a virtual necessity for MCMs of any complexity since identification of which IC is causing an MCM logic fault is very difficult with the lack of electromechanical access. BSCAN circuits are readily designed through most ASIC suppliers as standard cell implementations. Various pad register cell designs, instruction register cells, and the test access port (TAP) controller are available as circuit design menu items and can easily be added to most ASICs. The TAP controller is an approximate equivalent eighty 2-NAND subcircuit located on every IC. The TAP controller converts the serial test control information from the

TMS line to a test or diagnostic action. ASICs offer a unique opportunity to add test and diagnostic intelligence to the MCM in contrast to chips that are necessary, but may have no DFT such as commercial microprocessors, microcontrollers, memories, and glue chips. Intelligent ASICs provide a minimum capability to augment testing of the non-DFT chips.

It is desirable to have all die on the MCM provided with boundary scan, but the reality is that this will not be possible for a few years. One alternative is to add boundary scan capability to several locations in the MCM in the form of glue chips that have this capability. The SCOAP octal product family of ICs from Texas Instruments are bus line buffers having full boundary scan on each die.[38] These octal chips can be placed at strategic locations to improve the boundary scan communication within the MCM. The background currents on these ICs are in the tens of mA and therefore thermal and I_{DDQ} test considerations must be made.

The four boundary scan TAP signals are driven usually by an external tester, but in some products, it is efficient to consider using an intermediate control element to coordinate clusters of boards or modules. Inexpensive, special purpose ICs called boundary scan masters (BSMs) have been designed to perform these tasks.[39,40] BSMs are a consideration in MCMs since they can automatically generate and evaluate test patterns applied at a regional level. BSMs can convert parallel test vectors to a serial form required for the boundary scan signals.

Another alternative for providing BSCAN capability in the absence of BSCAN die is to build BSCAN into the substrate. This concept is known as the smart substrate design.[41] Those using silicon as a substrate have an option of placing some of these added values within the substrate. Of course, putting devices in the substrate multiplies the cost of the substrate by a factor of four, an extravagance not possible in commercial endeavors. There is a tension created by asking IC designers to add more features to a chip that are of greatest use in MCMs; for example, boundary scan, agile impedance, and DC power conditioning. Boundary scan within the substrate allows complete self-testing of the MCM substrate interconnections before chips have been committed to the unit. These same boundary scan cells can also be used to test the individual ICs. A substrate with a "sea of boundary scan cells" would only need one customized level of metal to accommodate a specific chip placement pattern.

There is a compromise position where smart chip carriers are used. With chip carriers, any substrate technology is possible including traditional laminates for COB. The carrier can also allow the handling for pretesting of the IC. In this way, commercial ICs can have MCM technology dictated features added for thermal, power input, signal intensity, and testability. Micromachining of the smart carrier such as thinning, vias, cooling channels, microconnectors, and optical applications can add further value.

For example, silicon carriers thinned to 5 mils add little thermal resistance and conductive vias can allow flip-chip attachment of carriers to boards.

Another IEEE test and diagnostic standard under development deals with mixed signal test problems. A working group was formed in 1991 under an IEEE P1149.4 standards designation that seeks to define an analog bus standard that would serve the same analog intent as the 1149.1 bus does for digital circuits.[5] Their mission statement is "to define, document, and promote the use of a standard mixed-signal test bus that can be used at the device and assembly levels to improve the controllability and observability of mixed-signal designs and to support mixed-signal built-in test structures in order to reduce test development time, testing costs, and improve test quality." P1149.4 seeks to define how an analog signal bus can be connected selectively to any internal node within the mixed-signal module. The intent of P1149.4 is compatibility with the other 1149.x standards such as 1149.1 and 1149.5. The problems of analog test standard definition are difficult, but some analog test structures have been proposed.[42] Analog ICs or mixed signal ICs on MCMs require parametric values to be measured, such as absolute voltages and currents, bandwidths, and noise levels. Automobile MCM applications require the ability to monitor continuously parametric drifts and diagnose accurately and rapidly the system electronics.[43]

System (MCM) DFT tools have not historically had the intense coordinated focus of ICs and boards. However, there is recent progress in an overall systems test and diagnostic standard referred to as the IEEE P1149.5 "Standard Backplane Module Test and Maintenance (MTM) Bus Protocol".[4] This methodology applies to complex systems (MCMs) in which the modules of the final product are sufficiently complex to warrant a higher test and diagnostic structure. The P1149.5 standard defines a hierarchy of test and maintenance buses so that test and diagnostic access extends to the system, subsystem, module (or board), and chip level. The standard is compatible with the 1149.1 boundary scan standard so that all tests defined by that structure can be performed at the systems level of control. Figure 2-7 is taken from the P1149.5 draft and shows the organizational goal of the standard.

At this point, there are few systems implemented with the P1149.5 systems test proposal, but alternative designs have been reported that have a common objective to link the BSCAN capability at the board (substrate) level to control at the backplane level. Bhavsar proposed a design for simple backplane systems[44] and Hilla described the linking of BSCAN on a system of 12 MCMs.[45]

General proposals for MCM testing have appeared. Posse outlined a proposal for MCM DFT features within modules and the architecture of tests for these modules[46] and Keezer discussed challenges and possible solutions to MCM testing.[47] Maierhofer described a design for hierarchical

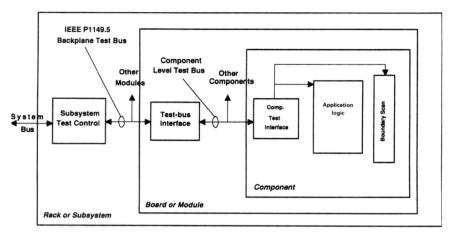

Figure 2-7 Hierarchy of test and maintenance buses in P1149.5 System Testability Standard.[4]

self-testing that applies to system, subsystem, modules, and components.[48] The concept is based upon a hierarchy of controllers that control the self-test at each level. Test buses transport test information between the hierarchical levels and the scheme is intended for 1149.1 boundary scan interface. Maunder proposed a structured or managed approach to using built-in self-test at a system level that would include to the level of field testing.[49] Zorian and Jarwala have provided information on the incorporation of built-in test for MCM testing.[50]

2.3.3 Defect Analysis and Test Strategies

Zero defects in electronic products requires knowledge of the real defects known to occur in the process assembly. Each hierarchical level creates different types of defects and defect distributions. The electrical properties of CMOS IC defects from fabrication are well known[7,30] and test processes must target these real defects with test methodologies matched to the defects.[51] CMOS defects are most often not detected by simply monitoring the logic zero and one at a primary output, but care must be taken of their effect of weakening node voltages or causing sequence-dependent failure mechanisms.[52,53] Recent data show that I_{DDQ} monitoring and at-speed testing of functional vectors are the most effective test methods for CMOS ICs.[31-33]

The dominant defects expected when components are mounted on an MCM substrate are related to the bonding process for the ICs (interconnection opens and bridges) and ESD damage. Interconnection defects can be difficult to detect. Bridging defects require a precise test vector that places the electrical nodes of the postulated bridge at opposite logic levels and then detects the defect either by voltage or current sensing methods. The 1149.1 boundary scan structures are designed to detect interconnect bridges when they cause low frequency logic upset. However, defect bridges can be resistive and not affect low frequency functionality, but rather act as timing or delay faults. This form of bridging defect may be detected by either high frequency tests or I_{DDQ} monitoring (where possible). Open circuit defects have variable parametric responses depending upon the size of the open and its local topological conditions.[54]

EOS/ESD are expected, and the test detection of that damage may be difficult. ESD takes a variety of effects such as gate oxide ruptures, lines opened, or lines fused. The I/O circuitry is the most common location for the handling damage to occur, but failure analysis reports have also found ESD deep in the inner core logic of ICs as the ESD energy is channeled down power rails.[55] When die are placed on MCM substrates, the likelihood of ESD is significant. The detection of that damage would ideally be done by a quiescent current measurement since EOS/ESD often escapes voltage sensing tests, but later causes functional failure due to progressive reliability degradation. One suggestion to monitor assembly damage on MCM substrates would use built-in current (BIC) monitors which would be specialized small die placed at locations in the substrate that bypass large DC background currents.

2.3.4 Concurrent Engineering

The organization of the people involved in design, test, diagnostics, assembly, and field servicing is crucial for success. Older, traditional styles used a sequential or "over the wall" organization in which it was assumed that all problems experienced at next levels of assembly were fixable. Often that was true for products in which all components were mechanically accessible. Resistors could be unsoldered or rewired with jumper wires. The total number of components on these systems could often allow a checklist procedure for diagnostics, and repair was a matter of disassembling the product, replacing the identified bad part, and then reassembling. Although this style seemed to work for those products, significant cost increases could be observed for engineering changes that were undertaken simply because there was no coordination between manufacturing and design.

A common complaint as systems became more complex was that insufficient coordination was taking place in the dynamics of all persons involved from product definition through design, test, assembly, field service, sales, and customer satisfaction. The successful total quality management (TQM) movements over the past 40 years have emphasized an integration of design and manufacturing.[56] This has recently evolved to the project management style called concurrent engineering.[57] Concurrent engineering has been defined as "A systematic approach to the integrated concurrent design of products and their related processes, including manufacturing and support. This approach is intended to cause developers, from the outset, to consider all elements of the product life cycle from conception through disposal, including quality, cost, schedule, and user requirements."–Institute for Defense Analysis (IDA) Report R-338, 1988.

Experience with the sequential or "over the wall" project development indicates that inattention to manufacturing details at the design phase leads to expensive engineering changes. MCMs are more difficult to redesign, particularly since the test and diagnostic solutions usually involve the design of special DFT structures at the IC level. Hence, test and diagnostics should be addressed in parallel with design.

The central issue is the commitment of a corporation to quality practices that have short-term costs, but long-term gain. Education of manufacturing personnel is a critical investment in higher quality return. Additional investments must be made in Type I error root cause identification and correction. A concurrent engineering team must devise designs that anticipate the problems of high density electronics and uniquely design-in rapid and accurate diagnostics. Design for testability must be coordinated at all levels: MCM system, MCM substrate (board), and component.[58]

ACKNOWLEDGMENT

This work performed by Sandia National Laboratories supported by the U.S. Department of Energy under contract DE-AC04-76DP00789.

REFERENCES

1. IEEE 1149.1 Std., IEEE Inc., 345 East 47th St., New York, NY 10017.

2. Maunder, C., and R. Tulloss, *The Test Access Port and Boundary Scan Architecture*, IEEE Computer Society Press, 1990.

3. Parker, K. P., *The Boundary Scan Handbook*, Kluwer Academic Pub., Norwell, MA, 1992.

4. IEEE P1149.5 Std. Draft, "Standard Backplane Module Test and Maintenance (MTM) Bus Protocol," Pat McHugh, US Army LABCOM, Ft. Monmouth, NJ.

5. IEEE P1149.4 Working Group on Mixed Signal Test Bus Standards, contact Prof. Mani Soma, Univ. of Washington, Dept. of Electrical Engr., Seattle, WA 98195, USA.

6. Hawkins, C. F., J. M. Soden, R. R. Fritzemeier, and L. K. Horning, "Quiescent Power Supply Current Measurement for CMOS IC Defect Detection," *IEEE Trans. Industrial Electronics*, pp. 211-218, May 1989.

7. Soden, J. M., C. F. Hawkins, R. K. Gulati, and W. Mao, "I_{DDQ} Testing: A Review," *J. of Electronic Test: Theory and Applications (JETTA)*, pp. 291-303, Vol. 3, Dec. 1992.

8. Orr, D., MCC, Austin TX, and M. Orr, of Intel, Chandler AZ, are main drivers for this effort.

9. McEuen, S. D., "Reliability Benefits of I_{DDQ}," *J. of Electronic Test: Theory and Applications (JETTA)*, pp. 328-335, Vol. 3, no. 4, Dec. 1992.

10. Hawkins, C. F., and J. M. Soden, "Reliability and Electrical Properties of Gate Oxide Shorts in CMOS ICs," *Int. Test Cont.*, pp. 443-451, Sept. 1986.

11. Chu, D., C. A. Reber, B. L. Draper, J. N. Sweet, and D. W. Palmer, "Multichip Module Enablers for High Reliability Applications," *Proceedings 1992 IEEE Multi-Chip Module Conference*, pp. 102-105, April 1992.

12. Chu, D., C. A. Reber, and D. W. Palmer, "Screening ICs on the Bare Chip Level: Temporary Packaging," 1992 *Proceedings of IEMT/IEEE* published September 1992, pp. 223-226.

13. Forman, G. A., J. A. Nieznanski, and J. Rose, "Die for MCMs: IC Preparation for Testing, Analysis, and Assembly," *Proceedings 1992 IEEE Multi-Chip Module Conference*, pp. 32-35, April 1992.

14. Newsnote in supplement to *Electronic Packaging & Production*, August 1992, p. 18.

15. Nakatsuka, Y., T. Okabayashi, T. Yoshizawa, and H. Kondo, "A feasibility study on Bonding Technique for a high-density Interconnection system," pp. 343-347, *Proceedings ISHM*, 1991.

16. Sekulic, Z., "The MCM Dilemma," *Advanced Packaging*, Summer 1992, pp. 42-45.

17. Scannell, R. K., "TAB Technology for Miniature Multiprocessor High Speed Interceptors," *ITAB 1992 Proceedings*, Feb. 1992.

18. Inpyn, B. "Overview of Surface Mount Technology," supplement to *Electronic Packaging & Production*, pp. 8-12, Aug. 1992.

19. Mood, A. M., *Introduction to the Theory of Statistics*, McGraw-Hill Book Co. Inc., 1950.

20. Williams, R. H., *Electrical Engineering Probability*, West Publishing Company, St. Paul, 1991.

21. Simpson, W. R., and J. W. Sheppard, "System Complexity and Integrated Diagnostics," *IEEE Design and Test of Computers*, pp. 16-30, Sept. 1991.

22. Henderson, C. L., R. H. Williams, and C. F. Hawkins, "Economic Impact of Type I Errors at System and Board Levels," *Proc. Int. Test. Conf.*, Sept. 1992.

23. Williams, R. H., R. G. Wagner, and C. F. Hawkins, "Testing Errors: Data and Calculations in an IC Manufacturing Process," *Int. Test Conf.*, Sept. 1992.

24. Williams, R. H., and C. F. Hawkins, "Errors in Testing," *Int. Test. Conf.*, pp. 1018-1027, Sept. 1990.

25. Williams, T. W., and K. P. Parker, "Design for Testability–A Survey," *Proc. IEEE*, Vol. 71, no. 1, pp. 98-112, Jan. 1984.

26. Eichelberger, E. B., and T. W. Williams, "A Logic Design Structure for LSI Testability," *Proc. 14th Des. Auto. Conf.*, pp. 462-468, June 1977.

27. McCluskey, E. J., *Logic Design Principles with Emphasis on Testable Semicustom Circuits*, Prentice-Hall, Englewood Cliffs, NJ, 1986.

28. Daniels, R. G., and W. B. Bruce, "Built-in Self-test Trends in Motorola Microprocessors," *IEEE Design and Test of Computers*, pp. 64-71, April 1985.

29. Konnemann, B., J. Mucha, and G. Zwiehof, "Built-In Logic Block Techniques," *Int. Test Conf.*, pp. 37-41, 1979.

30. Soden, J. M., and C. F. Hawkins, "Electrical Properties and Detection Methods for CMOS IC Defects," *1st European Test Conf.*, pp. 159-167, April 1989.

31. Maxwell, P. M., R. Aitken, V. Johansen, and I. Chiang, *J. of Electronic Test: Theory and Applications (JETTA)*, Vol. 3, pp. 305-316, Dec. 1992.

32. Perry, R. J., "I_{DDQ} Testing in CMOS Digital ASICs," *J. of Electronics Test: Theory and Applications (JETTA)*, pp. 317-325, Vol. 3, Dec. 1992.

33. Storey, T., W. Maly, J. Andrews, and M. Miske, "Stuck Fault and Current Test Comparison Using CMOS Chip Test," *Int. Test Conf.*, pp. 311-318, Sept. 1991.

34. Aitken, R., "Diagnosis of Leakage Faults with I_{DDQ}', *J. of Electronic Test: Theory and Applications (JETTA)*, pp. 367-375, Vol. 3, Dec. 1992.

35. Feltham, D., P. Nigh, L. R. Carley, and W. Maly, "Current Sensing for Built-in Testing of CMOS Circuits," *Int. Conf. on Computer Des.*, pp. 454-457, Rye Brook, NY, Oct. 1988.

36. Rius, J., and J. Figueras, "Proportional Testing for Current Testing," *J. of Electronic Test: Theory and Applications (JETTA)*, pp. 387-396, Vol. 3, Dec. 1992.

37. Josephson, D. D., D. J. Dixon, and B. J. Arnold, "Test Features of the HP PA7100LC Processor," *Int. Test Conf.*, pp. 764-772, Oct. 1993.

38. *Testability Primer*, Texas Instruments Inc., Semiconductor Group, 1991.

39. Vining, S., "Tradeoff Decisions Made for a P1149.1 Controller Design," *Int. Test Conf.*, pp. 47-54., Aug. 1989.

40. Yau, C. W., and N. Jarwala, "The Boundary-Scan Master: Target Applications and Functional Requirements," *Int. Test Conf.*, pp. 311-315, Sept. 1990.

41. Treece, R. K., "A Strategy for Obtaining KGD for MCMs," *Application Solutions for KGD Symposium*, Dec. 1992, sponsored by DARPA and MCC.

42. Parker, K., S. Orejo, and J. E. McDermid, "Structure and Methodology for an Analog Testability Bus," *Int. Test Conf.*, pp. 309-322, Oct. 1993.

43. Thatcher, C. W., and R. E. Tulloss, "Testing Mixed-Signal Interconnects in Boards & Systems," *Int. Test Conf.*, pp. 300-308, Oct. 1993.

44. Bhavsar, D., "An Architecture for Extending the IEEE Standard 1149.1 Test Access Port to System Backplanes," *Int. Test Conf.*, pp. 768-776, Oct. 1991.

45. Hilla, S. C., "Boundary Scan Testing for Multichip Modules," *Int. Test Conf.*, pp. 224-230, Sept. 1992.

46. Posse, K. E., "A Design-for-Testability Architecture for Multichip Modules," *Int. Test Conf.*, pp. 113-121, Oct. 1991.

47. Keezer, D. C., "MCM Test Using Available Technology," *Int. Test Conf.*, p. 253, Oct. 1992.

48. Maierhofer, J., "Hierarchical Self-Test Concept Based on the JTAG Standard," *Int. Test Conf.*, pp. 127-132, Sept. 1990.

49. Maunder, C. M., "A Universal Framework for Managed Built-In Test," *Int. Test Conf.*, pp. 21-29, Oct. 1993.

50. Zorian, Y., and N. Jarwala, Tutorial on MCM Testing given at *Int. Test Conf.*, Oct. 1993.

51. Soden, J. M., R. R. Fritzemeier, and C. F. Hawkins, "Zero Defects or Zero Stuck-At Faults–CMOS IC Process Improvement with I_{DDQ}," *Int. Test Conf.*, pp. 255-256, Sept. 1990.

52. Wadsack, R. L., "Fault Modeling and Logic Simulation of CMOS and MOS Integrated Circuits," *Bell Systems Tech. J.*, Vol. 57, no. 5, pp. 1449-1488, May-June 1978.

53. Soden, J. M., R. K. Treece, M. R. Taylor, and C. F. Hawkins, "CMOS IC Stuck-Open Fault Electrical Effects and Design Considerations," *Int. Test Conf.*, pp. 423-430, Aug. 1989.

54. Henderson, C. L., J. M. Soden, and C. F. Hawkins, "The Behavior and Testing Implications of CMOS IC Open Circuits," *Int. Test Conf.*, pp. 302-310, Oct. 1991.

55. Personal Comm. J. M. Soden, Sandia Failure Analysis Dept., Oct. 1989.

56. Dobyns, L., and C. Crawford-Mason, *Quality or Else*, Houghton Mifflin Co., 1991.

57. Shina, S. G., *Concurrent Engineering and Design for Manufacturing of Electronic Products*, Van Nostrand Reinhold, New York, 1991.

58. Hawkins, C. F., "Systems Testing: The Future for All of Us," *Int. Test Conf.*, p. 548, Sept. 1992.

3

Chip Level Interconnect: Wire Bonding for Multichip Modules

Harry K. Charles, Jr.

3.1 INTRODUCTION

Complex, high density integrated circuits have placed major demands on packaging structures and electronic interconnects. As clock speeds exceed 100 MHz and rise times fall below 100 ps, the need for short interconnect length and properly terminated transmission lines, which preserve pulse shape and minimize propagation delay, is paramount. The increased number of inputs and outputs (in some cases, several hundred) of the integrated circuits, coupled with the exponential rise in chip density[1] and exponential decline in device feature size,[2,3] have produced significant increases in packaging and interconnect density and the need for high conductivity, low capacitance (and inductance) interconnects.

Currently, there are several packaging methods which can accommodate the increased density and complexity of today's modern systems while providing the high electrical and thermal performance necessary for advanced integrated circuits. While some fully integrated techniques such as wafer scale integration (see Section 3.2.5 below) can support the required input/output (I/O) density and electrical performance requirements, multichip modules are the mainstay of modern, high density packaging. Multichip modules[4,5] combine several high performance silicon or gallium arsenide integrated circuits with a custom designed substrate structure. These complex "hybrid" substrates are fabricated from various materials (e.g., ceramics, semiconductors, organic materials, and sometimes metals) using several different technologies including thin film, cofired, and/or printed wiring board methods. The various multichip module types (C, D, and L) and their salient characteristics are listed in Table 3-1.

Several chip interconnection methods are being used to meet the performance and high density interconnect requirements of multichip modules. These methods include wafer scale interconnection, laser pantography, wire bonding, tape automated bonding, and flip chip (or inverted reflow). Of all the techniques used for multichip modules to date, wire bonding is by far the most common interconnect method despite its performance, reliability, and density limitations as described below.

3.2 WIRE BONDING PRINCIPLES

As mentioned above, wire bonding is the most widely used technique for electrical chip interconnection in the microelectronics industry.[6] The estimated number of wire bonds fabricated per year exceeds 10^{12} according to Harman.[7] Historically, wire bonding has been used with all chip and package styles ranging from the first transistors in single chip packages during the late 1950s[8] to today's large, high density integrated circuits packaged in multichip modules.[9]

To begin the wire bonding process, the back side of the transistor or integrated circuit chip must be firmly attached to the appropriate substrate location or package bottom using organic adhesives, low melting point glasses, or a metal-alloy reflow process.[10] Suitably attached chips are shown schematically in Figure 3-1. The wires are then bonded or welded, one at a time between the chip bonding pads and the appropriate package or substrate interconnection points. Similar bonds can be made between interconnection points on the substrate or between the substrate and the package. This interconnect process uses a special tool (wedge or capillary)

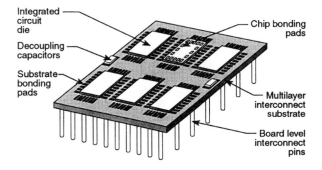

Figure 3-1   Generic multichip module structure without wire bonds.

Table 3-1 Multichip Module Parameters

Parameters	MCM-C		MCM-D		MCM-L
	Cofired Ceramic	Low-K Ceramic (Thick Film)	Silicon-on-Silicon	Low-K (Organic Dielectrics on Ceramic, Metal, etc.)	Printed Wiring Boards
Line density (cm/cm²/layer)	20	40	400	200	30
Line width/gap (µm)	125/(125–375)	125/(125–375)	10/(10–30)	15/(35–75)	750/(750–2250)
Dielectric constant	9	5	3.6–4.0	2.4–4.0	3.5–5.0
Maximum substrate area (cm²)	225	225	100	100	500+
I/O (×1000) Peripheral	1.6–6.4	1.6–6.4	0.8–3.2	0.8–3.2	1.6–3.2
Substrate costs ($/layer/cm²)	15–30	5–10	30–50	10–30	0.10–0.15
Terminating resistors	Built-in (acceptable)	Built-in or Surface Mount	Built-in Difficult	Surface Mount	Surface Mount
Decoupling capacitance	Built-in Surface Mount	Built-in Surface Mount	Built-in	Built-in Surface Mount	Surface Mount
Transmission lines	Stripline µstrip	Stripline µstrip	Stripline µstrip 50 Ω	Stripline µstrip 50 Ω	Stripline µstrip

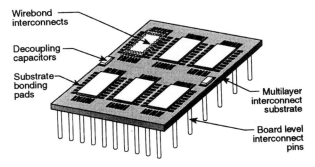

Figure 3-2   Generic multichip module structure with wire bonds added.

and a combination of heat, pressure, and/or ultrasonic energy to effect the bond. Figure 3-2 illustrates the circuit of Figure 3-1 with the wires in place.

Depending on the tool type and the choice of energy source (heat, ultrasonic energy, etc.), three major wire bonding methods can be defined: thermocompression bonding, ultrasonic bonding, and thermosonic bonding. Thermocompression and thermosonic bonding typically use a round capillary and produce a ball-wedge bond (first bond-second bond), where the wedge or second bond lies on an arc about the first or ball bond. Top and side views of a ball-wedge bond are shown schematically in Figure 3.3.

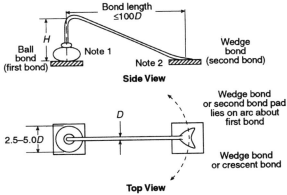

Note 1   Ball shape dependent on capillary and whether thermocompression or thermosonic bonding process was used

Note 2   Wedge shape somewhat dependent on capillary type and shape and whether thermocompression or thermosonic bonding process was used

H = Bond height or loop height

D = Wire diameter

Figure 3-3   Side and top schematic views of ball-wedge bonds. Schematic views applicable to both thermocompression and thermosonic wire bonding processes.

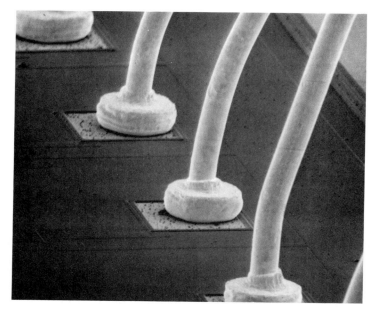

Figure 3-4   Photomicrograph of 25.4 μm diameter gold wire thermosonically bonded to an integrated circuit chip.

Figure 3-4 is a photomicrograph of several thermosonic (ball-wedge) bonds on an integrated circuit. Ultrasonic bonding uses a rectangular wedge and produces a wedge-wedge bond (first bond-second bond), where the second bond lies on the centerline of the first bond as shown schematically in Figure 3-5. Figure 3-6 is a photomicrograph of ultrasonic bonds on an integrated circuit chip.

3.2.1 Thermocompression Bonding

Thermocompression bonding results when two metal surfaces (e.g., bonding wire and a metallic integrated circuit bonding pad) are brought into intimate contact during a controlled temperature-pressure-time cycle. During this cycle, the wire and the underlying metallization undergo plastic deformation and atomic interdiffusion. This interdiffusion can result in a uniform gold-gold weld, if both a gold wire and a gold pad (or substrate metallization) are used. If a gold wire and an aluminum pad (or vice versa) are used, the result and wire pad interface is composed of aluminum-gold intermetallics.[11] The plastic deformation at the bonding interface ensures intimate surface contact, an increase in bonding area, and a breakdown of thin interfacial film layers. Surface roughness (and voids), oxide formation (or aluminum

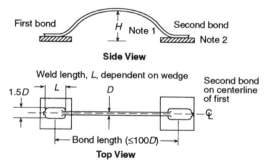

Side View

Top View

Note 1 Loop height, H, is generally lower than in the ball–wedge processes

Note 2 Flatness of weld dependent on wedge shape and ultrasonic bonding parameters

H = Bond height

D = Wire diameter

Figure 3-5 Side and top schematic views of ultrasonic wedge-wedge bonds.

Figure 3-6 Photomicrograph of 25.4 µm diameter aluminum (1% silicon) wire ultrasonically bonded to an integrated circuit chip.

pads), and absorbed chemical species or moisture layers can prevent the necessary metal-to-metal contact, thus totally or partially inhibiting the welding and thus producing no bond or a weakened bond.[12]

During thermocompression bonding the temperature typically ranges from 300 to 400°C and the bonding cycle (exclusive of positioning) lasts no more than a fraction of a second. During bonding the required heat for interface formation is provided by either a heated capillary through which the wire feeds or the package or substrate is mounted on a heated stage (column). In most thermocompression bonding operations a combination of capillary and column heat is used. The capillary or bonding tool is made of ceramic (Al_2O_3), tungsten carbide, ruby, or other refractory materials.

A typical ball bonding cycle is shown in Figure 3-7. First, gold wire is threaded through a heated capillary ($\sim 350°C$). The ball is formed by using an electronic discharge or a hydrogen torch flame (flame-off) to cut the wire. All modern bonding machines use an electronic flame-off. Because of capillary action (surface tension), the cut end of the wire hanging from the capillary forms a ball. Then, the still plastic ball is brought in contact with the heated bonding pad (nominally 150 to 250°C), and the weld is effected by applying vertical pressure with the capillary as the ball is deformed. Next, the capillary is raised, allowing the wire to play out from the spool as it is repositioned over the site of the second bond. The second bond is made by deforming the wire with capillary pressure against the bonding pad. Then a wire clamp is closed and the capillary is raised, thereby breaking the wire at the heel of the second bond. Another flame-off starts the process again.

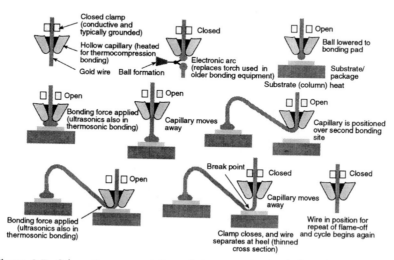

Figure 3-7 Schematic representation of the steps in the ball-wedge bonding cycle. Schematic views applicable to both thermocompression and thermosonic wire bonding.

Gold wire is used in almost all thermocompression operations because it easily deforms under pressure and does not oxidize at elevated temperatures. Aluminum has been used in thermocompression bonding[13] but typically with a wedge tool rather than a capillary because of the difficulty in forming aluminum balls during flame-off.[13]

3.2.2 Ultrasonic Bonding

Ultrasonic bonding is a lower temperature bonding process in which the energy for the weld formation is supplied from an ultrasonic transducer vibrating (20 to 120 kHz) the bonding tool or wedge. The ultrasonic wedge bonding process is shown in Figure 3-8. Aluminum or aluminum alloy wire is threaded through a hole in the wedge and trailed under the bonding tip and secured with a wire clamp. With the clamp closed, the bonding tool is positioned over the site of the first bond with the wire protruding under and

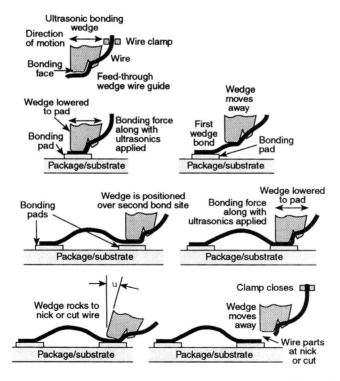

Figure 3-8 Schematic representation of the steps in the ultrasonic wedge-wedge bonding cycle.

somewhat beyond the front of the wedge. The wedge is lowered and the wire is pressed tightly between the wedge and the bonding pad. A burst of ultrasonic energy is applied and the combination of wedge pressure and vibrational energy forms a metallurgical weld at the wire-pad interface. Substrate or column heat can be added to help bond difficult materials or when using large diameter wire. Next the wire clamp is opened, and the wedge is raised and moved into position over the second bonding site. The wedge is then lowered, and the wire is clamped tightly against the bonding pad. Another burst of ultrasonic energy forms the second bond. A slight rocking motion of the wedge partially cuts the wire at the heel of the second bond. The clamp then closes, and the wedge is raised separating the wire.

The tip of the wedge vibrates parallel to the bonding pad. Ultrasonic bonds are typically formed with aluminum or aluminum alloy wire on either aluminum or gold pads. Gold wire ultrasonic bonding has been performed along with copper and other metals in special applications.[14] Wedges can be shaped to handle flat ribbons as well as the more widely used round wires.

3.2.3 Thermosonic Bonding

The thermosonic bonding process combines ultrasonic energy with the ball bonding-capillary technique of thermocompression bonding. It is analogous to thermocompression bonding except that ultrasonic energy is used to replace some or all of the capillary heat. The substrates (columns) are typically held between 100 and 150°C and ultrasonic energy bursts (20 to 50 ms in duration) applied through (by) the capillary are used to effect the bond. Because the integrated circuit chips are maintained at temperatures less than or equal to 150°C, they may be attached with epoxy or other organic adhesives without fear of adhesive degradation. Organic adhesives change properties when raised above their glass transition temperatures. At these low column temperatures there is also less risk of uncontrolled intermetallic growth (see Section 3.6.2 below).

Thermosonic wire bonding is done primarily with gold wire, but aluminum,[15] copper,[16] and palladium[17] have been used. Thermosonic bonding has been particularly useful in attaching wires to the rough surfaces of thick film metallizations, which have historically been hard to bond.[18] Because of the close pad spacings and high I/O densities of multichip modules, automatic wire bonding is a necessity. Even with automatic bonding, however, the relatively long round wire interconnect necessary in most large multichip modules will ultimately limit the performance of integrated circuits and modules due to its high inductance (see Table 3-2).

Table 3-2 Current Size and Performance Criteria for Various Chip Interconnection Methods

Method	Diameter (μm)	Length (mm)	I/O Pitch (μm)	I/O Number	Lead Inductance (nH)	Mutal Inductance (pH)
Wirebond	25.4	1	200	320	1–2	100
TAB	50[a]	1	100	400	1	5
Flip-chip	100	0.1	400	625[b]	0.05–1	1
Laser pantography	5.5[c]	1	25[d]	1600	0.25	1

[a] Effective flatwire or ribbon "diameter" = (thickness + width)/2. Thus, a 50 μm × 50 μm square lead has an effective diameter of 50 μm.
[b] Peripheral I/O only. Area interconnects can raise this number to 16,000.
[c] A 10 μm wide × 1 μm thick laser line yields an effective diameter of 5.5 μm.
[d] Laser pantography lines can be written to any practical length because they are entirely supported by a substrate.

3.2.4 Automatic Wire Bonding

Wire bonding is currently a fully automated process. Both automatic thermosonic and ultrasonic bonding machines are in worldwide use.[19] Automatic wire bonding uses pattern recognition to locate reference marks (fiducial marks) on both the chip and the package. The position of the reference marks relative to the bonding pads is stored in the machine memory and, once aligned, the machine automatically bonds all connections according to a pre-programmed sequence at a rate up to 10 bonds per second. Today's shrinking bond pad size and the increasing number of input/output connections on current integrated circuits, have forced wire bonding to keep pace in order to maintain its position as the dominant chip interconnection method. Chips with I/Os greater than 300 have been interconnected using wire bond techniques. High I/O chips typically use two rows of alternating bonding pads along their perimeter with pad sizes as small as $50\,\mu m \times 50\,\mu m$ with $100\,\mu m$ between on row pad centers. These close spacings have dictated the greater use of wedge-wedge bonding or wire with diameters below $25\,\mu m$ for the ball-wedge process. Wires as small as $15\mu m$ in diameter have been bonded using the thermosonic process. In the ball bonding process, Mil-Std-883C requires the ball size to be within 2.5 to 5 times the wire diameter. In the ultrasonic process, the bond foot width (deformation) is typically 1.5 times the wire diameter. Thus, for the same wire diameter, wedge-wedge bonds can be placed closer together provided the required bonding pattern (height, length, location, etc.) can be accomplished with the ultrasonic bonders in-line step (see Figure 3-8). Other limiting geometric factors for wire bonding are shown in Figure 3-9.

3.2.5 Other Wire-Like Interconnects

Wire bonding is the most widely used chip interconnection technique for many reasons including cost, flexibility (any chip, any bond pattern), no chip preprocessing (except die attach), etc. Wire bonding, however, is relatively slow even with automated processing producing up to 10 bonds a second. Other techniques are potentially faster including tape automated bonding (TAB). TAB is a "gang" bonding technique in which all bonds (on a single chip or lead frame) are formed simultaneously. The TAB process involves the use of prefabricated metallic interconnection patterns (either single or multiple level) on an organic film carrier–typically polyimide. The film carrier is usually one circuit interconnection pattern (or lead frame) wide and several hundred patterns long, making it suitable for winding on reels or spools for use in automated assembly equipment.[10] In order to connect a chip to these film-mounted lead frames, the die or tape must

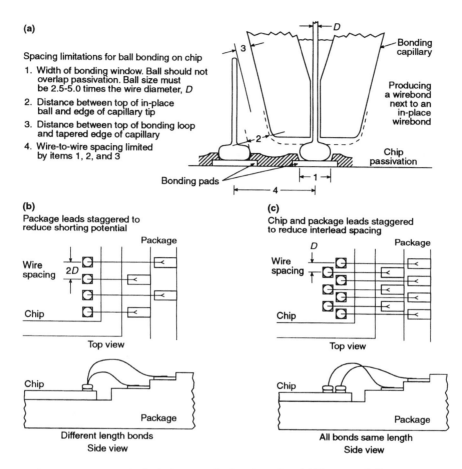

(a)

Spacing limitations for ball bonding on chip

1. Width of bonding window. Ball should not overlap passivation. Ball size must be 2.5–5.0 times the wire diameter, *D*
2. Distance between top of in-place ball and edge of capillary tip
3. Distance between top of bonding loop and tapered edge of capillary
4. Wire-to-wire spacing limited by items 1, 2, and 3

D

Bonding capillary

Producing a wirebond next to an in-place wirebond

Chip passivation

Bonding pads

(b)

Package leads staggered to reduce shorting potential

Package

Wire spacing

2D

Chip

Top view

Chip

Package

Different length bonds
Side view

(c)

Chip and package leads staggered to reduce interlead spacing

D

Package

Wire spacing

Chip

Top view

Chip

Package

All bonds same length
Side view

Figure 3-9 Geometric limitations to wire bond spacing. (a) Thermosonic/thermocompression capillary spacing requirements; (b) staggered package leads to minimize shorting potential; (c) staggered package and chip leads to minimize interlead spacing.

undergo additional processing, which involves the plating of gold or solder interconnection materials in the form of a spheroidal bump on either the die bonding pads or the tape lead ends. If the integrated circuit has aluminum bonding pads and gold is the interconnection material, a gold diffusion barrier such as titanium-palladium must be used. Since the integrated circuits for multichip modules may come from different vendors who use inherently different processes, it might be difficult to modify the bonding pad and incorporate this integral plating. Thus, plated tape (bumped tape) will be the most attractive module option. Tape automated bonding offers inherent testability since the chip is premounted to the tape lead frame and can easily be handled with a full function tester. It offers improved electrical performance with lower inductance (because of its planar structure and the ability with multilayer tape to bring ground planes close to the chip bond) and high densities (compared to wire bonding), including 50 μm leads on 100 μm centers. The major limitation to the density of these interconnects is the location of the chip bonding pads. Up to 300 inputs/outputs can reasonably be addressed with multilayer peripheral tape structures. Area bonding will have to be used for chips with more than 300 inputs/outputs, requiring the full development of techniques for placing chip bonding pads through the active area as well as on the die periphery.

The major disadvantage of tape automated bonding compared to, for example, wire bonding is that high density, multilayer tape automated bonding systems are expensive and have long production lead times. A specific interconnect pattern is required for each different chip; thus, new masking and tooling are needed for each chip (and/or lead connection pattern) to be mounted. Accordingly, this method has been considered practical only for high volume production. Because of its performance advantages (see Table 3-2), it is now being considered as an important interconnection method for current and future high performance chips.[20]

Laser pantography is an experimental technique which utilizes a laser to deposit "wire-like" interconnect (metal lines) from gaseous components in a reaction cell.[21] Extremely fine lines (640/cm) and, thus, high density interconnects can be "written" in this manner. Because the wires (lines) are fabricated by a deposition-type process, the lines must be written on an underlying support structure. Thus, to interconnect silicon integrated circuit chips, the side of each chip must be beveled (to avoid the lines trying to be written on a vertical step) and passivated after die attach to ensure electrical isolation. Prototype multichip modules have been fabricated using this technique,[22] but no production units have been made.

Wafer scale integration (WSI)[23] is another planar wire-like interconnection scheme in which all the system building block chips or integrated circuits are fabricated on a single wafer, tested, and then overlaid with a thin film metallization scheme or interconnect structure (using physical vapor deposition for example) that only links the functioning integrated circuits on the wafer. This approach has great potential for high density, high speed, and good thermal management,[24] but requires redundancy since the devices and interconnects do not have 100 percent yield and cannot be tested separately.[23] Even more important, different chip technologies cannot be interrconnected since all devices must be fabricated on a common wafer. The alternative approach is to fabricate integrated circuits and interconnects separately, and then test and assemble them. This approach allows maximum density (no redundancy) since only functional chips and interconnects are assembled–as in today's multichip modules.

Another wire-like interconnection scheme uses extremely short, fat wires (solder pillars) to form an extremely high performance interconnect. In this method, solder metallization is attached to the chip bonding pads and/or the substrate lead pattern. The chip is then inverted over the substrate pattern and the bond is formed by solder reflow. The solder metallization melts and gravity, coupled with surface tension, causes a perfectly shaped and aligned solder joint to be formed. Inverted chip solder reflow provides very short, low resistance interconnections that minimize inductance and capacitance–especially important for a high frequency operation such as that encountered in multichip modules. Inverted chip solder reflow is also amenable to full area attachment (i.e., bonding pads over the full active device area, not only on the perimeter) since no force is used in the bonding process. It has three disadvantages: the inverted geometry prevents inspection of the interconnection, heat sinking is poor (if no thermal post is brought into contact with the back of the chip) because the only thermal escape path is through the solder joints, and the devices are difficult to replace if they fail.

The electrical performance of the three major multichip interconnection schemes (wire bonding, tape automated bonding, and inverted reflow) is compared in Table 3-2 along with that of a typical laser pantography line. Wafer-scale integration is not included because interconnect performance varies depending upon known good die location and routing channel availability–although one might expect it to have performance similar to the flat rectangular cross-section metallizations found in both TAB and laser pantography methods.

3.3 BONDING PROCESS

The wire bonding process outlined above must be carefully controlled in order to produce reliable, strong, and repeatable bonds. Key to bonding process optimization are the machine set-up (bonding parameters), the wire and pad metallizations, the presence (or absence) of contamination and the influence of external environmental parameters.

3.3.1 Bonding Parameters

The technique for varying thermosonic ball bonding parameters to obtain reliable ball bonds on aluminum metallization is still often considered to be an art, and prior to the development of ball shear testing (see Section 3.5 below) the only method available to ascertain the effectiveness of the bond was the wire bond pull test. This, however, was ineffective in providing much information about the strength of the bond interface in the case of gold ball bonds, where the tensile strength of the wire is usually much less than the strength of the bond interface. With the availability of ball shear testing it is now possible to provide specific guidelines for bonding parameter adjustment, thus eliminating the wide variations in bond-to-bond reproducibility. In the following discussion the effect of bonding parameters will be examined utilizing the factorial approach to experimental analysis of parameter variation. This technique can be found in many excellent texts on statistics and experimental design.[25-27]

The bonding parameters that will be considered are the first bond power (P), the substrate temperature (T), and the first bond dwell parameter (D). The bonding force is not considered so that the analysis and subsequent interpretation of the data will not be overly complicated. Also, the bonding force is usually adjusted so that the ceramic or other capillaries do not become worn too quickly and thus are not subjected to day-to-day alteration. The other variational parameters considered are the substrate type and metallization thickness.

Bond Test Samples

Aluminum metallized test substrates were prepared on silicon, alumina and sapphire substrates. The 75 mm diameter silicon wafers were n- on p-type with a (100) orientation and a thickness of 380 μm.

The polycrystalline alumina substrates were 2.54 cm square with a thickness of 635 μm while the single crystal 2.54 cm square Al_2O_3 substrates were polished optically flat to a thickness of 380 μm.

The silicon wafers were coated with photoresist, and then diced into 5.08 cm square wafers. After dicing, the substrates were cleaned using the following procedure: (1) three, 3-minute immersions in boiling acetone (semiconductor grade); (2) a 3-minute hot sulfuric acid immersion; and (3) a hydrofluoric acid etch (30-40% low ion grade) to remove oxides. The alumina and sapphire substrates were solvent cleaned prior to aluminum deposition. In order to eliminate possible run-to-run variations, all substrates used for a particular experimental test were metallized at the same time. The same is true for all subsequent fabrication steps. Substrates were placed symmetrically into the vacuum system and the specific amount of aluminum was evaporated at a mean pressure of 5×10^{-7} torr, from a resistance heated boron nitride crucible.

The 5.08 cm square aluminized wafers were photolithographically patterned into sixteen sector substrates each section containing 44 metallized bonding pads. A quadrant of this pattern is shown schematically in Figure 3-10. This 16-sector pattern was chosen to provide the necessary

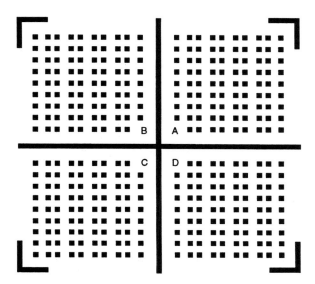

Figure 3.10 Schematic representation of a quadrant (four sectors) of a 16-sector bonding test pattern. Each of the sectors (i.e., A through D) contains 44 bonding pad pairs spaced approximately 1.25 mm apart. Each pad is approximately 0.25 × 0.25 mm.

statistics for a complete single block, four parameter factorial experiment. The 2.54 cm square alumina and sapphire substrates were patterned with a similar test array consisting of 4 sectors or quadrants each containing 44 bonding sites. After etching, the substrates were alloyed in vacuum $(5 \times 10^{-7}$ torr) at 550°C for 15 minutes. The substrates were then given a standard solvent cleaning followed by an additional 10 minutes in UV-ozone. Once cleaned, the substrates were stored in a dry nitrogen box to await bonding.

Wire Bond Fabrication

Wire bond test samples were fabricated using either semiautomatic or full automatic techniques. The semiautomatic bonder was a Mech-El Model 827 thermosonic ball bonding machine equipped with a UTHE 10E ultrasonic generator. The bonding wire was 25.4 μm diameter, pure gold (99.999%), with a specific elongation of 4-6% and a nominal tensile strength of 6 to 8 grams (force). The first bond force setting was 25-30 grams (force) and the second bond force was set at 70-80 grams (force). Ceramic capillaries were used and removed for cleaning and or replacement every 1500 bonds to ensure high quality bonding throughout. A single operator was used in all semiautomatic bonding to eliminate any multi operator effects.

Automatic wire bonding was performed using a Kulicke-Soffa Model 1419 automatic wire bonder with similar wire and a first bond force setting in the range of 30-40 grams (force).

Test samples were bonded in a random manner to avoid any systematic introduction of a geometrical or placement preference.

Bond Testing

The ball bond shear strength and the wire bond pull strength were used as response parameters for all the bonding parameter investigations. The ball shear test (as shown in Table 3-3 and discussed in Section 3.5.3 below) can provide more information on the first bond than can the standard pull test and thus is used as the primary response parameter in our bonding optimization process. The ball bond shear data presented in this study were obtained using a Dage-Precima MCT 20 microtester. This is a semiautomatic microprocessor-controlled unit which can perform both wire bond pull and ball (or die) shear tests. Various load cells can be used with the machine. For these tests a 0-200 gram (force) load cell was used. The shear tool face at the tip was approximately 0.015 cm wide.

Table 3-3 Possible Information Obtained from Wire Bond Tests

Information Area	Wire Bond Pull Test	Ball Bond Shear Test
Hybrid geometry	Yes	No
Wirebond geometry	Yes	No
Wire quality (bondability, defects, etc.)	Yes	No[a]
Second bond	Yes[b]	No
Bonding machine parameters (optimization, etc.)	No[c]	Yes
Process development	No[c]	Yes
Substrate, bonding pads	No[c]	Yes

[a] Sensitive to impurities, insensitive to mechanical defects.
[b] Extremely geometry dependent.
[c] Provides no information unless the bonds are catastrophically weak on the pads lift.

3.3.2 Process Optimization

Experimental Design

The bonding parameter experiments consisted of simple 2^3 factorial arrangements as shown in Table 3-4. This experimental design in most cases was unreplicated due to the substantial number of samples per treatment.[26] For all of the various treatments the order of execution was random. This experimental design required the use of approximately half of the bonding sites available on the 5.08 cm square silicon substrate. The unused bonding sites were also bonded and subsequently exposed to a high temperature bake at 125°C for 200 hours and 400 hours. The effect of this high temperature exposure will be discussed below.

The values for the first bond power, the substrate temperature and the dwell time for the −1 and +1 levels are shown in Table 3-5. The dial settings for the power are calibrated in units of millivolts and correspond to the voltage applied to the piezoelectric crystal driver for the ultrasonic

Table 3-4 2^3 Factorial Design Table

Power (P)	Temperature (T)	Dwell (D)	Response Parameter Mean Shear Strength grams (force)
−1	−1	−1	S_1
−1	−1	+1	S_2
−1	+1	−1	S_3
−1	+1	+1	S_4
+1	−1	−1	S_5
+1	−1	+1	S_6
+1	+1	−1	S_7
+1	+1	+1	S_8

Table 3-5 Bonding Machine Parameter Settings for the Low (−1) and High (+1) Factorial Levels

Bonding Machine Parameters	Factorial Levels	
	(−1)	(+1)
P. First bond power (millivolts)	260	290
T. Substrate temperature (°C)	100	150
D. Dwell time (milliseconds)	26	50

Table 3-6 Complete 2^3 Factorial Design Table Indicating Primary, Secondary and Tertiary Effects

Primary			Secondary			Tertiary	Response
P	T	D	$P \times D$	$P \times D$	$T \times D$	$P \times T \times D$	Parameter
-1	-1	-1	$+1$	$+1$	$+1$	-1	S_1
-1	-1	$+1$	$+1$	-1	-1	$+1$	S_2
-1	$+1$	-1	-1	$+1$	-1	$+1$	S_3
-1	$+1$	$+1$	-1	-1	$+1$	-1	S_4
$+1$	-1	-1	-1	-1	$+1$	$+1$	S_5
$+1$	-1	$+1$	-1	$+1$	-1	-1	S_6
$+1$	$+1$	-1	$+1$	-1	-1	-1	S_7
$+1$	$+1$	$+1$	$+1$	$+1$	$+1$	$+1$	S_8

generator. An increase in the power setting from 260 mV to 290 mV represents approximately, a 25% power gain. The responses denoted as S_i are the mean shear strengths for each of the treatments. A complete table of effects including second and third order effects is shown in Table 3-6. The calculation of any one of the effects is simply the sum of the products of each level with the corresponding response, all divided by $2^{(k-1)}$ where k in our case is 3. For example, the effect of first bond power is

$$P = (-S_1 - S_2 - S_3 - S_4 + S_5 + S_6 + S_7 + S_8)/4 \qquad (3\text{-}1)$$

In order to determine the statistical significance of a particular effect for an unreplicated design we need to estimate the sample variance, S^2. This can be done by making the assumption that all higher order interactions are negligible and then by pooling together all of the sum of squares of these higher order effects. The 95% confidence interval for testing whether the effect is significant or not can then be computed from the standard equation.[25]

$$\text{Estimate of the effect} = \pm t_{n,0.05}\sqrt{\left(\frac{1}{n-1}\right)S^2}$$

where $t_{n,0.05}$ is the appropriate value for a two-tailed test at a 95% confidence level and for the n-degrees of freedom used in the estimation of S^2.

Power-Dwell-Temperature Effects

In Table 3-7 all the effects are listed for the various thicknesses of aluminum on silicon. The main effects (shaded portion of Table 3-7) are significant to a 95% confidence level with the exception of the effect of power for the 6 μm film. This result would indicate that an increase of power setting by 30 mV (25% increase in power) has no significant effect on increasing the bond strength for this particular case. The mean response shown in Table 3-7 as $\Sigma s_i/8$ is the simple average of the mean shear strength for all the treatments and represents the center point of the factorial design, i.e., $P = 275$ mV, $T = 125°C$, and $D = 38$ ms. A curve representing the mean response for the four metallization thicknesses is shown in Figure 3-11. The linear plot clearly indicates a reduction in the net shear strength as a function of film thickness for constant bonding parameters with a slope of about 1.9 gf per μm of aluminum.

Table 3-7   Effects of Power, Temperature and Dwell Time on Mean Shear Strength in Grams (force) for Thin Film Aluminum on Silicon

Effects	Metallization Thickness			
	1 μm	2 μm	3 μm	6 μm
P(ower)	9.4	6.0	9.6	-0.7^a
	(3.1/10 mV)	(2.0/10 mV)	(3.2/10 mV)	
T(emperature)	11.5	5.9	7.7	9.6
	(2.3/10°C)	(1.2/10°C)	(1.5/10°C)	(1.9/10°C)
D(well time)	7.3	9.1	16.0	12.3
	(3.0/10 ms)	(6.1/10 ms)	(6.7/10 ms)	(5.1/10 ms)
P × T	0.8^a	-1.7^a	-1.3^a	1.3^a
P × D	-2.2^a	0.6^a	0.3^a	-0.3^a
T × D	1.5^a	3.6^a	-0.8^a	2.2
P × T × D	1.0^a	0.2^a	1.3^a	-0.9^a
$\Sigma S_i/8^b$	36.9	34.9	33.3	27.6
Sample variance	4.4	8.1	1.3	3.2
95% confidence interval	±4.1	±5.6	±2.2	±2.0

[a] Not significant at the level tested.
[b] Average response: corresponds to the center of the experimental design ($P = 275$ mV, $T = 125°C$, $D = 38$ ms).

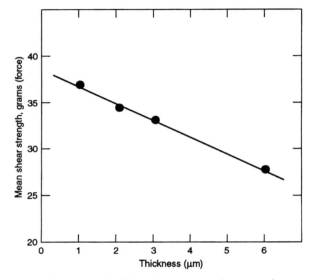

Figure 3-11 Mean thermosonic ball bond shear strength versus aluminum (on silicon) metallization thickness.

Since it is plausible that the area of the ball contacting the bonding pad is proportional to the shear strength, optical measurements of the mean diameter on the gold balls after bonding were made for the 3 µm thick aluminum on silicon group. Factorial analysis was employed using both the area and the shear strength per unit area as the responses to the various treatments. This analysis would indicate which bonding parameters were responsible for the ball deformation and which parameters give the most significant increases in the shear strength per unit area. The experimental design and results are displayed in Table 3-8. The results indicate that the temperature has a significant effect on the shear strength with no significant effect on ball deformation. This is a useful result in cases where the ball diameter is limited by the size of the bonding pad. Likewise, increasing the first bond power setting provides a greater gain in shear strength per unit increase in the bond area than does extending the dwell time. Increasing the dwell time by 26 ms yielded the largest gain in ball diameter. However, since the effect of dwell on shear strength per unit area is relatively large, we can conclude that the increase in shear strength cannot be attributed solely to the observed increase in ball diameter.

In Figure 3.12 the mean shear strength as a function of the dwell time is plotted for various substrate temperatures and power settings. For substrate temperatures of about 100°C, the mean shear strength is a linear function of the dwell time for the range of 26-50 ms with an increase of approximately 0.25 gf/ms. At 150°C this linear relationship breaks down. The shear

Table 3-8 Effect of Power, Temperature, and Dwell on Gold Ball Bond Deformation for Aluminum Metallization on Silicon

Parameters			Responses	
P (MV)	T (°C)	D (MS)	Area $\mu m^2 \times 10^3$	Strength/Area grams (force)/ $\mu m^2 \times 10^{-3}$
260	100	26	3.81	5.56
260	100	50	4.90	6.81
260	150	26	4.00	6.35
260	150	50	5.03	7.75
290	100	26	4.32	6.20
290	100	50	5.55	7.75
290	150	26	4.45	7.60
290	150	50	5.81	8.67
	Main Effects			
	Power		0.58	1.24
	Temperature		0.19	1.42
	Dwell		1.16	1.71

strength increases rapidly as the dwell is increased to about 40 ms with little or no change for further increases. At a power setting of 290 mV at this temperature, very strong bonds were initiated for dwell times less than or equal to 30 ms. Thus the effect of increasing the substrate temperature permits the making of strong bonds for reduced dwell times (less ball deformation).

Dwell Time and Wire Tensile Strength

The above results indicate that the efficacy for initiating strong bonds can be increased significantly at lower substrate temperatures by extending the dwell time. Since it was not quite clear how this increase in ultrasonic scrubbing time would affect the results of a wire bond pull test, an experiment was conducted to study the effect of dwell time versus wire tensile strength. Four groups of 44 bonds each were bonded with dwell times of 30, 40, 50 and 60 ms respectively. Half of the bonds were pulled immediately after bonding whereas the other half were pull tested after 250 minutes at 250°C. The results for this test are presented in Table 3-9. No significant differences were observed between the four groups prior to the

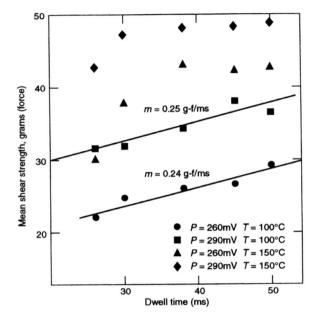

Figure 3-12 Mean thermosonic ball bond shear strength versus dwell time for various first bond powers and substrate temperature settings.

Table 3-9 Wire Bond Pull Strength for Various Dwell times[a]

	Pull Strength, grams (force)			
Parameters	30 ms Dwell	40 ms Dwell	50 ms Dwell	60 ms Dwell
Pre-bake	6.6±1.2	6.7±0.9	7.2±0.9	6.7±1.2
Post-bake (250°C for 250 min.)	7.1±0.7	7.2±0.8	7.5±0.9	7.8±0.7

[a] First bond power setting = 275 mV.
Bonding temperature = 125°C.

high temperature bake. As is usually observed for thermosonic gold balls bonded on aluminum metallization, over 90% of the wire breaks occurred in the neck region just above the ball. After the temperature bake, the pull test results improved significantly for all the groups. More interesting was the observed ordering of the wire bond pull strength which indicated larger values for those bonds with longer dwell times. This increase in net pull strength was a direct result of fewer wire breaks in the neck region. This effect is probably due to the process of recrystallization at elevated temperatures after coldworking (ultrasonic scrubbing). During the ball formation (flame-off) prior to initiating a bond, the grain size in the ball and neck area is known to increase by as much as a factor of 100.[28] Subsequent coldworking during the ultrasonic scrubbing process renders the ball and neck portions of the bond susceptible to recrystallization if it is subjected to post-heating at sufficiently high temperatures (200°C). Recrystallization reduces the grain size and thus improves the wire tensile properties.

Substrate Bonding Parameters

The effects of substrate parameters such as surface roughness (morphology), metallization thickness and substrate material which may influence ball bond shear strength were investigated. Silicon, alumina (99.6% pure) and single crystal sapphire substrates were used for this comparative study for two aluminum film thicknesses (1 and 2 μm). The results of this analysis are shown in Table 3-10. The treatments consisted of strong bond and weak bond parameter settings which confounded the effects of the parameters' long dwell-low power ($P = 290$ mV, $D = 45$ ms)

Table 3-10 Experimental Design and Response Values for Linear Ball Shear Model. Thermosonically Bonded 25.4 μm Diameter Gold Wire on 1 μm Thick Aluminum Pad Metallization (on Silicon)

Bonding Parameters			Response Parameter Mean Shear Strength grams (force)
Power[a] (P)	Temperature[b] (T)	Dwell[c] (D)	
−1	−1	−1	24.7
−1	−1	+1	26.2
−1	+1	−1	37.8
−1	+1	+1	42.0
+1	−1	−1	31.8
+1	−1	+1	37.9
+1	+1	−1	47.0
+1	+1	+1	48.4
0	0	0	37.6
0	0	0	36.9
0	0	0	37.4
0	0	0	35.1

[a] P: +1 = 290 mV; 0 = 275 mV; −1 = 260 mV.
[b] T: +1 = 150°C; 0 = 125°C; −1 = 100°C.
[c] D: +1 = 45 ms; 0 = 38 ms; −1 = 30 ms.

and short dwell-low power ($P = 260\,\text{mV}$, $D = 30\,\text{ms}$) respectively. Two substrate temperatures, $100°C$ and $150°C$ were used. The effect of temperature was found to be similar for all three substrate types. The effect of thickness was the only significant pre-bake effect that differed for the three substrate types. This effect consisted of an improved mean shear strength for the alumina as the film thickness increased, whereas the effect for the smoother silicon and sapphire was a decrease in mean shear strength. This effect was noted previously for the silicon.

Linear Ball Shear Model

Using the same 2^3 factorial design concept with replicated center points, a linear model for the ball bond mean shear strength in terms of P, D, and T was constructed. The resultant ball shear equation simplifies the understanding of how the bonding parameters influence bond strength, eliminating the need for complex three dimensional plots. In addition, the linear factorial design provides an efficient means for generating new models should different substrates and/or substrate metallizations be required. A typical example of this linear factorial design and its response functions for $1\,\mu\text{m}$ thick aluminum metallization on silicon is shown in Table 3-10. The resultant linear model for the shear strength takes the form:

$$\text{Shear Strength} = B_0 + B_1 P + B_2 T + B_3 D$$

where the coefficients for the responses shown in Table 3-10, are given by $B_0 = -104.3$, $B_1 = 0.29$, $B_2 = 0.27$, and $B_3 = 0.22$. The linear equation's goodness of fit to the experimental data was estimated and found to be quite acceptable (i.e., within 95%) for the range of bonding parameters used in this study. If a wider range of bonding parameters is used it may be necessary to fit the data to a second order (quadratic) model.[25] An optimized histogram for gold bonds to aluminum metallization is shown in Figure 3.13.

Prior Testing Influences

A series of experiments was conducted to determine if prior wire bond testing (i.e., non-destructive ball shear (NDBS) and destructive wire bond pulling) would have a significant influence on subsequent destructive ball shear testing. Bonds to gold and aluminum substrate metallizations with various combinations of wire bond pulling and non-destructive ball shear were made. The results of the ball shear tests are shown in Table 3-11. For

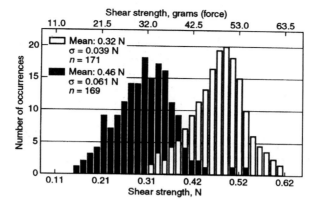

Figure 3-13 Histograms of gold thermosonic ball bond shear strength on aluminum metallization (on silicon). The black histogram resulted from bonding machine set-up using the wire bond pull test. The white histogram illustrates the improvements obtained in ball bond shear strength by optimizing the bonding cycle using the ball shear test.

Table 3-11 Influence of Prior Testing on Ball Bond Shear Strength. Thermosonically Bonded 25.4 μm Diameter Gold Wire. Gold Pads on Ceramic Substrates. Aluminum Pads on Silicon Substrates

Wire-Pad	Ball Bond Shear Strength[a]	Wire Bond Pull Strength[a]	Non-Destructive Ball Shear (NDBS)	$\Delta \bar{x}$ (Significance)
Gold–gold	44.3 ± 3.3	8.6 ± 1.1	N/A	+0.2
Gold–gold	44.5 ± 6.5	not pulled	N/A	(None)
Gold–aluminum	34.8 ± 3.3	not pulled	N/A	−3.0
Gold–aluminum	31.8 ± 4.1	7.1 ± 0.7	N/A	(Yes, highly)
Gold–gold	42.7 ± 3.2	N/A	15.0	−0.8
Gold–gold	43.5 ± 3.6	N/A	No NDBS	(None)
Gold–aluminum	35.4 ± 3.9	N/A	15.0	+0.3
Gold–aluminum	35.1 ± 4.0	N/A	No NDBS	(None)

[a] All strength data in grams (force) based on 44 bond samples.

the gold ball bonds on gold metallization there appeared to be no influence of prior testing (either NDBS or destructive wire bond pull) on the resultant shear strength distributions. Similarly NDBS produced no significant effect on the gold ball bond to aluminum bonding pad situation. The mean shear strength of the gold-aluminum sample which contained previously pulled bonds was reduced by three grams (force). This reduction was determined to be significant based on a Student's T-test analysis.[25] Although this shear strength reduction due to destructive wire bond pulling is statistically significant, the effect is relatively small (e.g., 10% or less of the mean shear strength).

Temperature Effects

Thermal exposure studies (up to 1000 hours) have been conducted on various test specimens over the temperature range from 125°C to 300°C. Full details of these studies have been published previously.[29] Some interesting results are described below. Extended thermal exposure (1000 hours) at 125°C produced increases in mean shear strength (when compared to unexposed controls) on the order of 25% (i.e., 10 grams (force)) for aluminum metallization pads directly on the silicon substrates. Decreases of the same order of magnitude were observed[30] for aluminum metallization on silicon dioxide under similar conditions. An explanation for this behavior might be linked to the rate of diffusion of aluminum depending on whether it has been alloyed with silicon or with SiO_2.

Activation Energy   Activation energy for the initial thermal annealing shear strength changes was calculated at 30 kcal/mole for 1 μm thick aluminum metallization on silicon. In later growth states, especially at high temperature (250°C and 300°C), a much higher activation energy (nominally 18 kcal/mole) was exhibited for the change in shear strength over time. The possible reason for this was the relative absence of the strong purple intermetallic ($AuAl_2$) in the low temperature (125°C) bond group as compared with the high temperature situation. Thus, there appear to be (at least) two primary activation energies that govern ball shear strength at different temperatures. Such an explanation may explain the differences in activation energies reported by several authors including Philofsky[11] (15.9 kcal/mole) and Chen[31] (12.8 kcal/mole). From our studies, based on parabolic growth laws and observation of $AuAl_2$ growth patterns, a lower bound for the activation energy of the $AuAl_2$ phase is 18 kcal/mole.

Cratering   The cratering of the silicon has been observed to occur only when $AuAl_2$ has formed in and around the weld region. A plausible explanation suggests that the strong, inelastic and brittle nature of this intermetallic and its good adhesion to the silicon, allows the shearing ram to transmit most of its energy directly through the intermetallic to the substrate. Thus, this effect should not occur when little or no intermetallic is present since most of the energy would easily be absorbed by the ball or at a relatively weak bond interface. In strong bonds where no cratering appears there are (typically) large areas of intermetallic, thus permitting the transmission of a greater amount of energy to the silicon prior to cratering. This is probably the reason for the predominance of cratering in the weaker bonds. A full study of the influence of bonding parameters on cratering is presented in a recent article by Clatterbaugh and Charles.[32] A summary of these results is discussed in Section 3.6.2 below.

Parametric Analysis

The ball bond shear test can be used as a response parameter to study the influences of hybrid processing procedures. One particularly interesting study considered the influences of contamination on the strength of the ball bond-bonding pad interface and the efficacy of various hybrid cleaning processes. Full details of this contamination-interface cleaning study have been presented previously.[33] In this study two basic forms of contamination (photoresist and epoxy off-gas products) were considered along with four basic cleaning methods (hydrocarbon solvents, oxidizing acid, oxygen plasma and UV-ozone). Results of the cleaning evaluation tests depended largely on the ability to reduce significantly ball bond strengths through the introduction of surface contamination prior to bonding. Surface contamination with photoresist showed a significant reduction in ball bond shear strength (approximately 20% for both aluminum and gold substrate metallizations). Epoxy off-gas contamination was less successful with the results highly dependent on metallization type. Bonds made to aluminum metallized substrates were significantly affected (approximately 18%) by epoxy off-gas contaminants, while those with gold metallization were not. The cleaning procedures used to remove the contaminants were found to vary in effectiveness as measured by the amount of shear strength restoration (and were dependent on both metallization and contamination type). Typical ball bond shear test data is shown in Table 3-12.

Solvent cleaning[33] was found to be the least effective in restoring ball shear strength to uncontaminated levels. Cleaning both contaminants with the sulfuric acid/potassium persulfate solution[34] did restore the mean shear strength to the level of bonds made on uncontaminated substrates (of either type). Unfortunately, this type of cleaning process is incompatible with many organic materials used in microcircuit assembly, including the epoxy resins used in die and substrate attach. Thus, the acid mixture may only be viable for empty packages, bare substrates or unmounted chips. Oxygen plasma cleaning was also found to be an acceptable technique for restoring shear strengths for both contaminants and substrate types. Only in the case of combined off-gas products from two epoxies, did the oxygen plasma fail to improve the mean shear strength after cleaning; however, after thermal

Table 3-12 Influence of Contamination and Cleaning on Ball Bond Shear Strength. Thermosonically Bonded 25.4 μm Diameter Gold Wire on 1 μm Thick Aluminum Bonding Pad Metallization (on Silicon)

Contaminant	Cleaning Method	Thermal Condition	Control Shear Strength, gf	After Contamination Shear Strength, gf	After Cleaning Shear Strength, gf
Photoresist	Solvent	As bonded	40.0 ± 6.2	30.0 ± 4.9	29.4 ± 7.7
		post burn-in	38.9 ± 8.7	30.9 ± 9.0	35.8 ± 8.0
	Acid	As bonded	39.9 ± 5.3	32.9 ± 5.4	37.4 ± 6.5
		post burn-in	43.7 ± 4.1	35.3 ± 9.4	47.3 ± 5.2
	Plasma	As bonded	34.7 ± 6.0	29.1 ± 6.9	32.4 ± 5.7
		post burn-in	41.3 ± 7.3	33.4 ± 8.1	42.8 ± 6.8
	UV-ozone	As bonded	34.8 ± 5.4	23.4 ± 5.4	27.6 ± 5.4
		post burn-in	44.5 ± 5.7	32.6 ± 9.9	40.3 ± 6.9
Epoxy Offgassing (517/550)	Solvent	As bonded	37.1 ± 10.6	31.9 ± 7.6	30.0 ± 10.0
		post burn-in	45.2 ± 4.8	35.3 ± 7.5	41.9 ± 6.1
	Acid	As bonded	40.8 ± 6.9	24.0 ± 10.1	37.1 ± 7.0
		post burn-in	47.7 ± 4.9	36.3 ± 7.6	42.5 ± 6.4
	Plasma	As bonded	40.5 ± 6.0	36.4 ± 5.8	36.4 ± 4.9
		post burn-in	40.4 ± 9.7	39.6 ± 10.4	47.9 ± 6.7
	UV-ozone	As bonded	40.9 ± 5.3	35.0 ± 5.8	36.8 ± 5.8
		post burn-in	47.8 ± 7.6	42.0 ± 7.8	48.2 ± 6.9

processing the shear strengths of the cleaned samples were much stronger than uncleaned contaminated samples. UV-ozone cleaning yielded the most significant improvement on gold metallization. For the aluminum on silicon test samples, although there was some improvement, the shear strength was not restored to uncontaminated level. Cleaning techniques based on strong oxidants could conceivably have a harmful effect on gold ball bonding to aluminum due to increased surface oxidation. To test this, three uncontaminated aluminum on silicon substrates were subjected to UV-ozone, oxygen plasma and acid cleaning techniques. No significant effect on shear strength was observed when compared to control substrates. Advanced materials analysis techniques including Auger electron spectroscopy (AES) and secondary ion mass spectrometry (SIMS) were applied to typical aluminum on silicon test specimens with the result that UV-ozone techniques produced the cleanest surfaces. AES spectra and a SIMS profile have been shown in a previous publication.[35]

3.4 MATERIALS

A wide variety of materials are used in the fabrication of integrated circuit die and their host substrates in today's multichip modules. Of these materials, the ones most important to the wire bonding process include (1) the chip metallization and its associated passivation layers, (2) the bonding wire itself, (3) the lead frame or package bonding pad, (4) the substrate metallization and its associated passivation, and (5) module or wire encapsulants. Different materials and their method of application can greatly influence both the strength and reliability of the wire bonding pad interfaces and, hence, the wire bond itself.

3.4.1 Chip Metallizations and Dielectrics

Silicon devices are typically metallized or interconnected (both bonding pads and conductor traces) using aluminum metal because of its high electrical conductivity and its ability to form a protective oxide layer on its top surface. For bonding pads the metal contacts the silicon through openings in the protective oxide layer. Although aluminum is the dominant metallization used by most integrated circuit manufacturers, it has one significant problem. Silicon is soluble in aluminum (typically 0.5 to 1% in the temperature range from 450 to 500°C) and during post-deposition (physical vapor deposition[36] or sputtering[37]) heat treatment or alloying, silicon diffuses into the aluminum, leaving holes or voids in the silicon which are filled by aluminum metallization "spikes". Spikes can be minimized by reducing processing times and rapidly increasing temperatures.[38] As device junctions became shallower due to the rapid shrinkage in device size, the aluminum spikes were deep enough to cause junction shorting. To reduce spiking to a minimum, aluminum alloys with 1 to 5 atomic percent silicon are routinely deposited.

Aluminum with a percentage of silicon also has its problems, because if there is too much silicon in the aluminum, it will precipitate and form silicon crystallites or nodules on the bonding pad surface during heat treatment. These nodules can cause poor bonding as well as potentially upset the interconnects' atomic behavior. The crystallites are typically p-type (aluminum is a p-type dopant for silicon) and can form a rectifying function if they come in contact with an underlying n-type layer.

As interconnect dimensions shrink, current densities carried by the circuit traces increase. Once current densities reach 10^6 A/cm^2, the interconnects begin to fail due to a process called electromigration.[39] Pure aluminum interconnects will crack along conductors carrying currents of 10^6 A/cm^2 when held for periods of less than 100 hours at 220°C. The cracks are due to mass transport of the conductor materials. Atoms move through grain boundaries, creating voids at one end and hillocks on the other. Electromigration is reduced by decreasing grain boundaries (i.e., increasing grain size) and by reducing grain boundary diffusion by the introduction of a grain boundary pinning impurity such as copper. A typical aluminum alloy includes 4 wt% copper and 1 to 2 wt% silicon. Such alloys can increase the electromigration resistance by several percent. The addition of titanium can further improve the electromigration resistance of the aluminum-copper-silicon interconnect system.[40] These alloys can significantly alter the bondability of the chip and force the development of new or optimized bonding parameters to ensure interconnect reliability.

The metallization of gallium arsenide is a significantly more difficult problem. Metallization on GaAs can form either ohmic contacts or Schottky barriers.[41] Because GaAs decomposes into Ga and As$_2$ gas at 590°C, the formation of metal contacts is especially difficult. Similarly, the Schottky barrier height on n-type GaAs is large ($\sim$0.8 eV) and essentially independent of metallization type. GaAs is very difficult to dope highly, making ohmic doping levels of 5×10^{18} impurities/cm^3 almost impossible to reach by conventional gaseous diffusion methods. Implanted dopants must be annealed at 850°C to activate their conductivity which necessitates capping or high arsenic overpressures to prevent GaAs decomposition. AuGe$_2$ metallization schemes have formed the most stable GaAs contact material.

Bonding to GaAs is as difficult as producing ohmic contacts with the metallization. The different pad alloys as well as the brittle nature of the semiconductor material itself (including thinned down die for improved performance) make selection of bonding parameters, especially force, extremely critical for successful bonding (high strength, no cratering, etc.).

Oxides and nitrides can be used as insulating layers between metallization levels, as insulating protective covers and as edge protection around bonding pads. These layers are generally deposited by chemical vapor deposition (CVD) techniques in which silane (SiH_4) and oxygen are reacted to form SiO_2. The temperatures of these reactions typically occur at 450°C or below depending upon the gas mixture and whether plasma enhancements are used. These low temperature processes allow the SiO_2 films to be deposited over aluminum metallization. A very important property of deposited SiO_2 films is the film stress. Low stress in the as-deposited films improves their stability under thermal cycling. The addition of phosphorus pentoxide (P_2O_5) to the growing SiO_2 can significantly reduce the built-in stress. Silicon dioxide doped with phosphorus is called phosphosilicate glass (PSG). At a number of 20 wt% P_2O_5, the internal slice film stress is zero, but normally less than 8% is used because high P_2O_5 content causes the PSG film to be hydroscopic.

The coefficient of thermal expansion (CTE) for SiO_2 films with incorporated P_2O_5 increases with P_2O_5 content. PSG films can be tailored to meet the CTE of the underlying substrate, thus, significantly improving passivation layer stability under thermal cycling. The CTE of undoped SiO_2 can be increased from 6×10^{-7}/K to about 6×10^{-6}/K by the incorporation of approximately 22 wt% P_2O_5. The latter CTE matches that of GaAs, thus allowing PSG to act as a stable passivation for GaAs.

PSG is also an extremely effective diffusion barrier to such elements as zinc and tin because of its dense microstructure (compared to SiO_2). It also stops sodium ion transport through oxide layers in metal-oxide semiconductor field-effect transistors (MOSFETs).

Silicon nitride (Si_3N_4) can also be grown by chemical vapor deposition by reacting silane gas with ammonia. Si_3N_4 is typically deposited in thin layers (1 to 2 μm in thickness) by plasma assisted processes. Si_3N_4 forms excellent mobile-ion and moisture barriers with a performance that typically exceeds SiO_2 for equal thickness layers.

3.4.2 Bonding Wire

Bonding wire is key to the manufacture of high quality wire bonds. Bonding wire is typically specified by its material type, diameter, and mechanical properties, such as the breaking strength and the elongation. Historically, bonding wires have been made of aluminum, aluminum alloys, and gold. Since aluminum is typically too soft to be drawn into fine microelectronic interconnect wires, it is alloyed with other materials (notably silicon and magnesium) to toughen its behavior and improve its metallurgical stability. Table 3.13 contains information on various bonding wire types and properties. Other materials have also been used for bonding wire including copper,[16] silver,[13] and palladium.[17]

Although impurities in bonding wire and lubricants from its manufacture (die drawing) have been linked to wire bonding problems in the past, modern wire production practices lead to a stable wire metallurgy with repeatable mechanical properties, so reliability and consistency of the wire itself is quite high. Despite this quality, interface problems and the control of bonding parameters can still lead to situations which bring the quality of wire into question. Further implications of these factors will be discussed in Section 3.6.2 below.

Table 3-13 Bonding Wire for Thermocompression, Thermosonic, and Ultrasonic Wire Bonding

Wire Type	Comments
Aluminum	Pure wire typically too soft to be drawn into fine wire
Aluminum + 1% Silicon	Standard bonding wire. Because 1% Si exceeds room temperature solubility of silicon in aluminum by approximately a factor of 50, there is a tendency for silicon to participate at ordinary bonding temperature
Aluminum + Magnesium	Aluminum with 0.5 to 1% Mg exhibits properties (strength and elongation) similar to silicon (1%) alloying. Does not form a second phase (at room temperature Mg solubility in aluminum is about 2%). Has excellent fatigue resistance
Gold	Gold wire surface finish, cleanliness, and purity must be strictly controlled. 99.99% pure wire can usually be drawn with adequate breaking strength and proper elongation. Ultrapure gold is very soft, but even extremely small amounts of impurities (for example, 5 to 10 ppm Be by weight or 30 to 100 ppm Cu by weight) make the wire workable. Be-doped wire (Be-stabilized) is stronger than Cu-doped wire by about 10 to 20% under all conditions
Other wires	Silver,[13] copper,[16] and palladium[17] wires have all been bonded successfully using both thermocompression and thermosonic techniques. The ball bonding of aluminum[15] with special additives such as nickel has been demonstrated

3.4.3 Lead Frames and Packages

The packaging of integrated circuits falls into two generic classes based on the protection afforded by the package to the environment and in particular moisture. These classes are hermetic and non-hermetic (or plastic). Hermetic packages are made of metal, glass and/or ceramic. The hermetic package typically has a recessed die cavity in which the integrated circuit is attached to the bottom using either a metal alloy or an organic adhesive. In two piece ceramic DIPs (CERDIPs) the die cavity is replaced by a metal lead frame similar to that in plastic packages described below. Typical die attach materials for these packages are Au-Si or silver-filled glass or Pb- or Sn-base solders. Temperatures for these processes and the associated sealing operation (fusing of two ceramic halves together with low melting point glasses) range from 370 to 450°C. These temperatures are high enough to affect lead frame materials and care must be used in alloy selection to prevent softening and subsequent lead frame damage. Following die attach, the chip is wire bonded and then the die cavity is sealed using a metal or ceramic lid by either welding, soldering or fusing with a low melting point glass.

In non-hermetic or plastic packages the die is attched to a metal lead frame, wire bonded and then encapsulated typically using a hard rigid plastic placed around the die and lead frame by transfer molding. The lead frame is a patterned piece of sheet metal formed by stamping or chemical milling. There are three primary categories of metals used for lead frames: copper-based alloys, nickel-iron alloys, and certain clad metals. Copper is the ideal material for lead frames because of its high electrical and thermal conductivity. However, copper must be alloyed to improve its tensile strength. Materials such as iron, zirconium, zinc, tin and phosphorus have been used to improve its tensile strength and ductility. Copper alloy frames can be specified with tempers from half-hard to spring temper. Half-hard material is less likely to crack upon bending, but spring temper is more amenable for automatic insertion. The most widely used metal for lead frames is Alloy 42 (42% nickel, 58% iron). This material and the more familiar Kovar® (iron-nickel-cobalt alloy) were developed to match the expansion coefficient of glass for use in vacuum tubes and the making of glass-to-metal seals. Low thermal conductivity is the major drawback of Alloy 42. Since the lead frame is the major heat flow path in plastic packages, high power chips cannot be used with Alloy 42 lead frames.

Copper-clad stainless steel was developed to provide the required mechanical stability of Alloy 42 while improving the thermal conductivity. Table 3-14 presents physical, thermal and electrical properties of various lead frame materials illustrating a wide range of thermal conductivities.

Table 3-14 Typical Lead Frame Materials

Material	Composition	Thermal Conductivity W/m · K	Coefficient Thermal Expansion $\times 10^{-6}$/K	Electrical Conductivity % IACS[a]	Strength GPa
Alloy 42	Ni 42%, Fe 58%	15.9	4.3	3.0	0.64
CDA-151 (Olin)	Cu, 0.1% Zr	359.8	17.7	90.0	0.35
KPC (Hussey)	Cu, 0.1% Fe 0.058% P	435.1	17.0	92.0	0.39
410SS (TI)	12% Cr, 88% Fe	24.3	11.0	30.0	0.62
PMC-IC (PAN)	Cu, 1% Ni, 0.2% Si, 0.03% P	259.4	16.9	60.0	0.55
KLF-2	Cu, 2.09% Sn, 0.1% Fe, 0.03% P	133.9	16.5	35.0	0.54

[a] Percentage of the electrical conductivity of copper.

Wire bonding to lead frames has been studied by many authors.[42,43] The gold-copper alloy system produces three ductile intermetallics (Cu_3Au, $AuCu$, Au_3Cu) with an activation energy range for formation from 0.8 to 1.0 eV. These intermetallics, through the action of Kirkendall-like voiding, can be shown to decrease bond strength, but overall the strength over time should be adequate for commercial devices. Studies predict a lifetime of at least five years at 100°C according to Harman.[44] As with most bonds, the quality of the gold-copper bond is influenced by the copper microstructure, the weld formation process and bonding parameters, and the impurity content of the copper.

A clean, oxide-free, lead frame is of paramount importance. All hydrocarbon contamination must be removed including the residue from the curing of organic die attach adhesives. Removal must be done without oxidizing the lead frame. Similarly, the organic adhesive cure must be done in an inert atmosphere to avoid oxidizing the copper during the adhesive setting process.

Many manufacturers have gold-plated their lead frames to prevent copper gold intermetallics from forming. But these platings (usually with a nickel strike underplate) can be subject to several problems which can cause poor bonding or delamination. Problems such as inadequate film purity, hardness, plating uniformity and hydrogen content have all led to poor quality wire bonds. In particular iron-nickel alloys (or nickel understrikes) can be particularly susceptible to hydrogen embrittlement. Gold films for high quality wire bonds should contain no thallium and less than 50 ppm total of Ni, Cu and Pb.[44] The inclusion of hydrogen should be at a minimum and the resultant gold should be soft (< 70 HKN) with a continuous small grained nodular appearance. Annealing (heat treating at high temperature $\sim$ 200°C for several hours) can be used to remove hydrogen, but care must be taken not to allow underplatings of Ni, Cr, Ti or Cu to diffuse to the surface, oxidize and prevent the sticking of wire bonds.

Other aspects of bonding to lead frames include appropriate clamping to the frame in ultrasonic or thermosonic methods.

3.4.4 Substrate Metallizations and Materials

Substrate metallizations display many of the same features and effects as described above for chip metallizations and lead frames. Researchers have studied the bondability of both thick and thin film metallizations for many years. From all the studies several general results have been obtained.

1. Increased bonding temperature produces stronger thermosonic bonds. In fact temperature has a more significant impact than bonding power (i.e., ultrasonic energy).
2. For both gold thick[45] and thin films, bond strength after reaching a threshold essentially remained constant (for a given thickness metallization) over a wide range of ultrasonic energy. A curve for thin film gold is shown in Figure 3-14.
3. Thermosonic bonding to aluminum metallization (with gold wire) shows a wider variation with a greater dependence on ultrasonic energy. A curve for aluminum thin film is shown in Figure 3-14.
4. For a given diameter bond the expected shear strength of a gold-gold bond (gold wire-gold (thin-film) metallization) would be greater than a similar gold wire on appropriate aluminum metallization. Thick film gold metallizations tend to produce strong bonds but with greater variability.[18]

Most underlying ceramic type substrates seem to have little effect on metallization bondability unless they contribute to pad roughness or impurity contamination of the bonding surface. The location of bonding pads on silicon (and polysilicon) versus SiO_2 can have a profound effect on bond strength, bond reliability and the cratering phenomena.[32] Aluminum silicon alloys and the formation of aluminum gold intermetallics also play a major role on bondability and the ultimate bond strength. GaAs is very

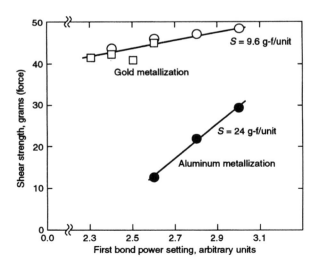

Figure 3-14 Gold thermosonic ball bond shear strength versus first bond power (ultrasonic energy) for both gold and aluminum thin film metallizations.

prone to cratering and mechanically induced defects. To minimize cratering one follows the same rules as applied to silicon: minimum ultrasonic energy, minimum tool bounce and thick (greater than 1μm) pad metallization. Preheating and avoidance of rapid thermal shock are also good measures to ensure GaAs bondability.

3.5 WIRE BOND TESTING

The destructive wire bond pull test, since its introduction in the 1960s, is the most widely accepted technique for the evaluation and control of wire bond quality and the associated bonding process.[46,47,48] The destructive wire bond pull test, however, has limitations in at least two significant areas. First, since the pull test by its very nature is destructive, it can provide information only on a lot sample basis or in a post-mortem fashion. Thus, it does not provide a measure of quality for each bond. Second, the destructive wire bond pull test provides very little information on the strength and overall quality of the first and second bonds themselves (i.e., the wire bonding-bond pad interfaces). Obviously, only in cases of catastrophic bond failures, such as impurity-driven intermetallic growth,[12] will the destructive wire bond pull test yield information other than the relative breaking strength of the bonding wire. This point is especially significant in the case of gold ball bonding, such as thermocompression or thermosonic bonding, where a ball of relatively large diameter (nominally 2.5 to 5 times the wire diameter) forms an effective bonding pad attachment which is many times stronger than the wire's breaking strength. Although typically stronger than the nominal wire bond pull forces,[45,49] the strength of the ball-to-bonding pad attachment can vary substantially due to the influences of bonding parameters, interfacial material composition, and environmental stresses. Although aluminum ball bonding is only in experimental development stages, the results presented here for gold ball bonding should be directly applicable when aluminum ball bonding is introduced into hybrid production. Similar comments can be made about silver and copper ball bonding.

These factors have led to the development of two major complementary wire bond tests: the 100% non-destructive pull test[50] which provides a degree of confidence in the quality of each bond and the ball shear test[51,52] which can be used to investigate not only the integrity of the ball bond-bonding pad interface, but also the influence of both pre- and post-bond processing factors. Non-destructive ball shear testing[53] has also been investigated, but the difficulty of its implementation as a 100% test in most practical integrated circuit, hybrid, and multichip module geometries has limited its use. Typical experimental configurations for both the wire bond pull test and the ball shear test are shown in Figure 3-15.

3.5.1 Wire Bond Testing Equipment

With the advent of the widespread application of micoprocessors to measurement equipment, wire bond testing has been turned into a repeatable, routine operation. Modern, semi-automatic wire bond pull testers are marketed by several firms. These units, once manually positioned, will automatically perform destructive or non-destructive wire bond pull tests, digitally display the pull test results (or the non-destructive pull test (NDPT) limit), and provide a permanent record either through their own integral printer or some external output device. These systems can be programmed (or are pre-programmed to calculate the mean ($\bar{X}$), standard deviation (σ) and other desired parameters of interest such as ($\bar{X} - 3\sigma$) and ($\sigma/\bar{X}$)). The machines automatically detect and record NDPT wire bond failures (i.e., pull forces that do not reach the pre-set limit) which occur during the test, as well as excessive wire bond loop height. Because of the confidence produced in wire bonds by the 100% non-destructive pull tests, manufacturers have created wire bonders which automatically pull each wire (to a pre-set limit) after the completion of each wire attachment operation. These machines were cumbersome with added hardware in the vicinity of the bonding head. Current trends are leading to separate fully automated wire bond testers which can be programmed in a manner similar to automatic wire bonders to locate the center of each wire and pull it non-destructively. In these units, once the location and pull parameters are programmed into the machine, the operator loads the samples, locates some initial fiducial marks and starts the machine. When complete, the operator removes the sample and repeats the process.

Ball shear testers have not quite reached the sophistication of some of the latest wire bond pull testers. In most cases, they are either precision manual machines with analog force gauges similar to many die shear machines used in the semiconductor and hybrid industries for years or semi-automatic machines with digital readouts obtained by reconfiguring wire bond pull testers. The shear can be conducted with the sample held either horizontally or vertically as shown in Figure 3-15. Shearing rams with sizes on the order of 0.152 mm (6 mils) have been shown to be most effective for balls formed on 25.4 μm (1 mil) diameter gold wire.[33]

Because of the accuracy and precision built into modern microprocessor-based pull and shear test machines, it is possible (assuming machines are properly calibrated and in working order) not only to believe the

(a) Wirebond pull testing

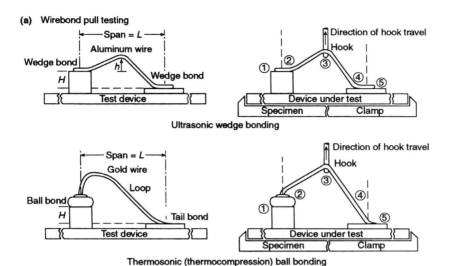

Numbers 1-5 correspond to the standard wirebond failure codes. (1) First bond lift; (2) break at first bond transition; (3) mid-span break; (4) second bond transition; (5) second bond lift.

(b) Ball shear testing

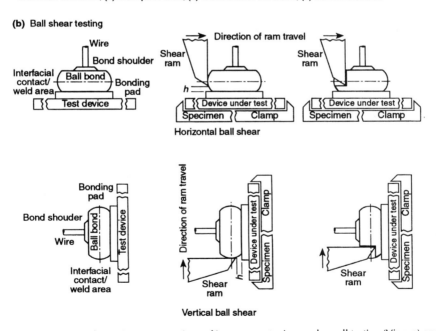

Figure 3-15 Schematic representations of interconnect wires under pull testing (View a) and ball bonds under ball shear testing (View b).

validity of data collected at different sites and on different machines, but also to combine data sets if they have a common sample base. Typical destructive pull test and ball shear experimental data at various sites and machines with common sample bases are shown in Tables 3-15 and 3-16 respectively.

3.5.2 Wire Bond Pull Tests

Destructive Wire Bond Pulls

The destructive wire bond pull test has been used by many researchers as a measure of wire bond strength, quality, and hence, long-term reliability. It is also considered by many as a standard by which to judge the bondability of a metallized surface, the quality of bonding wire and as a parametric response parameter for optimizing bonding machines and establishing hybrid processing procedures (epoxy cures, cleaning methods) and test methods. However, because of the previously mentioned insensitivity to the factors influencing the quality of the bonding pad-wire interface, the destructive wire bond test is severely limited in shedding meaningful insight into many of the tests and evaluations described above. Table 3-3 gives a list of the possible information that can be supplied by the destructive wire bond pull test.

Wire bond geometry is one of the most significant factors in determining overall wire bond pull strength. As an example, special test substrates were constructed with photolithographically patterned bonding pad pairs spaced apart in multiples of 0.51 mm (20 mils) from 0.51 mm to 2.54 mm (100 mils). Gold wires, both 18 μm (0.7 mil) and 25.4 μm (1.0 mil) in diameter, were thermosonically bonded to these pad pairs and subsequently pulled. The pull data is shown in Table 3-17. Upon analysis by a resolution of force technique the data was shown to be consistent with the geometrical interpretations of Harman and Cannon.[48]

To study the geometrical influences on wire bond strength in actual multichip situations, multichip module layouts were reviewed to determine average bond length and type. Typical bond lengths ranged from 1.02 mm (40 mils) to 1.52 mm (60 mils) with some as short as 0.51 mm (20 mils) and a few over 2.54 mm (100 mils) in length. Table 3-18 summarizes bond pull strength data for representative multichip populations. Three different multichip types (I, II, III) were considered. The Type I multichip modules consisted of a few large integrated circuit die, relatively uniform in size and height, with closely spaced but uniform length wire bonds. Type II multichip modules contained a mixture of chip types (i.e., GaAs, Si, and

Table 3-15 Typical 25.4 μm Diameter Thermosonic Gold Wire Bond Testing Comparisons Between Various Pull Test Machines and Measurement Sites

| Bond Geometry (Metallization/Substrate) | Ball Bond Pull Strength (grams (force)) Wire Bond Pull Tester Model (Site) | | | | |
	Tester Model #1 (Site A)	Tester Model #2 (Site A)	Tester Model #1 (Site B)	Tester Model #2 (Site B)	Tester Model #3 (Site C)
Chip-Substrate (Al/Ceramic)	9.5 (0.9)[a]	9.3 (0.7)	9.8 (1.1)	9.0 (0.7)	9.5 (0.9)
Substrate-Substrate (Al/Ceramic)	8.3 (0.6)	8.0 (0.5)	8.0 (0.5)	8.3 (1.6)	8.1 (0.7)
Chip/Substrate (Au/Ceramic)	9.8 (1.0)	9.6 (0.8)	9.5 (0.8)	9.3 (1.2)	9.4 (0.9)
Substrate-Substrate (Au/Ceramic)	8.1 (0.8)	8.3 (0.9)	8.2 (0.8)	8.2 (0.9)	8.1 (0.8)

[a] 9.5 (0.9) = 9.5 ± 0.9 grams (force).
Each data point represents the average of 100 samples.

Table 3-16 Typical 25.4 μm Diameter Gold Wire Bond Testing Comparisons Between Various Shear Test Machines and Measurement Sites

Metallization (Substrate)	Ball Bond Shear Strength (grams (force)) Ball Shear Tester Model (Site)					
	Tester Model #1 (Site A)	Tester Model #2 (Site A)	Tester Model #1 (Site B)	Tester Model #2 (Site B)	Tester Model #2 (Site C)	
Gold (2.5 μm thick) (Ceramic)	49.4 (6.5)[a]	47.5 (7.1)	48.5 (6.8)	46.8 (6.9)	45.2 (7.3)	
Aluminum (1 μm thick) (Silicon)	48.8 (6.7)	49.5 (7.5)	47.6 (6.9)	48.8 (7.3)	—	
Aluminum (1 μm thick) (SiO$_2$ on Silicon)	48.5 (6.3)	46.5 (6.8)	—	47.1 (7.0)	47.2 (6.5)	

[a] 49.4 (6.5) = 49.4 ± 6.5 grams (force).
Each data point represents the average of 100 samples.

Table 3-17 The Effects of Bond Length on Wire Bond Pull Strength (Substrate to Substrate Bonds). Thermosonically Bonded Gold Wire on Gold Metallized Ceramic Substrates

Bonding Wire Diameter	25.4 μm (1-mil)		18 μm (0.7-mil)	
Bond Length	Pull Statistic[a]	Min.–Max Pulls[a]	Pull Statistic[a]	Min.–Max Pulls[a]
0.51 mm (20-mils)	10.5 ± 0.8	8.8 – 11.8	4.5 ± 0.4	3.8 – 5.5
1.02 mm (40-mils)	9.2 ± 0.9	6.9 – 10.8	3.2 ± 0.6	2.4 – 4.3
1.52 mm (60-mils)	7.4 ± 1.0	5.7 – 9.1	2.3 ± 0.4	1.8 – 3.0
2.03 mm (80-mils)	6.6 ± 0.7	5.2 – 7.9	2.0 ± 0.3	1.4 – 2.3
2.54 mm (100-mils)	5.8 ± 0.5	4.8 – 6.6	1.8 ± 0.3	1.5 – 2.7

[a] Pull strength in grams (force) for a nominal 50 bond sample.

both digital and analog) with varying heights, but relatively uniform size. Type III multichip modules had a complete mixture of chip sizes and heights as well as a variety of chip attachment modes from eutectic to epoxy. This type, although designated a multichip module, was more representative of the conventional hybrids of yesteryear. Both Type I and II multichip modules used epoxy die attach. All multichip modules used gold metallization. The bond populations in Table 3-18 were segregated according to bond type (i.e., chip to module substrate, module substrate to enclosure pin, and substrate to substrate). It is interesting to note the increase in the standard deviation (for chip to substrate bonds) as chip sizes and heights begin to vary widely. Thus, the regular geometry in the Type I multichip module would allow them to pass routinely any statistical bond screens (e.g., $\bar{X} - 3\sigma \geq 3$ grams (force)) while the varied geometry of the Type III multichip modules would not pass such a screen even though the minimum bond strength was 3.2 grams (force).

Non-destructive Pull Testing

Non-destructive pull testing[50] is used to check the validity of each wire bond by stressing the bond to a preset (non-harmful) limit. The non-destructive pull test (NDPT) limit is based, obviously, on the wire size, type, and ultimte wire breaking strength. Generalized guidelines are contained in the standardized ASTM Test Method.[50] NDPT has been shown by many authors to be effective in improving the reliability of microcircuit wire bonds. An example of the improvement in destructive wire bond pull test results after the introduction of NDPT is shown in Figure 3-16.

NDPT can be introduced into almost all integrated circuit, hybrid, and multichip module microelectronic application areas, provided a proper training, qualification, and screening program is followed. The elements of this program may vary but should include a pre-introduction operator training and testing program, a controlled evaluation of NDPT procedures on representative circuits in the production line (to identify problem areas), and the establishment of geometrical screening guidelines (determined by your equipment and operators) for all future circuits designs. With skilled operators and the current semi-automated testers, minimum applicable wire spacing appears to be 0.51 mm (20 mils). This criterion is based on two hook lengths (plus a 50% safety margin). Fully automated testers may reduce this minimum spacing limit.

Table 3-18 The Effects of Multichip Module Bond Geometry and Type on Wire Bond Pull Strength. Thermosonically Bonded 25.4 μm Diameter Gold Wire. Gold Metallized Ceramic Test Substrates

MCM Type	Bond Type	Number of Bonds[a]	Pull Test Statistic[b]	Min. – Max. Pull Strength[b]
Type I	Chip–substrate	2376	8.2 ± 1.1	4.5 – 10.3
(9, 132 I/O Chips)	Pin–substrate	400	7.1 ± 1.2	5.2 – 12.0
	Substrate–substrate	72	8.5 ± 2.1	5.1 – 11.2
Type II	Chip–substrate	1560	9.2 ± 1.6	5.0 – 12.4
(9 Chips 48-132 I/O)	Pin–substrate	320	7.8 ± 0.7	5.9 – 10.9
	Substrate–substrate	80	8.5 ± 2.4	6.5 – 11.1
Type III	Chip–substrate	992	8.9 ± 2.2	3.2 – 11.2
(25 Chips 2-84 I/Os)	Pin–substrate	192	7.7 ± 1.4	6.2 – 12.6
	Substrate–substrate	32	7.5 ± 0.9	6.3 – 9.1

[a] Bond totals represent two test substrates each.
[b] Pull strengths in grams (force).

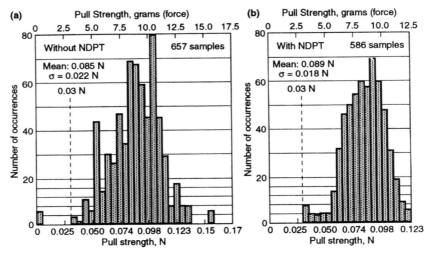

Figure 3-16 Wirebond pull strength histograms for 25 μm diameter, thermosonically bonded gold wire on gold thin film metallization (on ceramic). Histogram (a): after burn-in without NDPT. Histogram (b): after burn-in with NDPT applied post bonding.

3.5.3 Ball Shear Testing
Destructive Ball Bond Shearing

Using the generic configuration illustrated in Figure 3-15 and the equipment described above, ball shear data can be collected on a variety of circuits and material sample configurations. Ball shear strength data for various testers, tester sites, and substrate metallization combinations is presented in Table 3-16. Since the ball bond shear test provides a direct measure of the ball bond-bonding pad interface strength, it can be used to optimize the bonding process as well as parametrically study the effects of changes in wire and pad metallurgy and thickness, pad cleaning, processing, and subsequent thermal processing. Such parametric studies have been performed by several authors leading to such conclusions as: the gold ball bond shear strength decreases with increasing thickness of aluminum metallization,[33] both UV-ozone and oxygen plasma appear to be effective pre-bonding cleaning methods,[35] prior testing influences (such as destructive pull testing and non-destructive ball shear) are minimal,[35] and gold ball bond-aluminum bonding pad intermetallic growth (bond aging) is dependent on many factors and can display multiple activation energies.[30]

A typical ball shear histogram for gold ball bonds on aluminum (on silicon) is shown in Figure 3-13 along with a subsequent histogram after bonder optimization. These histograms contain only true ball shear data as defined in the immediately following failure mode discussion. In complete analogy with the wire bond pull test (e.g., see Fig. 3-15), there are several modes of failure associated with ball shear testing. These modes are defined in Table 3-19. In addition to the failure modes encountered during the ball shear operation, there are many interference modes. Rams that are too high or angled upward result in lower than normal shear strength values. Rams that are angled downward or positioned too low will strike the bonding pad and/or the substrate (chip) and cause inordinately high shear strength as well as potentially damage either the specimen or the shearing ram. In bonding systems in which excessive intermetallic growth has formed around the ball bond, the shearing ram may contact the intermetallic rather than the ball bond, and thus, the shear readings can be in error (i.e., weak ball bond shear is masked by the shear strength of the strong intermetallic wreath surrounding it).

Shearing gold ball bonds on gold metallization pads or substrates can lead to another interference (depending on relative size of pad in relationship to the ball) called friction rewelding.[35] As a strongly welded gold ball bond is sheared, the ball tends to tip away from the ram and contact the substrate as it moves. The ball smears against the pad metallization and rewelds itself often several times before it finally clears the metallization.

Non-destructive Ball Bond Shearing

Non-destructive ball shear (NDBS)[53] is possible in direct analogy with the non-destructive wire bond pull test. In NDBS the ram loads the ball to a preset value (nominally less than the envisioned shearing strength of the system under test) and then if a failure does not occur the ram is retracted and moved to the next bond. Experiments have shown that NDBS does not affect ultimate destructive shear value, at least with balls formed from 25.4 μm diameter wire and an NDBS limit of 15 grams (force).[54] Other experiments[35] have shown that the ball can be loaded non-destructively up to 60% of its shearing strength without influencing the subsequent destructive shear.

3.6 WIRE BOND QUALITY ASSURANCE/RELIABILITY

3.6.1 Quality Assurance (QA)

Wire bond QA is accomplished by establishing base line standards for wire bond quality and then assuring that product meets or exceeds these requirements on a daily basis. Two principal QA tools are the wire bond pull

Table 3-19 Ball Shear Failure Modes

Failure Mode No.	Failure Terminology	Failure Description
1	Ball lift	Separation of the ball bond at the bonding pad interface with little or no (less than 25% of the bond deformation area) residual ball metallization remaining on intact bonding pad. Specifically for gold ball bonds on aluminum metallization, a ball lift is defined as a separation of the ball bond at the bonding pad interface with little or no intermetallic formation either present or remaining (intermetallic area less than 25% of the bond deformation area)
2	Ball shear	Separation of the ball bond at or immediately above the bonding pad with appreciable intermetallic (in the case of an aluminum–gold system) and/or ball metallization (in the case of the gold-gold system) remains on the bonding pad (area of remaining metal or intermetallic greater than 25% of the bond deformation area)
3	Bonding pad lift	Separation between the bonding pad and the underlying substrate. The interface between the ball bond and the residual pad metallization attached to the ball remains intact
4	Cratering	The bonding pad lifts taking a portion of the underlying substrate material with it. Residual pad and substrate material remain attached to the ball. The interface between the ball and this residual material remains intact
5	Wire shear	The ball sheared too high or off line. Only a minor fragment of the ball attached to the wire. The major portion of the ball remains on the pad with the bonding pad-ball interface region intact. This may be regarded as one of the interferences

test and the ball bond shear test. These test methods can be used to ensure that the strength of initial bonds as well as bonds after processing and other forms of stress meet established standards. Section 3.5 above discusses these tests in detail while Section 3.2 describes how the ball bond shear test can be used to optimize bonding machine parameters for a given die-package configuration and metallization scheme.

Since most wire bond tests are practiced in a destructive mode, 100% testing is not possible, and lot sampling and pre- and post-run qualifier techniques must be used. Pre- and post-run qualifiers are used to ensure that on a standard test sample (representative of product), bonding machine wire and operator performance has remained constant from the beginning of a run to the end. If any change is made during a product run an additional qualifier must be used and tested. If no changes occurred during the day, usually beginning of day and end of day qualifiers are sufficient. Pull test and/or shear test data are recorded for each qualifier and used to determine bond strength history over time. Run charts are usually produced and previous data samples are used to set control limits for the process. An example of a wire bond strength run chart is shown in Figure 3-17. All deviations from the controlled process should be noted and their root causes explored. Samples should consist of at least 32 bonds so that large number statistics can be employed and the strength distribution averages and standard deviations have their conventional meanings.

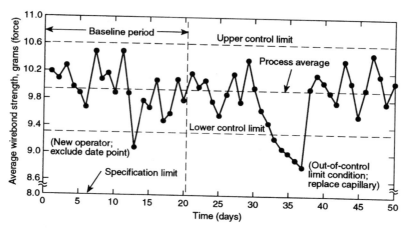

Figure 3-17   Run chart for destructive wire bond pull test for thermosonically bonded 25.4 μm diameter gold wire on gold metallization (on ceramic). Process monitoring is used to develop control limits for the process during a baseline period. These limits are much tighter than the specification limit of 8 grams (force) generated from $\bar{X} - 3\sigma = 3$ grams (force) (Reference 54). Here the run chart allowed the identification of capillary wear before the product exceeded the specification limit.

In addition to plotting the average and standard deviations on control charts, strength histograms should be produced on a regular basis to identify any bimodality (e.g., due to different underlying failure mechanisms, etc.). Single mode underlying distributions are required to use adequately and appropriately the statistical mean and standard deviations. Most test methods put specification limits on the mean and standard deviation. For example, with 25 μm diameter gold wire these limites are: $\bar{X} - 3\sigma \geq 3$ grams (force), and $\sigma/\bar{X} \leq 0.25$, where $\bar{X}$ equals the distribution mean and σ is the standard deviation.

Full information on the use of the wire bond tests and how specification limits may be set are given in References 35 and 48. Another quality assurance measure involves working with wire vendors to ensure a continued supply of high quality wire of appropriate strength and temper with a minimum amount of lubricants. Bonding wire has been linked to several failure mechanisms over the years.[55] Wire bond geometry[49] has a significant influence on the apparent strength and quality of wire bonds. Careful control of pad geometry (as described in Section 3.7 below) as well as bond length and height is necessary to ensure high quality bonds. In low profile bonds, wire bond pull testing can indicate low strength (due to resolution of force issues) while the bonds and wires are actually strong (see Table 3-7).

Lot sample wire bond testing is usually done on the finished product as part of a destructive physical analysis program to ensure that strength has been maintained in the product after prolonged testing or thermal aging.

The wire bond test and in particular the wire bond pull test can be done in a non-destructive mode (NDPT). NDPT can assure that all bonds possess a minimum level of strength. One hundred percent NDPT is time consuming but allows significant improvement in product quality as discussed above.

3.6.2 Reliability

Wire bonds have been shown to be a highly reliable flexible interconnect scheme for decades. In fact, wire bonded products have been in continuous use in space for over 25 years.[56] Wire bond, like all complex physical and chemical processes, can be fraught with reliability detractors if proper cautions and controls are not exercised and if the phenomenological factors are not well understood. An entire text book by Harman[44] has been devoted to reliability issues and yield problems. Of all the issues two particular ones deserve further description: intermetallics and cratering.

Intermetallics

Many analyses of aluminum-gold intermetallics and their formation kinetics were made during the 1960s and early 1970s. Initial studies indicated that high temperatures (300 to 400°C) and large time-temperature products were needed to form excessive amounts of intermetallic and produce bond failure. Many questions were raised during this period including: intermetallic phase structure and properties, the growth kinetics of each intermetallic phase, bond strength and the actual intermetallic bond failure mechanism, the role of silicon, the role of impurities (both organic and inorganic) and the influence of the ambient package gas. Many of these questions were answered or resolved by the work of Philofsky in 1970,[11] but several questions still remain, even after the extensive studies during the 1980s[12,30,32,33] especially concerning impurities, package gas ambient, interfacial stress and automated, high-speed bonding.

The basic intermetallic failure mechanism associated with wire bonds involves embrittlement of the gold-aluminum interconnection by intermetallic formation followed by eventual degradation of the bond strength by Kirkendall voiding.[11] Various workers[57,58] have shown that void-free intermetallics are stronger than either of the constituent metals–gold or aluminum. During the initial stages of intermetallic formation, the interface is relatively continuous and the bond strength is high. As the bond is stressed (e.g., subjected to large temperature-time products), vacancies or Kirkendall voids form in the intermetallics due to the different mutual interdiffusional rates of gold and aluminum. Prolonged stress (or thermal exposure) can cause these voids to coalesce and form a continuous void layer. Such a layer results in an electrical open circuit and an eventual lifted bond.

Table 3-20 presents the formation temperature of the five aluminum-gold intermetallics as determined from experiments with gold-aluminum reaction couples.[11,30] At high temperatures (400 to 450°C) such as encountered in ceramic package sealing, thermocompression bonding, and other processes, a certain amount of additional intermetallic growth (over that formed initially during the wire bonding operation) is unavoidable. However, it can be controlled if the time at high temperature is minimized and if proper materials and cleaning procedures are used. Several authors[30] have presented design rules for minimizing intermetallic failures by controlling film layer composition and thickness. In several cases,[55,59] impurities in the metallization, wire, or at the wire bond-pad interface have been shown to cause rapid intermetallic growth and voiding at lower temperatures than normally associated with intermetallic growth. Activation energy estimates for this impurity-driven intermetallic growth have been as low as 12.8 kcal/mole (0.5 eV).

Table 3-20 Aluminum-Gold Intermetallics: Basic Information and Observations

Au-Al Intermetallics[a]	Formation Temperature °C	Activation energy kcal/mol (eV)	Comments
$Au_5 Al_2$	100 23 (Ref. 55)	14.2 (0.68)[b]	Tan in color
$Au_2 Al$	50 80 (Ref. 55)	23.5 (1.02)	Metallic gray in color, orthorhombic randomly oriented crystals, parabolic growth observed at 50–80°C once Au_5Al_2 was begun. The boundary of Au_2Al advances into Au. Dark gray color (Ref. 11) begins at 80°C and Au_2Al becomes growing phase (Ref. 55).
$Au Al_2$	150	27.6 (1.2)	Deep purple color, resistivity = 8 micro-ohm centimeters
$Au_4 Al$	~ 150	—	Tan in color (Ref. 11)
$Au Al$	~ 250	—	White in color (Ref. 11)

[a] The intermetallic phases form in the order listed $Au_5Al_2 \rightarrow Au\ Al$.
[b] Value usually given as 13 kcal/mol (Ref. 11). Ref. 55 quoted 12.8 kcal/mol.

Initially, in most aluminum-gold wire bonds, a tan phase (typically Au_5Al_2) is present along with a trace of the purple phase ($AuAl_2$). As post-bond heating occurs the amount of $AuAl_2$ growth follows a parabolic type growth law with temperature. A lower bound for the activation energy of this phase has been estimated at 18 kcal/mole (0.78 eV). The role of the underlying substrate in contribution to the bond failure mechanisms has been studied by several authors.[32] Both silicon and silicon dioxide produce different ball bond shear failure mechanisms upon thermal aging. Initial un-aged bonds produced similar strength results. For short-term high temperature aging (200~250°C for 3 hours), bonds made to aluminum on silicon were very strong while those on SiO_2 were becoming weaker and exhibiting failures due to loss of adhesion on the SiO_2. After prolonged aging (≥ 1000 hours) at these temperatures, bonds made to aluminum on both silicon and SiO_2 were significantly weaker with similar reductions in shear strength. Thus, mechanically, neither contact system is significantly superior to the other after prolonged thermal aging.

Results for SiO_2 samples point to loss of bond adhesion at the SiO_2 interface as the primary reason for a reduction in shear strength after prolonged thermal aging at both temperatures. This phenomenon of aluminum depletion under the gold ball bond is unfortunately dominated by the low temperature intermetallic phases (Au_2Al and Au_5Al_2), both of which proceed at temperatures as low as 80°C (see Table 3-20). The ultimate loss of adhesion at the SiO_2 interface should therefore depend on the initial amount of aluminum present and the operating temperature at the bonding site. Prolonged burn-in or high temperature baking will accelerate the eventual demise of the Au-Al-SiO_2 contact system.

For the case of the Au-Al-Si contact system, the primary failure mode for prolonged high temperature annealing appears to be the continuous formation of gold-rich intermetallic phases. SEM analysis of ball sheared samples indicates that the net shear area (i.e., the smeared area that results from ball shearing) for these specimens is actually decreasing with time at elevated temperatures. This is probably due to the movement of the gold-rich intermetallics up into the gold ball and the formation of a void layer at the Au_5Al_2-Au_4Al interface. For this contact system, no aluminum depletion failures are found to occur. This is believed to result from the maintenance of good adhesion to the underlying silicon substrate, possibly due to the formation of a ternary alloy[32] at the silicon-intermetallic interface. In this case increasing the film thickness is not desirable since a good adhesion layer is maintained in spite of any prolonged thermal processing.

The cratering effect seen in silicon samples is probably due to the film stresses developed during the formation of Au_2Al at the bond interface. Film thickness measurements with a Sloan Dektak IIA stylus profilometer have indicated large volume increases when this particular phase was prepared from thin film diffusion couples. Measurements of diffusion induced strains for Au_5Al and Au_2Al by Takei and Francombe[58] have indicated film stresses which would correspond to about 2×10^9 dyne/cm^2 (approximately 90 grams (force) for an average ball area). They indicate that plastic flow and strain relief (which occur during recrystallization as the annealing proceeds) would limit the maximum strain observed. Subjecting a highly strained sample to a shear test could aid in the mechanical failure of the underlying silicon along slip planes. After the formation of intermetallics under the ball bond at the silicon interface is complete, strain relief occurring via recrystallization would proceed upon further annealing. This would explain the disappearance of cratering in samples that were thermally processed for long times.[32]

Cratering

As mentioned previously, cratering can be a significant problem associated with the bonding and subsequent shearing of ball bonds from

silicon integrated circuits. Intermetallic formation, induced stress, metallization thickness, bonding parameters, and underlying dielectric layers have all
been noted to have an effect.[32] To help separate these phenomena, a series
of cratering-related experiments (bonding, etching, metallizations, etc.) and
finite elements analysis (FEM) have been performed. The results of these
studies show:

1. the effects of gold-aluminum diffusion-induced strains within the weld
   region are negligible compared to those introduced by shear testing;
2. the smaller the weld region, the more likely the underlying silicon will
   crater when shear tested;
3. the taller the ball bond, the more likely the underlying silicon will
   crater when shear tested; and
4. the stress field for an angular type weld is similar to that for a
   continuous circular type of the same radius.

Thus, a flatter bond with a larger weld area (or large annulus) is less prone
to produce silicon cratering when shear tested.

The results of the etching experiments indicated that the occurrence of
cracks on bare silicon or SiO_2 on silicon due to improper bonding
parameters was not likely for parameters falling within the bonding window
used for this investigation. This rules out the requirement that initial
substrate damage due to improper bonding parameters is necessary for ball
shear-induced cratering to occur.

For the case of the BPSG on SiO_2, damage of the thin oxide coating was
prevalent for almost all bonding conditions. The best set of bonding
conditions to resist cracking in BPSG films was found to be the combination
of low power, short dwell, high stage temperature, and high force. This is
consistent with the results reported by Koch and his co-investigators[60] who
studied the effects of bonding parameters for thermosonic gold ball bonding
on aluminum pads over phosphosilicate glass (PSG).

The results from the metallization thickness experiment do indicate a
significant reduction in the incidence of ball shear-induced cratering as the
metallization thickness is increased. However, since the etching analysis of
untreated wire bonded samples showed no initial substrate damage as
mentioned above, a cushioning effect of the additional metallization was
ruled out. A more plausible explanation would be that the additional metal
would prevent the alloying of the gold ball and aluminum to the underlying
silicon substrate during the thermosonic scrub. This would prevent a rigid
link to the substrate available to transfer shear energy to the underlying
silicon substrate.

Results from the experiment constructed to study the effect of bonding
conditions on ball shear-induced cratering indicated that the greater the

force and the power parameters or, equivalently, the stronger the bond, the less likely the substrate was to crater when ball shear tested. This is equivalent to stating that the larger the weld (a flatter bond), the less likely that cratering will occur. Once again, this is consistent with the results from finite element modeling and previous data indicating that the manufacture of larger, more robust bonds is less likely to cause cratering.[32,58]

Some conclusions can be drawn from the information above concerning the effects of the gold-aluminum intermetallic on the cratering effect in silicon and thermally grown SiO_2 reported in References 32 and 33. As stated above, the effect of strains introduced due to structural misfit of intermetallic phases is of second order compared to those introduced by the ball shear ram. Also, no damage was observed for either the silicon or SiO_2 for the full range of bonding parameters used in this study. Therefore, it must be concluded that the cratering effect observed here is not a result of initial substrate damage. Several factors point to the rigid intermetallic bond between the ball and the substrate as the sufficient cause for ball shear-induced cratering. These factors include:

1. the absence of cratering in samples which had not been thermally annealed (i.e., there is little intermetallic formation, and, thus, the aluminum metallization will yield before much energy can be transmitted to the underlying substrate);
2. the delay in the onset of cratering for thicker metallizations (a longer period of time would be required to penetrate the thicker metallization and alloy with the underlying silicon substrate);
3. the effect is significantly less for pads over SiO_2 (Au-Al intermetallic does not form as good a bond to SiO_2 as it will to bare silicon); and
4. as the weld area increases with additional intermetallic formation, the estimated stress concentration factors become significantly reduced, thus explaining the reduction in cratering in strong bonds after prolonged thermal aging.

In summarizing this cratering effect, a thicker bond pad metallization, a larger weld area (high ultrasonic energy and stage temperatures), and flatter bonds (high force) will produce bonds that are more resistant to any form of shear-induced cratering for bond pads on SiO_2. Weak bonds are to be avoided since they can crater at much lower shear values. Preliminary information suggests that for bonding above multilevel oxides such as PSG, BPSG, and other low-strength CVD deposited oxides, thicker pad metallization and the combination of low power, short dwell, hot stage temperature, and high force will be required to produce reasonably strong bonds and reduce damage to the underlying dielectric layers.

3.7 MCM WIRE BONDING DESIGN

Proper MCM wire bonding design is critical for today's high density, high performance modules. In particular, the spacing of bonding pads both on and off chip is critical to ensure successful high reliability wire bonds. Critical spacing is determined by the shape of the wire bonds and the shape and motion of the wire bonding tool as well as the chip-substrate-package structure and the ability to perform bonding on a high yield (repeatable) basis.

From a bonded wire point of view, the type of bond has an influence on bonding pad design rules and their subsequent spacing. Ultrasonic wire bonds tend to have a smaller width (nominally 1.5 wire diameters) than either thermocompression or thermosonic wire bonds with ball bond widths ranging from 2.5 to 5.0 wire diameters. For 25.4 μm diameter gold wire these dimensions tend to limit pad width to 50 μm and 75 μm (assuming minimum ball size), respectively. Pad pitch (pad width + gap width) would be typically double the pad width dimension. Staggered chip bonding pads would allow this pitch to be reduced to the width of the bonding pad. Both these possibilities are illustrated in Figure 3-9.

In ball bonding, the size and shape of the capillary (as shown in Figure 3-9) limit the fineness of interconnection pitch. With thinned neck (bottleneck) capillaries, test structures down to 110 μm pitch have been bonded.[61] Current integrated circuit pitch limits in volume production are approximately 115 μm. Because of the fragility of the thinned capillary, further reductions in pitch below 100 μm would seem unlikely.

In ultrasonic bonding with narrow, extended wedges, a bonding pitch of between 75 and 90 μm has been demonstrated with 25.4 μm diameter wire.[62] On double row (staggered bonding pads) integrated circuits with 10 μm diameter wire, a pitch of 40 μm has been demonstrated. All these pitches assume that the die (first bond) and substrate or package (second bond) are in perfect alignment. Non-parallelism between die and package edges (or substrate land patterns) reduces the distance between adjacent wires and in extreme cases could lead to shorting.

At such small dimensions, the bonding wire covers the entire bonding pad, thus probe marks and other pad defects could inhibit yield and detract from overall bond reliability. At smaller size pad cleanliness becomes increasingly important and most MCM bonding is done in cleanrooms (class 10,000 or better). Pad cleaning with oxygen plasma and/or UV-ozone will be necessary to remove certain organic and inorganic contamination. Metallizations on both die and substrates will not be uniform and thus will have to be individually characterized to ensure high bonding yields.

3.8 SUMMARY

Wire bonding continues to be a mainstay for interconnect in the modern packaging world. More wire bonds are made on a daily basis than any other form of first level interconnect. Wire bonding is flexible, reliable and extremely low cost. As multichip modules begin to dominate the packaging scheme wire bonding faces many challenges such as pitch limitations, performance limitations, and assembly costs involving speed versus yield considerations.

ACKNOWLEDGMENTS

The author wishes to acknowledge the support of APL's Microelectronic Group personnel over the last ten years in their continuing efforts to make wire bonding an exact science rather than an art. Special words of praise are given to Ms. S. Lynn Hoff and Ms. Heather L. Jackson for help in preparation of this manuscript.

REFERENCES

1. Moore, G., "VLSI, What does the Future Hold?," *Electron. Aust.*, Vol. 42, p. 14, 1980.
2. Sze, S. M., *VLSI Technology*, McGraw-Hill, New York, 1983.
3. Howland, F. L., "Trends for Future Single and Multichip Packages," *Abstract Book ISHM 1991 Joint Technology Conference*, Dallas, TX, 3 pages, 1991.
4. Messner, G., "Price/Density Tradeoffs of Multi-Chip Modules," *Proc. 1988 International Microelectronics Symposium*, Seattle, WA, pp. 28-36, 1988.
5. Charles, H. K., Jr., "Applications of Adhesives and Sealants in Electronic Packaging," *Proc. 1991 International Microelectronics Symposium*, Orlando, FL, pp. 139-147, 1991.
6. Tummala, R. R., and E. J. Rymaszewski, Eds., *Microelectronic Packaging Handbook*, Van Nostrand Reinhold, p. 391, 1989.
7. Harman, G. G., Private communication.
8. Andersen, O. L., *Bell Laboratories Record*, 1957.
9. Singer, P. H., "Hybrid Wire Bonding Advances," *Semiconductor International*, Vol. 7, no. 7, p. 66, 1984; also, G. Dehaine, K. Kurzweil, and P. Lewandowski, "Crossing the 200 Wire Barrier in VLSI Bonding," *Proceedings of the 1984 International Microelectronics Symposium*, International Society for Hybrid Microelectronics, p. 353, 1984.
10. Charles, H. K., Jr., "Electronic Packaging Applications for Adhesives and Sealants," *Engineering Materials Handbook: Adhesives and Sealants*, ASM International, Metals Park, OH, Vol. 3, pp. 579-603, 1990.

11. Philofsky, E., "Intermetallic Formation in Gold Aluminum Systems," *Solid State Electronics*, Vol. 13 (10), p. 1391, 1970.
12. Charles, H. K., Jr., B. M. Romenesko, G. D. Wagner, R. C. Benson, and O. M. Uy, "The Influence of Contamination on Aluminum-Gold Intermetallics," *Proc. 20th IEEE International Reliability Physics Symposium*, p. 128, 1982.
13. Gehman, B. L., "Bonding Wire for Microelectronic Interconnections," *IEEE Transactions on Components, Hybrids and Manufacturing Technology*, Vol. CHMT-3, no. 3, p. 375, 1980.
14. Ling, J., and C. E. Albright, "The Influence of Atmospheric Contamination on Copper to Copper Ultrasonic Welding," *Proc. 34th Electronic Components Conf.*, Institute of Electrical and Electronics Engineers, p. 209, 1984.
15. Onuki, J., M. Suwa, T. Iizuka, and S. Okikawa, "Study of Aluminum Ball Bonding for Semiconductors," *Proc. 34th Electronic Components Conf.*, Institute of Electrical and Electronics Engineers, p. 7, 1984.
16. Kurtz, J., D. Cousens, and M. Defour, "Ball Bonding with Copper Wire," *Proc. IEPS Conference, International Electronics Packaging Society*, p. 688, 1984.
17. Bischoff, A., F. Aldinger, and W. Heraeus, "Reliability Criteria of New Low Cost Materials for Bonding Wires and Substrates," *Proc. 34th Electronic Components Conf.*, Institute of Electrical and Electronics Engineers, p. 411, 1984.
18. Noble, R. P., "HMC Intraconnections: Some Observations, Problems and Solutions," *Proc. 1983 International Microelectronics Symposium*, Philadelphia, PA, p. 566, 1983.
19. Harman, G. G., and F. Oettinger, "Bonding and Packaging," *Solid State Technology*, p. 65, 1985.
20. Brown, S., M. Kressley, R. Natali, R. Rath, and P. Rima, "Multilayer Interconnects and TAB for VHSIC Chips," *Proc. 1987 VHSIC Packaging Conference*, pp. 169-181, 1987.
21. Ehrlich, O. J., and J. Y. Tsao, "Laser Direct Wiring for VLSI," in *VLSI Electronics: Microstructure Science 7*, Academic Press, NY, pp. 129-164, 1983.
22. Tuckerman, D. B., D. J. Ashkenas, E. Schmidt, and C. Smith, "Die Attach and Interconnection Technology for Hybrid VLSI," *Laser Pantography: 1986 Status Report for the VHSIC Program*, UCAR-10195, Lawrence Livermore Laboratory, p. 45, 1986.
23. Neugenbauer, C. A., "Comparison of VLSI Packaging Approaches to Wafer Scale Integration," *Proc. IEEE 1985 Custom Integrated Circuit Conference*, pp. 32-37, 1985.
24. Winkler, E. R., *Semiconductor International*, p. 350, 1985.
25. Guttman, I., S. S. Wilks, and J. S. Hunter, *Intro. Engineering Statistics*, John Wiley & Sons, Inc., New York, p. 403, 1971.
26. Box, C. E., and J. S. Hunter, "The 2^{k-p} Fractional Factorial Designs, Part I," *Technometrics*, Vol. 3, p. 311, August 1961.
27. Fonney, D. J., *PROBIT Analysis*, Cambridge University Press, London, 1947.
28. Ramsey, T. H., "Metallurgical Behavior of Gold Wire in Thermal Compression Bonding," *Solid State Technology*, p. 44, 1973.

29. Charles, H. K., Jr., and G. V. Clatterbaugh, "Ball Bond Shearing–A Complement to the Wire Bond Pull Test," *International Journal of Hybrid Microelectronics*, Vol. 6 (1), p. 171, 1983.
30. Clatterbaugh, G. V., J. A. Weiner, and H. K. Charles, Jr., "Gold-Aluminum Intermetallics: Ball Bond Shear Testing and Thin Film Reaction Couples," *Proc. 34th Electronic Components Conf.*, Institute of Electrical and Electronics Engineers, p. 21, 1984. Also CHMT-7, no. 4, p. 349, 1984.
31. Chen, C., "On the Physics of Purple-Plaque Formation and the Observation of Purple-Plaque in Ultrasonically-Joined Gold Aluminum Bonds," *IEEE Transactions on Parts, Materials, and Packaging*, Vol. PMP-3(4), p. 149, 1967.
32. Clatterbaugh, G. V., and H. K. Charles, Jr., "The Effect of High-Temperature Intermetallic Growth on Ball Shear Induced Cratering," *IEEE Transactions on Components, Hybrids and Manufacturing Technology*, Vol. 13 (1), p. 167, 1990.
33. Weiner, J. A., G. V. Clatterbaugh, H. K. Charles, Jr., and B. M. Romenesko, "Gold Ball Bond Shear Strength–Effects of Cleaning, Metallization, and Bonding Parameters," *Proc. 33rd Electronic Components Conf.*, Institute of Electrical and Electronics Engineers, p. 208, 1983.
34. Kaplan, H. L., and B. K. Bergin, "Residues from Wet Processing of Positive Photoresists," *J. Electrochem. Soc.*, Vol. 127 (2), p. 386, 1980.
35. Charles, H. K., Jr., G. F. Clatterbaugh, and J. A. Weiner, "The Ball Bond Shear Test: Its Methodology and Applications," *Semiconductor Processing, ASTM STP 850*, Dinesh C. Gupta, Ed., American Society for Testing and Materials, p. 429, 1984.
36. Holland, L., *Vacuum Deposition of Thin Films*, Prentice-Hall, 1966.
37. Chapman, B., *Glow Discharge Processes*, John Wiley and Sons, 1980.
38. Mayer, J. W., and S. S. Lau, "Aluminum Metallization Scheme," in *Electronic Materials Science: For Integrated Circuits in Si and GaAs*, Macmillan Publishing Company, pp. 280-284, 1990.
39. Poate, J. M., K. N. Tu, and J. W. Mayer, Eds. *Thin Films–Interdiffusion and Reactions*, John Wiley and Sons, 1987.
40. Steppan, J. J., J. A. Roth, L. C. Hall, D. A. Jeannotte, and S. P. Carbone, "A Review of Corrosion Failure Mechanisms During Accelerated Tests: Electrolytic Metal Migration," *J. Electrochemical Soc.*, p. 175, 1983.
41. Einspruch, N. G., and G. B. Larrabee, Eds., *VLSI Electronics: Microstructure Science*, Vol. 6, Academic Press, 1983.
42. Pitt, V. A., and C. R. S. Needes, "Thermosonic Gold Wire Bonding to Copper Conductors," *IEEE Transactions on Components, Hybrids and Manufacturing Technology*, CHMT-5, no. 4, pp. 435-440, 1982.
43. Lang, B., and S. Pinamaneni, "Thermosonic Gold-Wire Bonding to Precious-Metal-Free Copper Lead Frames," *Proc. 38th IEEE Electronic Components Conference*, Los Angeles, CA, pp. 546-551, 1988.
44. Harman, G. G., "Reliability and Yield Problems of Wire Bonding in Microelectronics," *International Society For Hybrid Microelectronics*, 1989.
45. Jellison, J. L., "Kinetics of Thermocompression Bonding to Organic Contaminated Gold Surfaces," *IEEE Transactions on Parts, Hybrids and Packaging*, Vol. PHP-13, p. 132, 1977.

46. ASTM Standard: F459-78, "Standard Methods for Measuring Pull Strength of Microelectronic Wire Bonds," *Annual Book of ASTM Standards*, American Society for Testing and Materials, 1916 Race Street, Philadelphia, PA 19103, p. 838.
47. Schaff, H. A., "Testing and Fabrication of Wire Bond Electrical Connections: A Comprehensive Survey," *National Bureau of Standards Technical Note 726*, pp. 50-102, September 1972 and the references contained therein.
48. Harman, G. G., and C. A. Cannon, "The Microelectronic Wire Bond Pull Test– How to Use It, How to Abuse It," *IEEE Transactions on Components, Hybrids, and Manufacturing Technology*, CHMT-1, p. 203, 1978.
49. Charles, H. K., Jr., B. M. Romenesko, O. M. Uy, A. G. Bush, and R. Von Briesen, "Hybrid Wire Bond Testing Variables Influencing Bond Strength and Reliability," *International Journal of Hybrid Microelectronics*, Vol. 5 (2), p. 260, 1982.
50. ASTM Standard: F458-78, "Standard Recommended Practice for Non-Destructive Pull Testing of Wire Bonds," *Annual Book of ASTM Standards*, American Society for Testing and Materials, 1916 Race Street, Philadelphia, PA 19103, p. 883.
51. Arleth, J. A., and R. D. Demenus, "A New Test for Thermocompression Microbonds," *Electronic Products*, Vol. 9, p. 208, 1967.
52. Harman, G. G., "The Microelectronic Ball Bond Shear Test–A Critical Review and Comprehensive Guide to Its Use," *International Journal of Hybrid Microelectronics*, Vol. 6 (1), p. 127, 1983.
53. Panousis, N. T., and M. K. W. Fischer, "Non-Destructive Shear Testing of Ball Bonds," *International Journal of Hybrid Microelectronics*, Vol. 6 (1), p. 142, 1983.
54. Charles, H. K., Jr., "Ball Bond Shear Testing: An Interlaboratory Comparison," *Proc. 1986 Int. Microelectronic Symposium*, Atlanta, GA, pp. 265-274, 1986.
55. Horsting, C. W., "Purple Plaque and Gold Purity," *Proc. 1972 Int. Rel. Phys. Symposium*, p. 155.
56. Charles, H. K., Jr., and G. D. Wagner, "Applied Physics Laboratory: 30 Years of Development and Innovation," *Inside ISHM*, pp. 11-17, Nov.-Dec. 1989.
57. Footner, P. K., B. P. Richards, C. E. Stephens, and C. T. Amos, "A Study of Gold Ball Bond Intermetallic Formation in PEDs Using Infrared Microscopy," *Proc. 24th IEEE Int. Rel. Phys. Symposium*, pp. 102-108, 1986.
58. Takei, W. J., and M. H. Francombe, "Measurements of Diffusion Induced Strains at Metal Bond Interfaces," *Solid State Electronics*, Vol. 11, pp. 205-208, 1968.
59. Graves, J. F., and W. Gurany, "Reliability Effects of Aluminum Bonding Pads on Semiconductor Chips," *Proc. 32nd Electronic Components Conference*, p. 266, 1982.
60. Koch, T., W. Richling, J. Whitlock, and D. Hall, "A Bond Failure Mechanism," *Proc. 24th IEEE Int. Rel. Phys. Symposium*, pp. 55-60, 1986.
61. Tsuboi, T., K. Otsuka, Y. Shirai, T. Minwa and Utsumi, "Development of Gold Wire Wedge Bonding Technology for High Pin Count QPF," *Proc. 11th Int. Electronic Packaging Conference*, San Diego, CA, pp. 565-577, 1991.

62. Ohno, Y., Y. Ohzeki, T. Yamashita, Y. Iguchi, T. Kanamori and Y. Arao, "Development of Ultra Fine Wire for Fine Pitch Bending," *Proc. 41st IEEE Electronic Components and Technology Conference*, Atlanta, GA, pp. 519-523, 1991.

4

Chip Level Interconnect: Wafer Bumping and Inner Lead Bonding

Sung K. Kang, William T. Chen,
Richard B. Hammer, and Frank E. Andros

4.1 INTRODUCTION

Tape automated bonding (TAB) technology has been widely used in consumer products for some years. Products such as watches, cameras, calculators, hand held tools, displays, etc., requiring lightweight, thin profile, low cost silicon packaging have taken advantage of TAB's attributes.

Other sectors of the microelectronics industry, for example the computer industry, have been slow to accept this packaging technology. There have been applications of TAB on ceramic Multi Chip Modules (MCMs) as an alternative die attach technique to the conventional, back-bonded, wire bond interconnect methodology. This practice has grown and will continue to grow in popularity because TAB allows the die to be tested and burned in prior to attaching to the MCM substrate. The current trend toward portable computing systems–laptops, notebooks, palm tops, etc.–is driving demand for Single Chip Modules (SCMs) with requirements similar to consumer product needs–thinner, lighter, lower cost. Some manufacturers of the Personal Computer Memory Card International Association (PCMCIA) cards have chosen TAB for packaging memory devices over the more conventional Thin Small Outline Plastic Package (TSOP) due to the higher reliability of the Outer Lead Bond (OLB) interconnection. This is the result of thinner, more flexible conductors used for TAB, 1.4 mils vs. 5 mils for TSOPs. In addition, I/Os per die continue to increase, particularly for ASIC and processor devices.

Current practice for wire bond technology is a single or double staggered peripheral row of die bond pads, the latter having limited acceptance due to the potential for wire cross-over (shorting) and ease of inspectability. There are two limitations with wire bond technology: die bond pitch and length of wire or distance a wire can be "thrown'. Pad pitch is limited by the bond tip physical dimensions,[1,4] and typical pad pitch for wire bonding is 125 micron. TAB ILB (inner lead bond) pitch on the other hand has been demonstrated, using single point laser bonding, to achieve a 75 micron pitch.[2]

The wire length issue is related to lead frame packages, such as plastic quad flatpacks (PQFP) and results because the fine geometry required to get within 200 mils of the die (approximate wire length limitation) cannot be achieved on conventional, 125 micron thick lead frames. The solution here may be to use an interposer to bridge from the region near the device to a conventional lead frame. Ceramic, plastic TAB tape, and FR-4 materials, etc., are examples of interposers which have been proposed. Both of these limitations can also be overcome by physically increasing the size of the die; however, this results in fewer die per wafer and therefore higher die cost.

An alternative to the peripheral ILB attach techniques is to utilize the full area under the die for I/O, i.e., an area array interconnection scheme.[3] This technique is particularly beneficial for TAB when two metal layer tape is used because one layer can be used for power and ground, the second layer for signal I/O. This results in fewer peripheral I/O required on the die and provides the capability for enhanced electrical performance at both the die and package level. Improved power distribution is achieved at the die as well as reduced inductance on the tape.

The subject of TAB technology has been discussed extensively in recent publications,[4-8] including such topics as tape manufacture, wafer bumping, inner lead bonding, outer lead bonding, encapsulation, burn-in and testing, thermal management, reliability, and others. In this chapter, two key aspects of TAB technology, wafer bumping and inner lead bonding, are reviewed in conjunction with COB applications. Bumping is referred to as the process of adding raised metal contacts to bond pads on the die to provide both the necessary bonding metallurgy for die-to-lead bonding and a physical stand-off to prevent lead-die shorting. Bumping can be accomplished on the individual die on a wafer, called "wafer bumping" or on the TAB tape, specifically the tip of conductor leads, called "tape bumping." The bumping processes will be reviewed in terms of bump materials/processes, structure, geometry, bump properties and behavior during bonding. The success of TAB technology is largely determined by the success of inner lead bonding (ILB), the first assembly process step in the TAB process, which interconnects the die to the tape. ILB processes are reviewed in terms of process mechanics, bonding mechanism, bump deformation and impact on reliability.

4.2 TAB INNER LEAD BOND BUMP OPTIONS

TAB ILB options can be categorized by three factors; tape structure (planar vs. bumped tape), bump materials (gold, copper, gold/copper, aluminum, solder, etc.), and bonding tool (gang vs. single-point bonding). The bonding mechanism is determined by the combination of these factors.[9] Figure 4-1 describes four ILB options classified in terms of tape structure; (a) planar tape, (b) bumped tape, (c) transfer-bumped tape, and (d) balltape. Bumps, either on a die or on the tape, serve as a "stand-off" for bonding, which connect the die to the tape both mechanically and electrically. The planar tape option with bumps on the die is commonly used for high I/O count applications, while the bumped tape or the transfer-bumped tape options are usually used for low I/O applications because of planarity issues.

In the planar tape option, the bump deposition is performed on a wafer. Gold is the most common bump material. The gold may be electroplated on the wafer through a photoresist mask,[10,11] or vacuum deposited through a metal mask.[12] Other bump materials such as copper,[13-15] gold/copper,[16] aluminum,[17] and lead/tin[18] have also been demonstrated in TAB applications. In the following sections, the wafer bumping process for each bump material is discussed.

If bumped tape with a fine lead pitch is available, it is a better choice over the planar tape option, because it eliminates the wafer bumping process. It

TAB TAPE STRUCTURE

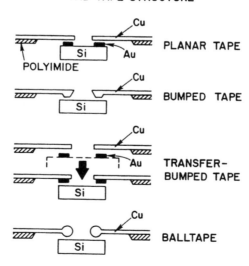

Figure 4-1   A schematic diagram showing four TAB ILB options classified acording to tape structure.

also provides the flexibility of packaging wire bond die with standard back-end-of-line metallization. Bumped tape can be manufactured either by etching a portion or section of copper from a beam to produce a bump structure[19] or by plating a bump onto a copper beam.[20] Both techniques require a lithographic process to define a window for bump production. This is a challenging problem, because the conventional tape materials are generally not rigid and therefore are susceptible to handling damage during processing. The result is alignment and planarity problems during ILB attach. This problem is more pronounced for fine pitch applications.

A third option shown in Figure 4-1 is a bumped-tape scheme, where gold bumps are electroplated on a "dummy" substrate, and then transferred onto a tape by bonding. The "transfer-bumped" tape is subsequently bonded to a die by another bonding step.[21] The advantage here, is the elimination of the wafer bumping process. However, two bonding steps are required for the die interconnection. The tape structure shown in the fourth option is called "balltape",[22,23] and is a variation of bumped tape. The balltape structure is produced from a planar tape (either two or three layer tape) by melting the ends of copper beams into spherical bumps. The melting of copper beams is possible by means of a laser beam or an arc process.

4.3 WAFER BUMPING

Wafer bumping can be classified by either bump materials (e.g., gold vs. solder) or bumping processes (e.g., wet vs. dry process). Since bumping processes are determined by the choice of bump material, we classify the wafer bumping process by: gold, copper, aluminum, and solder bumps. In the following, bump structure, material processing, geometry, properties and deformation during bonding are discussed for each material.

4.3.1 Gold Bump

Since its introduction in early 1970,[24] the gold bump has been the most popular metallurgy for TAB technology. Gold bump is usually made of two regions; thin film adhesion layers to aluminum metallization and the main body of the bump. The thin film structure consists of three layers: first an adhesion layer of Ti or Cr a few hundred ångstrom thick, second a diffusion barrier of Cu, Pd, W or Pt about 10K ångstrom thick, and third the top capping layer. The capping layer is commonly gold a few thousand ångstrom thick, which provides an easy surface for gold bump plating. Several examples of bump structures which have been used are listed in Table 4-1.

Table 4-1 Bump Structures Produced by Wafer Bumping

	Thin Films	Bump (Process)	Capping	Reference
Gold Bump				
(1)	Cr/Cu/Au	Au(plating)	—	GE, Sharp,
(2)	Ti/Pd/Au	Au(sputter)	—	Honeywell
(3)	Ti/W/Au	Au(plating)	—	
(4)	Ti/Pt/Au	Au(plating)	—	Signetics
Copper bump				
(5)	Cr/Cu/Au	Cu(plating)	Au(Ni)	Nat. Semi.
(6)	Al/Ni/Cu	Cu(plating)	Au	Nat. Semi.
Aluminum bump				
(7)	Ti (or Cr)	Al/(evapor'n)	Ti/Cu/Au	IBM
Solder bump				
(8)	Cr/Cu/Au	Pb–Sn(evapor'n)	—	IBM
(9)	Ni/Cu	Pb–Sn(plating)	—	Philips
(10)	Cr/Cu/Au	Pb–Sn(plating)	—	Honeywell

Figure 4-2 shows schematically a typical gold bumping process. To protect the aluminum metallization from corrosion and oxidation, one or two passivation layers of silicon oxide, silicon nitride or polyimide are applied before the thin film deposition. The via window in the passivation layers is opened by plasma or reactive ion etching. Blanket deposition of the thin film layers is performed by sequential evaporation or sputtering without breaking vacuum. The thin film layers play critical roles: as an adhesion layer between the aluminum metallization and the gold bump, and as a common ground for the subsequent electroplating of the gold bump. Before plating, a photo resist layer is applied and developed to define the window for the bumps. The photo resist material and thickness determine the geometry and shape of the plated bump. The earliest gold bumps had a mushroom shape as a result of the thin photo resist applied by wet processing.[25,26] Recently, bumps with straight walls were produced by using a thick, laminated resist,[27,28] as shown in Figure 4-3. The advantages of the straight-walled bumps over the mushroom-shaped bumps are better control of lateral dimensions, finer-pitch bumping, and less possibility of shorting between adjacent bumps after bonding. For plating, the most commonly used processes in the wafer bump industry are pulse[28] and rack or fountain plating. To produce a soft gold bump, high purity gold plating is required. Common plating solutions used are either gold cyanide or gold sulphite. By optimizing both chemical and electrical variables in pulse plating, gold

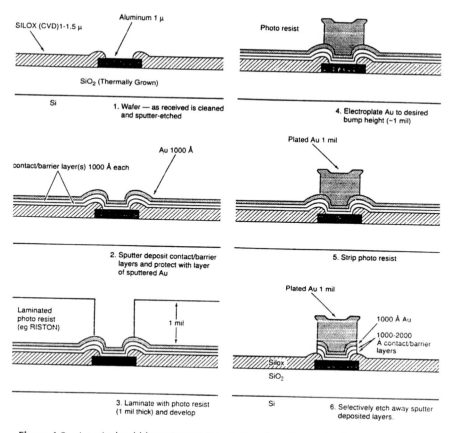

Figure 4-2 A typical gold bumping process starting from the passivation step to the bump anneal (reprinted from Ref. 27).

bumps with uniform bump height (good lateral dimension), consistent bump hardness, flat bump surface morphology, and roughened top surface[28] can be produced. Typical bump heights range from 17 to 25 μm. After gold bump plating, the photo-resist layer is stripped off and the thin film layers are selectively etched. The final process is annealing of the bumps at an elevated temperature, such as 300°C, to obtain the desired hardness. Table 4-2 shows the microhardness of gold bumps annealed under various conditions. The microhardness of a well annealed gold bump is in the range of 50-60 on the Vickers" scale (HVN), while that of the as-deposited bumps is about 120. Since the hardness of an electroplated copper beam is about 100, the annealed gold bumps are easily bondable, while the as-deposited gold bumps are hard to bond to the copper tape beams.

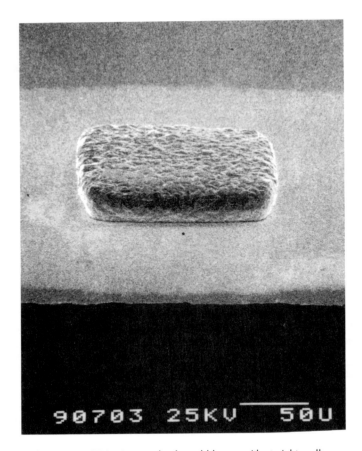

Figure 4-3   SEM micrograph of a gold bump with straight walls.

Table 4-2   Microhardness of Gold Bumps and Copper Tape

Samples	Hardness (HVN)	St'd Dev.	Max/Min
Au bumps			
As-deposited	123	4.0	127/117
200°C, 0.5 h	116	6.3	123/107
300°C, 0.5 h	52	1.1	54/51
Cu tape			
As-received	97	3.1	102/92

4.3.2 Copper Bump

Replacement of a gold bump with gold plated copper has been considered to reduce the high cost associated with gold plating. Since a copper surface is easily oxidized or corroded, for thermocompression bonding the copper bumps require a layer of gold or gold/nickel on top of the bump.[13-15] The thin film layers used are either Cr/Cu/Au or Al/Ni/Cu, as listed in Table 4-1. In the case where an aluminum contact layer is used, the adhesion to the aluminum bond pad is a critical factor in determining the fabrication yield and reliability of the copper bump.[29] A significant improvement in adhesion was achieved by a back-sputtering etch process in addition to wet acid etch prior to the deposition of the Al/Ni/Cu metallization.[29] For bump plating, a high purity copper bath, preferably copper sulfate solution, is required to produce soft copper bumps. The hardness of as-deposited copper bumps is expected to be in the range of 100 (HVN), which is similar to that of the copper beam listed in Table 4-2. Wafer annealing would reduce this hardness as in the case of gold. In general, low cost wafer bumping can be achieved with a copper bump instead of gold. However, difficulty in bonding and additional process steps for the protective layer have prevented the copper bump from gaining widespread use in TAB or COB technology.

4.3.3 Aluminum Bump

Gold and copper bumps are produced by means of electroplating or similar wet chemical processes for deposition of the metal onto the die. These wet chemical processes require the immersion of the entire wafer into the electroplating bath. The removal of the common plating electrode is accomplished by another wet chemical etching step, ion-beam etching or the equivalent. Wet chemical processes or ion-beam etching present a risk of damage or corrosion to the wafers, as well as processing complexity. The aluminum bump was designed to be manufacturable by evaporation using a metal mask, i.e., a dry physical process. Figure 4-4 shows a schematic of the aluminum bump structure.[30] The bump consists of metals evaporated sequentially; first a 500 ångstrom titanium adhesion layer, then the 120 to 180K ångstrom aluminum body, followed by a capping layer consisting of Ti/Cu/Au (5/10/2-5K ångstrom) (refer to Table 4.1). The thin adhesion layer consisting of titanium can be replaced by chromium if desired. To adjust the hardness of the aluminum bump, pure aluminum can be replaced by aluminum with a small percentage of Cu, Ni, Cr, Ti, Si or Fe. The fabrication of the bump starts with cleaning the wafers by means of plasma ashing, followed by a phosphoric/chromic acid clean. Next, the wafers are mounted against molybdenum masks with openings aligned to the die bump

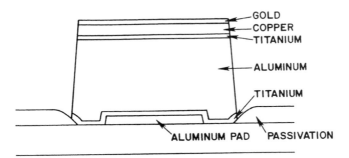

Figure 4-4 Schematic of aluminum bump structure produced by sequential evaporation of the metals with a molybdenum mask.[28]

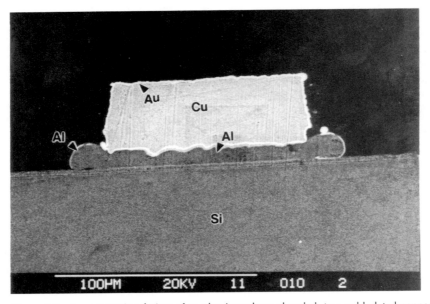

Figure 4-5 A cross-sectional view of an aluminum bump bonded to a gold-plated copper beam by thermocompression bonding.

pads. They are mounted in the evaporator which is pumped to the 10^{-6} Torr pressure range. The wafers are pre-heated to 150-200°C before evaporation of the metals. The resultant gold-aluminum bumps not only serve as a bonding platform, but also protect the pad below from corrosion by covering and isolating it from the environment. While it is known that aluminum is a good electrical conductor for the top surface metallurgy on a die, it is novel in pedestals for thermocompression bonding.[17] The relative mechanical softness of aluminum as contrasted to gold or copper can be beneficial because it permits good bonds to be made at the same time as the underlying die structure is being protected from transmission of excessive forces during bonding. The potential risk of cracking the die or passivation layer is therefore reduced. However, aluminum bumps may be subject to corrosion if not properly protected from environmental conditions, that is, encapsulation.

4.3.4 Solder Bump

The solder-bump interconnection of flip chips, the face-down soldering of silicon devices to alumina substrates, has been extensively practiced since the so-called Controlled Collapse Chip Connection (C4) technology was introduced in the early 1960s.[3] The major application of this technology has been for high performance, high I/O die interconnection. Solder bumps are evaporated on wettable metal terminals on the die through a molybdenum mask. High lead solders, 95% Pb-5% Sn, and 97%/3% have been most widely used for solder bumps to connect die to alumina ceramic substrates. Figure 4-6 is a schematic diagram of a solder bump structure used for flip-chip bonding with the interface structures on both the silicon side and the substrate side. The thin film structure of Cr/Cu/Au is one choice for the terminal connecting metallurgy on the die as shown in Table 4-1. The application of solder bumps for TAB interconnection was first reported by the Philips Research Laboratories in 1974.[31] The contact areas of aluminum metallization on the die are first coated by evaporation of a thin film of Ni. A relatively thick layer of Cu (about $10\,\mu m$) was electroplated onto the Ni layer. The copper layer served as a spacer between the die and the tape after soldering. The plating of Pb and Sn was performed

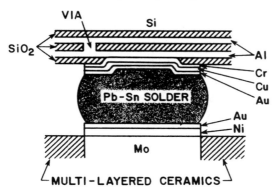

Figure 4-6 A schematic diagram of a solder bump structure used for flip-chip bonding with the interface structures at both the Si side and the substrate side.

by using a mask of a thick photographic lacquer to produce mushroom-shaped solder bumps. The final hemispherical form was obtained by "remelting" the solder bumps in a bath. The die with the solder bumps was connected to a TAB tape with Au/Ni plated copper beams.

Recently, the application of solder bumps for TAB interconnection has increased in popularity.[18,32,33] In a new application, Honeywell used the solder bump structure to interconnect TAB tape to high-performance CMOS die mounted on a ceramic substrate.[18] The ceramic carrier was then sealed with a Kovar lid and soldered to a multilayer printed circuit board. The solder bump structure was very similar to that used in the C4 technology (refer to Table 4-1). However, the deposition of lead and tin was done by electroplating rather than by evaporation. Figure 4-7 shows a schematic of the electroplated solder bump structure. The reflow bonding of the solder bump was achieved by a thermode in a manner similar to thermocompression bonding. To prevent collapse of the solder bumps during bonding, a clever polyimide support structure was built in the TAB design.[32]

A direct marriage of the C4 and TAB technology has been also demonstrated with the VLSI packaging application.[33] The process developed, called Solder Attach Tape Technology (SATT) utilizes a hot gas thermode to "reflow" the C4 solder bump. This process is described in more detail in Sections 4.7 and 4.8.

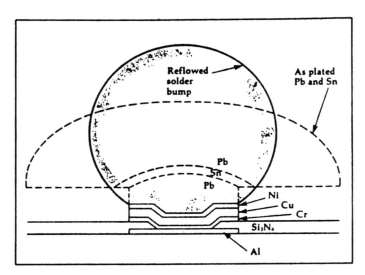

Figure 4-7 A schematic of an electroplated solder bump used for the interconnection of the TAB tape to high-performance CMOS chips.[15]

4.4 TAPE BUMPING

Although the bumped-chip or wafer approach is widely accepted for TAB applications, its inherent limitations have been well recognized.[34-36] The extra wafer-level processing (either wet or dry process) not only adds extra cost per die, but also risks yield reductions of the often expensive devices. Another limitation is the lack of flexibility in packaging die from various sources, because a device manufacturer may not be willing to supply his wafers to others for bumping. The bumped-tape or tape bumping approach can be an alternative. Here, three different schemes of the tape bumping technique are reviewed: bumped-TAB tape, transfer-bumped tape, and balltape.

4.4.1 Bumped-TAB Tape

The bumped-TAB (BTAB) tape approach was initially developed for hybrid microcircuit applications, for low volume wafer production.[35] In general, there are two methods of manufacturing the BTAB tape: subtractive and additive processes. In the subtractive process, the bump structure can be formed by selectively etching a thick copper foil as an integral part of the lead structure. In the additive process, the bump structure is plated in the photoresist windows on top of the lead structure. The simplest BTAB is one-layer, all metal tape, which is usually manufactured by selective photoetching a 2-ounce (0.0028 inch nominal thickness) piece of copper foil.[35] To facilitate metallurgical bonding, appropriate metallization (Au or Au/Ni, for example) is applied to the bump surface. Figure 4-8 shows two possible processes for the production of two-layer and three-layer BTAB tape by the additive process. For the two-layer tape, both the leads and bumps are plated, while for the three-layer tape only the bumps are plated. In either case, the bump material can be copper or gold.[20,37] A challenging problem in the additive process of the BTAB tape is the double photoresist process used to define the windows for the bump structure. To alleviate this problem, a method for the production of a self-aligned bump structure, employing the single expose double develop (SEDD) process, was proposed.[38] The SEDD process uses positive photoresist and a glass mask with patterned areas of high optical density and areas of intermediate optical density to enable two exposures to be made simultaneously. BTAB tape with fine pitch and high I/O count is not commercially available yet. Thermocompression bonding of the BTAB tape with copper bumps requires higher temperature and pressure than the eutectic bonding of gold-bumped die to tin-plated copper tape.[36] The difficulties in both bonding and tape

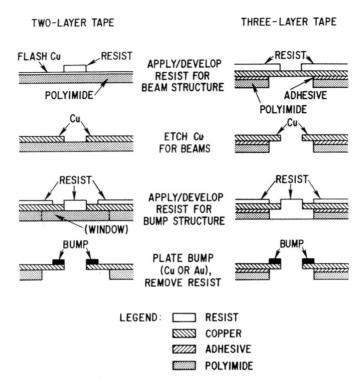

Figure 4-8 Schematic of bumped-TAB tape manufacturing steps by the additive process.

production are the principal reasons why the BTAB tape is not widely accepted for TAB or COB applications.

4.4.2 Transfer-Bumped Tape

This technology was developed in Japan and has been extensively used for consumer products, such as LCDs, TVs, smart cards, watches, personal computers, cameras, and others.[21,39] The major advantage of this technique is the elimination of wafer bumping. Gold bump plating is done on a glass substrate coated with a conductive layer which is used as the plating electrode. Photoresist is applied to the substrate and openings in the conductive layer are developed for plating. A mushroom-shaped bump is formed with a typical dimension of 75-80 μm diameter and 30 μm height. The gold bumps on the substrate are transferred onto the Sn-plated tape by eutectic bonding. The "transfer-bumped" tape is subsequently bonded to a die by another bonding step. Due to the limitation of bump size and mechanical accuracy involved with the two bonding steps, the transfer-bumped TAB has been used primarily for die having over 130 μm lead pitch.[40]

Recently, the need for the interconnection of fine pitch ICs has led to the development of the advanced transferred-bump technology with smaller dimensions and better shapes. The small transferred bumps have the dimensions of 40-60 μm diameter and 20-25 μm height, which can be used for the interconnection of ICs with 100 μm lead pitch or less.[40] The mushroom shape also has been modified to either T-shape or straight wall to improve bump transferability and alignment.

In the transferred-bump tape, where gold bumps are directly bonded to aluminum pads on a die, the formation of unwanted intermetallic phases can occur at the interconnection. The gold-aluminum system has been extensively investigated for the application of wire bonding.[41-44] An example of the gold-aluminum interface is shown in Figure 4-9, where a gold-plated copper beam was bonded to an aluminum pad on a silicon die at 500°C, 5 sec and 1800 psi.[9] Initially, both the gold and aluminum layer were about 2 μm thick. Figure 4-9 is a cross-sectional view of the as-bonded specimen, where a multilayer structure of the intermetallic phase is observed: Au, Au_2Al, $AuAl$, and $AuAl_2$. The identification of these layers was achieved by combining the information obtained from EMPA, XRD and RBS. As in the case of wire bonding, the formation and further growth of the intermetallics can raise serious concern of the interface reliability.

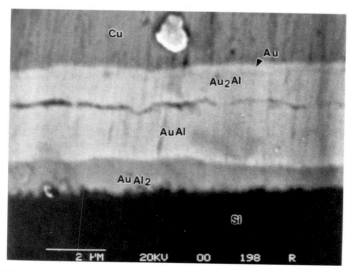

Figure 4-9 Growth of intermetallic phases during thermocompression bonding of Au-to-Al at 500°C, 5 sec, and 1800 psi: Au, Au_2Al, $AuAl$, and $AuAl_2$.

4.4.3 Balltape

A unique bumped TAB structure called "Balltape" has been developed by exposing the free standing inner leads to a focused yttrium-aluminum-garnet (YAG) laser to produce spherical bumps at the tip of the planar copper beams.[22,23] The laser-melting balltape process as well as a fully automated tool required for large quantity production, are described in detail in the above references. Figure 4-10 shows a typical balltape structure. The laser-formed ball has a very smooth surface with no indication of oxidation. The

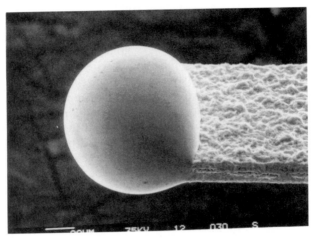

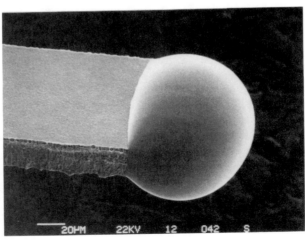

Figure 4-10   SEM micrographs of as-laser-melted copper balltape showing both the top side and the bottom side of the tape.

copper ball is significantly softer than the origianl copper beam, and therefore, thermocompression bonding can be achieved at a much lower force. To protect the copper ball from corrosion, the balltape is electroplated with gold. The thickness of the gold layer is in the range of 0.5 µm to 0.8 µm. With this balltape structure, thermocompression bonding to aluminum was demonstrated at (or above) 500°C, 5 sec, 15-30 Ksi. Because the bonding interface has a gold-aluminum metallurgy, as discussed in the previous section, this high temperature condition has raised reliability concerns regarding intermetallic formation. Recently, low temperature bonding was achieved by modifying the surface morphology of the gold layer on the balltape.[44]

4.5 TAB ILB PROCESSES AND TOOLS

With the completion of wafer bumping, the die are tested and diced. The next step in the tape automated bonding (TAB) technology is bonding the semiconductor die to the TAB tape using a suitable bonding process. Subsequent processing such as encapsulation, test and burn-in, can be carried out in reel to reel, strip, or singulated form. This is followed by excising the die tape subassembly and attaching it to the board by outer lead bonding (Figure 4-11). Inner lead bonding (ILB) is the first manufacturing process step which brings the die, bump, and tape together to create a functional product. The thermocompression bonding process for TAB technology has been reviewed.[4,5,8]

Thermocompression bonding is the primary bonding process being practiced in TAB production. In this process metal bonds between the bumps on the die and the conductor leads on the tape are achieved through mechanical pressure and elevated temperature. The critical bonding parameters are time, temperature, pressure, and energy. They vary depending upon the bonder, die/pad/lead size, tape, I/O, as well as the metallurgy used. The processes are categorized as single point or gang bonding depending upon whether the bonds are accomplished one at a time, or simultaneously. Single point bonding is similar to wire bonding and offers advantages over gang process, cf. Table 4-3. For high volume manufacturing, gang bonding is the preferred approach for high lead count packages because of higher throughput and lower cost.

There are different designs in ILB gang bonders and bonding stations. In the typical process TAB tape frames in roll, singulated, or strip form, are fed to the bonding station where they join the die. The die is placed on a heated anvil, and the tape frame is aligned so that the leads on the tape are positioned over the die bumps. A heated thermode, which may be a solid block or a bladed tool is actuated to apply a dynamic force against the tape

MODULE ASSEMBLY PROCESS

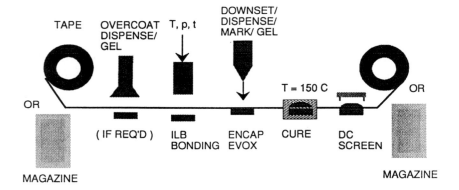

CARD ASSEMBLY PROCESS

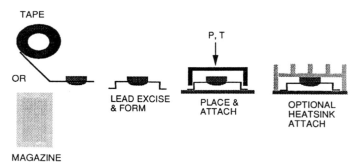

Figure 4-11   Module/card TAB assembly process.

Table 4-3   Advantages of Single Point Bonding

- No Planarity Problems
- Low Tooling Cost
- Low, Consistent Bond Force
- User Friendly, Easy to Install and Operate
- Allows Easy Repair of Individual Sites
- Eliminates Cratering Problems
- Not as Pad Design Limited as Gang Bonding

leads, which in turn is forced against the preheated die bump. Metallurgical bonding between the leads and the bumps takes place and the bonded tape frame is advanced to the next stage.

There are a number of ILB equipment manufacturers worldwide as shown in Table 4-4.[45]

TAB ILB is accomplished today with either thermocompression, ultrasonic, laser, thermosonic or the recently reported solder reflow bond process.

The thermocompression bond process typically utilizes gold plated leads with gold pads, or gold plated leads with gold coated aluminum bumps.[46] Synthetic diamond or diamond material is commonly used in thermode designs. Other applicable materials include: inconel, titanium, tungsten, nickel, molybdenum, etc. Thermocompression (TC) bonding is a slow (5-20 bonds/sec) process that requires high temperatures (> 500°C) and pressure that can result in die and/or passivation damage.

Unlike TC bonding, eutectic bonding typically uses a pulsed thermode and requires a gold/tin eutectic or solder configuration. Selection of this eutectic allows inner lead bond formation at lower temperatures and pressures than the gold to gold thermocompression bonding but less reliable metallurgy is a concern.[47]

Two single-point bonding methods have been developed for TAB ILB applications; thermosonic single-point bonding and laser joining process.

Table 4-4   ILB Equipment Manufacturers

Anorad (Gang and Single Point)
Dai-Nippon Screen (Gang)
Dynapert (ESI, Laser)
ESI (Laser)
Farco (Gang) (SMT Engineering)
GM-Hughes (Single Point and Laser)
Hybond
IMI (Gang and Single Point)
Jade (Gang)
K&S (Single Point)
Kaigo-Denke (Gang and Single Point)
Mat Sushita (Laser)
Orthodyne (Single Point)
Shinkawa (Gang and Single Point)
Toshiba
Universal Instruments (Single Point)

TAB thermosonic single-point[48] bonding is a modification of the well-established technique of thermosonic wire bonding in which heat and ultrasonic energy are applied together in addition to a small amount of force to produce solid-state diffusion bonding. Gold-plated TAB leads can be thermosonically bonded to gold bumps at relatively low temperature (150-200°C) as compared with thermocompression bonding.[49] As reported, a significant amount of force is often required to form successful bonds by thermosonic single-point bonding.[49-51] This introduces an excessive lead/bump deformation as well as stress in the underlying structure of the device. Silverberg[68] and Levine, et al.[69] have reported good bond formation with forces of 70-90 grams with 2-mil wide leads and 120-160 grams with 4-mil wide leads.

A laser joining process was developed to circumvent the high bonding force for thermocompression bonding; this is a non-contact process which is applicable to a high density TAB ILB.[51-53] Laser joining processes take advantage of the localized, concentrated energy in a focused yttrium-aluminum-garnet (YAG) laser with the wavelength of 1.06 µm. At this wavelength, the laser energy is well absorbed by tin ($\sim$20-50%), while only a few percent is absorbed by gold and copper. For this reason, most laser ILB development work is limited to bonding of tin-plated leads to gold bumps.[52-55] Recently, bonding of gold-plated leads to gold bumps was demonstrated by using the YAG laser at the half-wavelength, 0.53 µm, where the absorption of energy by gold and copper increases substantially.[56] Since the laser beam can be focused down to spot sizes less than 50 µm in diameter, a laser joining process to handle very high density ILB is potentially feasible.

Several investigations have been conducted to characterize the laser joining process, and thereby to improve the joint integrity and reliability.[54-56] A direct correlation has been reported between the amount of laser energy applied and the joint strength as well as the failure mode. At low energy levels, the ILB joint is formed by melting the tin layer and subsequent reaction with the gold bump. At high energy levels, the copper lead itself partially melts and reacts to form Cu-Sn-Au alloys, which generally promotes high joint strength. At very high energy levels, some indication of die thermal damage has been observed. Laser bonding has been demonstrated on tapes with three different metallizations (Sn, Sn/Ni, Au-plated leads) to Au or Sn-capped Au bumps.[55] A significant degradation in pull strength was reported for joints made with Sn-plated leads to Au bumps after thermal aging. This was attributed to the formation of Kirkendall voids at the contact interface due to the unidirectional diffusion of copper into the Au-Sn region. In order to avoid the void formation, Sn/Ni-plated leads were bonded to Au bumps and then again thermally aged. Although the void formation was not observed in the Sn/Ni-plated tape, the joint strength still degraded upon thermal aging. This resulted from the formation of brittle Au-Sn intermetallics during the thermal aging. In the third type of tape, Au-plated leads were bonded to Sn-capped gold bumps by melting the tin layer. This ILB joint showed remarkable thermal stability with thermal aging. The high stability was explained by the fact that the Au-rich phase formed at the contact interface acts as a diffusion barrier between the copper and the eutectic Au-Sn.[55]

Owing to recent advances, the laser joining process is now a feasible ILB technique. However, there are several important issues to be resolved before it is accepted widely in the production line. These include both the technical issues discussed earlier, such as control of bond temperature, minimizing thermal damage or stress to the device, improvement of bond reliability, and the non-technical issues such as development of low cost production equipment and laser safety concerns.

4.6 TAB INNER LEAD THERMOCOMPRESSION BONDING PROCESS

4.6.1 Thermocompression Bonding Considerations

The metallurgical interface mechanism for TAB thermocompression bonding is the same as for wire bonding, which has been used since the early days of semiconductor technology. In the simplest sense, it is a process for joining two material surfaces by interdiffusion across the boundaries, through the use of pressure and temperature. Whether each pair of the bump and lead surface metallurgies are gold to gold or gold to tin, the thermocompression bonding process is a complex physical phenomenon. This phenomenon involves the surfaces being brought into intimate contact, and heated to the appropriate temperature for an appropriate length of time in order for sufficient interdiffusion to form a strong metallurgical joint. Since there is no rubbing motion as in wire bond to create clean intimate contact, TAB ILB bonding usually requires higher temperature and pressure per lead than those for wire bonding.

TAB ILB thus requires a manufacturing bonding tool with high precision and very tight dimensional tolerance control of the tape and die to be bonded. Important bonding parameters for a given set of mating surface metallizations are thermode temperature and pressure, dwell time, and preheat temperature. Different mating metallurgies require different bonding parameters; for example, bonding tin-plated leads to electroplated gold bumps would require a lower thermode temperature than for gold-plated leads to gold bumps. The reason for the lower bonding temperature for tin-plated lead is that Sn melts at $232°C$ and forms a Cu-Sn-Au intermetallic compound rather easily, whereas gold-gold bonding depends only on solid state diffusion. Although high temperature and pressure appear to be preferable to achieve good bonding, other undesirable aspects should be considered. For example, high temperature may produce a change in device characteristics and will enhance oxidation of the thermode which shortens its life. Also, high temperature may cause dimensional instability in the tape, resulting in registration problems in the subsequent outer lead bond operation.

In many applications the gold bumps are selected for their inertness to environment and ductile properties. They are bonded to gold or tin plated copper leads. The advantage and disadvantage of the gold-gold metallurgical system and the gold-tin metallurgical system for TAB ILB gang bonding have been reviewed by Walshak.[8] In general the gold-gold system requires a relatively higher temperature and pressure. The tin-plated tapes have the well known tin whisker concern with shelf life. This problem has been minimized by tape manufacturers by heat treatment after plating. The gold plated tapes in the gold-gold system have the advantages of an inert surface and long shelf life, but are generally more costly. In either case, equivalent bond strength can be achieved with proper processing conditions, as was shown by Kawanobe.[57] Zakel, Leutenbauer, and Reichl[58] have studied the Sn-Cu, Cu-Au, and Cu-Ni-Au systems for the thermal aging of these metallurgies for their influence on inner lead bonding processing with gold bumps. For tape with 0.7 μm thick tin-plated copper leads they estimated a shelf life of two years at 20°C. They did not observe degradation in bondability after 2000 hours at 150°C for the Cu-Au and Cu-Ni-Au systems, and suggested at least a 150-year shelf life at 50°C.

4.6.2 The Mechanics of the Bonding Cycle

The TAB ILB bonding cycle can be divided into three stages as shown in Figure 4-12.

Stage 1. The thermode actuated by the bonder mechanism strikes the leads, which in turn strike the bumps on the die. The thermode and the leads oscillate and rebound, until the motion is damped. The thermode contacting the leads and the lead to bump contacts are heated through thermal conduction.

Stage 2. The lead and bump temperatures increase to the equilibrium temperature of the thermode. The actuator force is transmitted through the leads to the heated bumps, the bumps are deformed in compression until the reaction force on the bump is equal to the applied thermode actuator force. The temperature and the interface compressive pressure at each bump

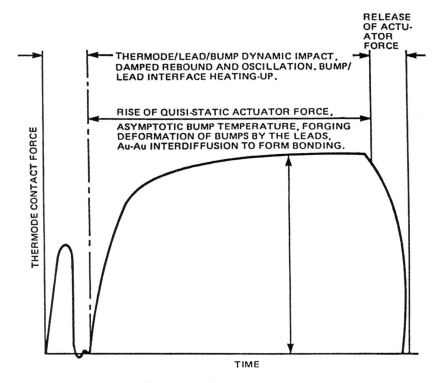

Figure 4-12   The TAB bonding cycle.

initiates the gold to gold interdiffusion mechanism for bonding to proceed over the total time of contact (dwell time).

Stage 3. The bonder actuator is released. The thermode starts to withdraw to reduce the contact force to zero. The bump and lead temperatures decrease to equilibriate to the environment.

Of these three stages, Stage 1 and Stage 2 are crucial for a successful bonding cycle. They will be discussed in further detail.

Stage 1: Thermal and Dynamic Models

In Stage 1, it is important to understand the time it takes for thermal conduction to heat the leads and the bumps, once contact has taken place. Next, the time taken for the thermode dynamic force to rise, rebound, and damp out to the desired actuator force would be quantified. It is desirable for the system (lead, bump and die) thermal heating time to be a small fraction of the dynamic thermode impact time, and for the time the thermode motion is damped out, to be a small fraction of the total bonder dwell time. This would insure that the bump and lead temperature and pressure parameters are the same as they would be if set statically for the process. Ramakrishna, Henderson, Chen, Lee, Eagle, and Hammer[59] have modelled the thermal and dynamic process for a commercially available bonder, both mathematically and experimentally, and verified it for a 47 and a 268 bump die.

The results showed that the die reaches the anvil preheat temperature in about 5 seconds, and occurs during the die/tape alignment and set-up stage. A range of thermode temperatures (550°C-570°C), with lead thicknesses of 35 μm to 45 μm, were used in the study. The bump heights were 17-25 μm. It was calculated that the 35 μm lead heats up in 20 μsec, and the 45 μm lead heats up in 40 μsec. The bump will heat up in 150 μsec from the preheat temperature of 250-570°C for the 25 μm bump height, assuming zero contact resistance, as shown in Figure 4-13. Contact resistance from surface roughness, and contamination will lengthen this heat-up time.

A lumped parameter dynamic model was developed to model the thermode, lead and bump system. It was estimated and confirmed with experiment (Figs 4-14 and 4-15), that the peak dynamic force occurred in 50 to 100 μsec. It was also noted that in some instances the model would predict a peak dynamic force higher than the preset actuator force, and that there was some dynamic rebound after the initial contact. High dynamic force at

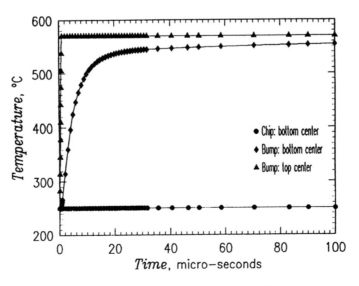

Figure 4-13 Bump heating: zero contact resistance.

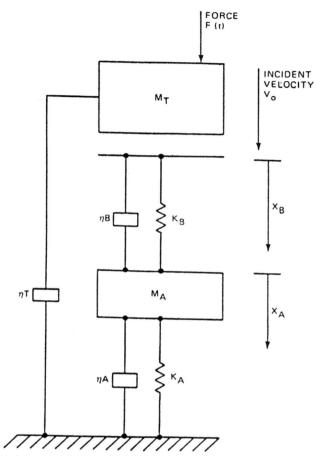

Figure 4-14   Schematic diagram of the spring mass model of the TAB ILB bonding process.

the time of impact could damage the die area underneath the bumps. In the system modeled, the thermode damped out to the actuator force value in tens of milliseconds. High peak dynamic force and dynamic rebound can be detrimental.

Stage 2: Gang Bonding Bump Forging Model

In the second stage the thermode has contacted the leads which in turn press against the bumps. Since the leads and bumps have thickness variations from the manufacturing processes, some lead-bump pairs will be

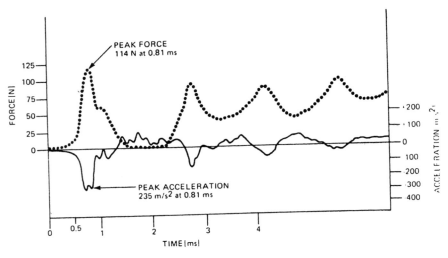

Figure 4-15 Force and acceleration traces of the thermode. Chip has 47 bumps.

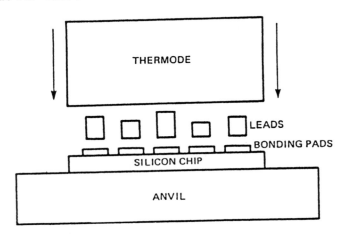

Figure 4-16 Schematic of thermocompression bonder.

contacted earlier than the others (Fig. 4-16). Chen, Raski, Young, and Jung[60] have employed a plasticity model for forging to apply to the bump deformation. They have derived the following equation

$$\left[\frac{P}{Y}\right]_{ave} = \left[1 + \alpha + \frac{b}{4h}\right]$$

to relate the bump material flow stress (Y) to the bump pressure (P), lead width (b), and bump height (h). The term α is a constant set either equal to zero or two depending upon whether the model line of sight is parallel or perpendicular to the lead length. Consider the graphs of tensile flow stress versus deformation of a pure metal and the same metal with some impurities (Fig. 4-17). The amount of impurities would cause the material to display a stiffer work-hardening characteristic. Applying the bump forging model to the set of tensile flow stress data, one obtains two significantly different pressure deformation characteristics as shown in Figure 4-18. Figure 4-18 also shows that as the amount of deformation increases, the pressure required to produce the deformation increases dramatically due to the influence of the stiff die substrate underlying the bump material.

If the compression of a single bump by a heated lead is considered as a forging process, then gang bonding can be considered to be the simultaneous forging of a set of individual bumps with the constraints of a compatibility condition satisfied by the thermode planarity, and the thickness tolerances of the individual bumps and leads. Newton's law requires that the sum of the forces on the individual bumps and leads is equal to the total thermode force. In this way, one can determine the penetration and thermode force based upon a known distribution of the lead and bump heights. In other words, the equilibrium condition is

$$F = \sum_{i=1}^{n} f_i$$

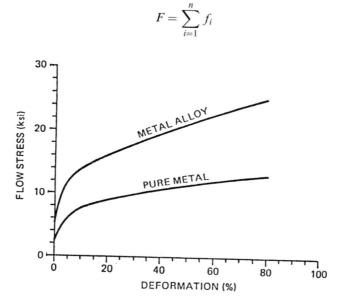

Figure 4-17 Flow stress versus deformation for two metal purities.

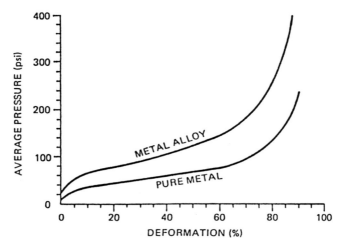

Figure 4-18 Average pressure versus compressive deformation of two metal purities between two rigid platens for $b/h = 5.7$.

where F is the total thermode force, f_i is the force in the "ith" bump, and n is the total number of I/Os. The compatibility condition is:

$$U = u_i + v_i + t_i + p_i \qquad \text{for } i = 1 \text{ to } n$$

where U is the thermode travel after the first bump contact, u_i is the deformation of the "ith" lead, v_i is the deformation of the "ith" bump, t_i is the tolerance for the bump and lead at the "ith" position, and p_i is the planarity for the "ith" position. The values of f_i and v_i are related by the forging (plasticity) characteristics of the lead material and the bump material described previously.

Figure 4-19 shows a normal distribution of lead heights within a tolerance range, and the effective widening of the distribution by thermode and die planarity. Using this normal distribution, the thermode force and thermode penetration have been computed, with the mean and 3 sigma limits of bump deformation. This included the effect of a thermode being out of plane by 0.01 degree over the diagonal of a 10 mm die. The results shown in Figure 4-20, include some actual measured mean, maximum, and minimum bump deformations at various levels of thermode force.

Within the distribution of lead thickness tolerances, the thickest lead will have the largest deformation corresponding to the highest local bump force. The thinnest lead will have the least amount of deformation corresponding to the lowest local bump force. Thermode planarity will have the effect of broadening the distribution. Figure 4-21 is an example of the probability density of the force distribution. This would allow one to estimate the risk

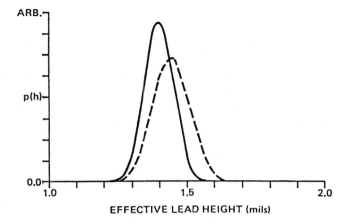

Figure 4-19 Probability density of lead heights. Mean lead height = 1.4 mils, $3\sigma = 0.15$ mils. Dashed line represents the change in effective lead heights when planarity is taken into account.

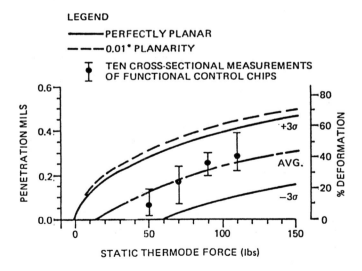

Figure 4-20 Total thermode force versus total thermode penetration. Variation in lead heights and thermode planarity are shown as changes in the amount of bump deformation.

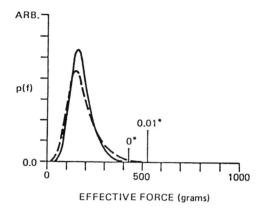

Figure 4-21   Probability density, $p,(f)$, of the forces on individual bumps for a total thermode force of 55 pounds.

for silicon crater, and the risk of a non-bond at the two ends of the distribution.

4.6.3 Thermocompression Bonding Mechanical Process Window

In the development of the ILB bonding process window, the process development engineer would usually design an experimental matrix on parts to understand the upper and lower bounds of the bonding parameter. The lower limits of pressure and temperature would have the risk of poor bonds, while the upper limits of the process window would lead to the risks of silicon crater, and passivation damage. The process engineer would set the process window well inside these two limits in order to obtain a high yield manufacturing process.

The modelling study showed that the bump deformation provides a critical indication of the upper and lower limits, and these limits are affected by the bump flow stresses and the lead width/thickness tolerances as described in the bump forging equation. The three graphs in Figure 4-22 represent the mean, lower and upper limits of the flow stress versus bump deformation as they are affected by the strain hardening and thickness. The ideal operating position would be the mid-point of the curve. To the left, the force will drop off steeply with the risk of poor bonds, and to the right the force will rise steeply with the risk of silicon crater or passivation damage. For effective control of the process window, the bump needs to be sufficiently high to accommodate the lead thickness variation and thermode planarity. The bump material compressive flow properties at the bonding temperature need to be well controlled by its process and composition. The

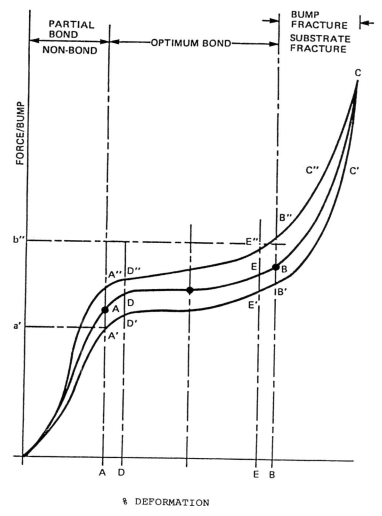

Figure 4-22 TAB thermocompression bonding mechanical process window.

lead tolerances need to be such that the bonding force on each bump will be in the mid-region of the graph.

While the simple forging model described previously is useful to study the overall deformation and for process optimization, a sophisticated finite element model simulating bump metal plastic flow would be necessary to study the bump strain behavior around the via connecting the metallurgy from the die to the bump, as well as the stress distribution in the passivation. Dawson, Lee, Young, and Chen[61] have simulated the deformation and flow

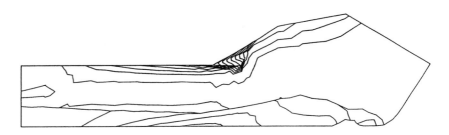

Figure 4-23   Strain contour after deformation.

of the bump and lead materials from initial contact to 50% compression. An example of the flow strain from the study is illustrated in Figure 4-23. It was found that the strain around the via was much higher than 50%, and that careful design of the via would be important to avoid excessive strain hardening that could lead to via fracture.

4.7 ILB SOLDER JOINING TOOL–HOT AIR THERMODE

The most recent advances in bonding center around solder reflow.[33,62,65,66] Solder bumped devices and electroless tin-plated tape reportedly form good, reliable inner lead bonds with a pulsed thermode.[70,71] This gang process is typically a longer cycle time than TC bonding, but planarity and die to tape stand-off is not as critical. Early solder reflow processes required flux which was applied to the die prior to bonding. This required cleaning, usually with solutions associated with environmental concerns, to assure reliability and eliminate corrosion concerns.

A fluxless reflow technology, solder attach tape technology (SATT) has been developed.[72-74] This technology utilizes a hot air thermode (HAT)[74,75] which can be used for most OLB TAB SMT assembly and is well suited for TAB applications. HAT uses a high velocity heated gas as the medium for solder melting. Because of the particular tool/process integration employed by the HAT system, the technique does not require ovens with tightly controlled atmospheres and produces reliable solder joints. Of course, elimination of flux and the cleaning steps during bonding, along with the solvents associated with the same, is a significant advancement. Other advantages of this process include minimized emissions during bonding and reduced bond corrosion since entrapped flux residue is eliminated.

The HAT process described was formulated starting from the design requirements listed above. The solder is melted with a heated jet of gas immediately adjacent to the joint to be reflowed. It is the jet of heated gas

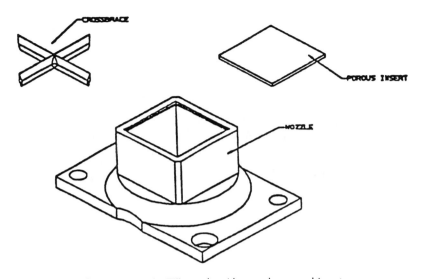

Figure 4-24   "SATT" nozzle with cross brace and insert.

and not the nozzle that conducts the required heat. Thus, extremely low bonding loads are required to overcome the surface tension of the melted solder.[75]

A key design features the use of a porous sintered metal insert in the tip of the nozzle, shown in Figure 4-24. With the porous insert, excellent contact between the leads and bumps was achieved, while still allowing the inert gas to flow through.

The tool for the HAT process consists of two basic parts. The heart of the tool is the heat exchanger which starts with a tool mount and contains gas chambers filled with ballast, cartridge heater wells, an integral vacuum tube, and control feedback thermocouples. The second part of the HAT tool is the nozzle. Each nozzle is custom designed for a specific TAB footprint. The basic nozzle components include a quick disconnect flange, pyramid gas deflector, gas jets, vacuum ports, and a gas dam to allow the die to be shielded from the heated gas.

The gas mixing chamber is filled with small stainless steel spheres which randomize the gas flow while providing a large surface area for efficient gas heat transfer.

The HAT process uses nitrogen gas which is regulated at a very low flow rate, typically about 2 SCFM. Other inert gases used (such as helium, forming gas, etc.) yielded equivalent results. Gas temperature, flow rate, velocity, and flow duration are the process variables used to establish the desired bonding results. It should be noted that the thermode remains in

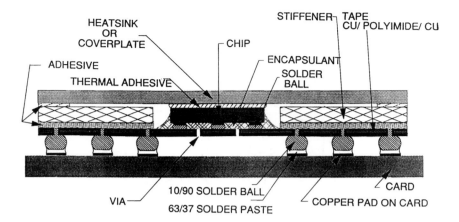

Figure 4-25 TBGA cross-section.

contact with the TAB frame for an additional 6 seconds, at which time the solder is well below the liquidus temperature. A total typical cycle time is < 10 sec.

To understand the mechanism that permits fluxless bonding to take place, assume that the TAB frame is now on the HAT thermode and is aligned on a footprint having individual pads coated with Sn-Pb solder. The HAT thermode body is kept at a constant temperature and then pulsed with nitrogen gas. The resultant stream of heated nitrogen, at 200-250°C, is sufficient to melt the solder on the circuit board. While the gas stream is hot enough to melt the solder, it must also be at a velocity above the critical velocity necessary to wipe the surface oxides off the solder pad. This "scrubbing" action is aided by the TAB leads which puncture and fracture the thin oxide layer on the molten solder pads as they are pushed into the solder. Such hydrodynamic scrubbing at the point where joining is to take place will ensure proper wetting, since the fresh solder surface has an affinity for the heated gold surface on the leads. Other surface metallurgies that do not have native surface oxides can also be used in place of gold plating.

4.8 ILB SOLDER ATTACH PROCESSES

In the solder attach process for inner lead bonding, a non-oxidizing gas such as nitrogen is heated by flowing through a heat exchanger assembly. The heated gas then flows through a porous sintered insert at the tip of the nozzle. The nozzle applies a force of about 8 grams per bump to assure good contact of the lead to the solder bump. The metallurgical bond is formed with the gold-plated copper beam by partially melting the solder bump. The encapsulation of the solder joint is performed to increase the joint strength, and to reduce stress on the joint as well as to protect the joints from handling damage. The SATT (solder attach tape technology) process has demonstrated several advantages of the use of solder bumps over the thermocompression bonding with gold or copper bumps. Because of the very small forces on the die, SATT can be used to bond over active circuitry. In addition, array chip footprints can be packaged where the conventional technology is limited to perimeter connections.[33]

Speer Schneider and Lee[65,66] reported a solder bump reflow process for TAB early in 1989. This initial solder bump TAB work was an extension of the C4 (controlled collapse chip connection) used in flip chip bonding.[64] They employed the 5/95 Sn/Pb with Cr-Cu interface metallurgy. A two row array of solder bumps on a 10 mil pitch was utilized. The bumps were 4 mil in diameter at the start of bonding and about 2 mil high after bonding. This demonstrates the fact that TAB is very forgiving in regard to planarity. For example, a gold bump TAB is limited to about 0.4 mil planarity variations.[67] When a leadframe was bonded or reflowed to these solder bumps, excellent wetting of the lead with solder was observed. A uniform stand-off from the die and excellent mechanical characteristics of each bond site were also recorded. The lead to bump solder bond was formed by localized heat applied over the die area to reflow the solder using a pulsed thermode.

More recent advances in inner lead bonding have been reported[61-64] and employ gold tape inner leads bonded to tin/lead solder bumps. The SATT process offers general advantages compared to the standard thermocompression process. For example, it allows bonding over active circuitry since forces are minimized, (i.e., < 10 times those of thermocompression bonding). This gang (mass) bonding process also allows area array die footprints to be assembled versus the more limited perimeter connections in thermocompression bonding. This means die size can be minimized while maximizing the total number of I/Os. The Area Array TAB (ATAB) package (Figure 4-26) utilizes the SATT technology for ILB bonding. The devices undergo solder bumping with 3/97 Sn/Pb composition and are typically about 4 mil in diameter and about 3.5 mil high after evaporation and reflow.

The first step in the inner lead bond process is to align the die to the tape. With the die on a preheat stage, with the bumps facing up under the tape, the nozzle and heater assembly is lowered into contact with the tape. The nozzle applies a force of approximately 8 grams per bump. This force assures good contact of the lead to the bump during the initial heat transfer stage of the bond, and settles the lead into the bump as the solder liquefies. In the SATT bonder, this force is generated by a motor which compresses springs located above the heater assembly. The amount of force applied can be varied by changing the displacement of the springs. Figure 4-27 shows the

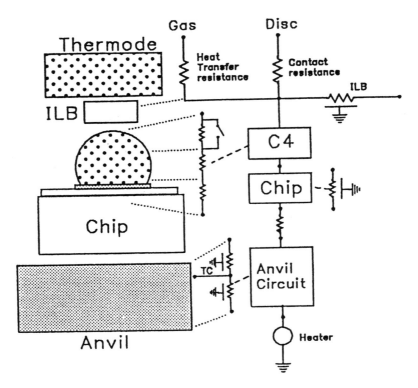

Figure 4-26 Heat transfer to the inner lead bonds.

force versus displacement curve for a particular configuration and spring stiffness.[33]

When the desired displacement is reached, the flow of heated nitrogen is started. The nitrogen provides the additional heat necessary to bring the tape leads and the preheated die to the necessary temperature to form the bond.

The bonding process is primarily Au/Pb eutectic which begins forming at 215°C. As the lead is heated the gold begins to combine with the Pb and Sn in the solder. This reaction becomes diffusion rate limited and begins to slow down once the gold is depleted from the lead surface. While the Au/Pb is liquid, the lead is pushed down to the unreacted solder and the Au/Pb wets up the edge and down the length of the lead. When the cool nitrogen is applied, the solder is solidified (Fig. 4-28) resulting in a reliable inner lead bond.

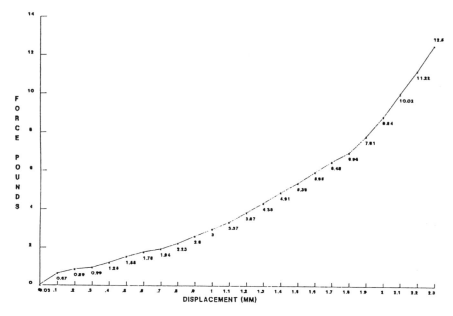

Figure 4-27   Bonder force versus displacement.

Figure 4-28   "SATT" bonded solder joint.

The diffusion continues with time and is accelerated by downstream processing at elevated temperature until the gold is uniformly distributed through the entire solder column. The once gold rich fillet on the side and along the length of the lead is now primarily lead. Although there is no direct evidence of interaction between the nickel and the Au/Sn or Pb, the strength of the joint is sufficient to produce primarily ductile fractures in the solder when a die pull is performed on a sample. Because of the inherently low tensile strength of the solder, encapsulation has been found to be critical to protect the bonds from handling damage and to minimize stress on the joint during temperature cycling.[33]

The parameters critical to the bonding process are the force applied during bonding, the heat input to form the bond, and the cooling necessary to solidify the joint. The criteria for bond optimization were to maximize the die pull force, while assuring a minimum wetting length, and to maintain a minimum die-tape stand-off to assure encapsulation will flow and fill the space between the die and tape. This processing has been successfully used to bond in excess of 700 I/O leads.

4.9 FUTURE TRENDS IN TAB TECHNOLOGY

TAB packaging technology is particularly suitable for applications with requirements for thin, lightweight, small packages. It will be used on COBs (MCMs) where device test and burn-in are required until a cost-effective solution to the "known good die" issue is resolved. In order to minimize silicon cost (smaller die), TAB (peripheral/area array) will be used for the high I/O packages because of its ability to achieve tighter ILB pitches than wire bond. Two-level metal layer TAB tape usage will grow as electrical performance requirements (number of drivers simultaneous switching) become more stringent. The ATAB package described earlier is an example of two-layer metal tape with bonds over active circuitry.

Significant advancements in the TAB packaging technology have occurred in recent years. Tape manufacturing capability and quality have improved dramatically–fewer lines and spaces, tape flatness and dielectric stability are a few examples. ILB bonding technology requirements and processes have become well understood allowing more consistent wafer bumping, repeatable/reliable interconnects, availability of tools and finer pitch bonding. A remaining manufacturing/assembly issue is bumping cost.

A significant barrier to TAB acceptance in the majority of applications is the OLB process. The difficulty in handling the "flimsy" fine pitched leads has resulted in an expensive attach process, requiring special tools and processes which are not compatible with existing SMT processing. Array

TAB is an alternative solution since it eliminates fine pitched OLB leads, and tool/process requirements. Most OLB attach is carried out using a hot bar thermode gang bonder as opposed to mass reflow using infrared solder reflow. An exception[76] developed and in practice with 0.5 mm OLB pitch, utilizes a unique outer lead reform to penetrate the solder paste and spring load the leads on the substrate; a plastic fixture then holds the TAB in place during standard SMT processing. The applicability of this process to finer OLB pitch is not known. Alternative approaches, currently in development, are the use of unidirectionally conductive, fast setting epoxies;[77] or, if the application allows, the Array TAB (ATAB) packaging concept.[78] The latter utilizes a mechanical stiffener which maintains tape planarity and an array (50 mil grid) of solder balls to interconnect the tape to the printed circuit card.

The TAB technology is sound, the challenge lies in the implementation for low cost manufacturability.

REFERENCES

1. Shu, Bill, "Fine Pitch Wire Bonding Development Using a New Multipurpose Multi-Pad Pitch Test Die, *41st Electronic Components and Technology Conference Proc.*, 1991.
2. Scheibner, J. B., and R. S. Dodsworth, "Manufacturing and Bonding of Fine Pitch TAB Tape," *Proc. 3rd International TAB Symposium*, February 1991, pp. 239-254.
3. Anderson, S. W., "Solder Attach Tape Technology (SATT) Inner Lead Bond Process Development," *Proc. 4th International TAB Symposium*, February 1992, pp. 158-172.
4. Tummala, R., and E. Rymaszewski (eds.), *Microelectronics Packaging Handbook*, Van Nostrand, 1989.
5. Lau, J., S. Erasmus, and D. Rice, *Electronic Materials Handbook–Packaging*, Vol. 1, ASM International, p. 274, 1989.
6. P. Burggroaf, *Semiconductor Int.*, p. 72, June 1988.
7. *An Introduction to TAB Fine Pitch Technology*, a Surface Mount Council publication, SMC-TR001, pp. 1-54, 1989.
8. Lau, J. H. (ed.), *Handbook of Tape Automated Bonding*, Van Nostrand, 1992.
9. Kang, S. K., "Inner Lead Bonding Mechanisms in TAB Technology," *Proc. 4th Elec. Mat. and Proc. Cong.*, ASM International, Montreal, Quebec, pp. 187-192, August 1991.
10. Liu, T., et. al., *Solid State Tech.*, March 1980, pp. 71-75.
11. Hyakawa, M., et al., *Solid State Tech.*, March 1979, pp. 52-55.
12. Liu, T., and H. Fraenkel, "Metallurgical Considerations in Tin-Gold Inner Lead Bonding Technology," *Int. J. Hybr. Microelec.*, Vol. 1(2), pp. 69-76, 1978.

13. Burns, C. D., "Copper-to-Gold Thermocompression Gang Bonding of Interconnect Leads to Semiconductive Devices," U.S. Patent No. 4,000,842, January 4, 1977.

14. Harris, J., and W. Gounin, "Method for Gold Plating of Metallic Layers on Semiconductive Devices," U.S. Patent No. 4,005,472, January 25, 1977.

15. Burns, C. D., "Antioxidant Coating of Copper Parts for Thermal Compression Gang Bonding of Semiconductive Devices," U.S. Patent No. 4,188,438, February 12, 1980.

16. Winkler, E., *Semiconductor Int.*, May 1985, p. 350.

17. Brady, M. J., et al., "Aluminum Bump, Reworkable Bump, and Titanium Nitride Structure for TAB Bonding," U.S. Patent No. 5,134,460, July 28, 1992.

18. Speidelberger, R., et al., *Gov. Microcir. Appl. Conf.*, 1985, p. 515.

19. Kanz, J., et al., *IEEE Trans. Comp. Hybr. Manuf. Tech.*, Vol. CHMT-2, 1979, p. 301.

20. Small, D., and A. Blain, *Br. Telecom. Tech. J.*, Vol. 3, No. 3, July 1985, p. 86.

21. Hatada, K., et al., *Nat. Tech. Rep. Japan*, Vol. 31, June 1985, p. 116.

22. Hodgson, R., et al., "Balltape Structure for Tape Automated Bonding, Multilayer Packaging, Universal Chip Interconnection and Energy Beam Processes for Manufacturing Balltape," U.S. Patent No. 4,814,855, March 21, 1989.

23. Ledermann, P. G., G. W. Johnson, and M. Ritter, "Laser Formed Structures to Facilitate TAB Bonding," *Proc. SPIE*, Vol. 1598, "Lasers in Microelectronic Manufacturing," pp. 160-163, December 1991.

24. Triggs, W. T., and C. J. Byrns, Jr., U.S. Patent No. 3,599,060, August 10, 1971.

25. Kawanobe, T., K. Miyamoto, and M. Hirano, "Tape Automated Bonding Process for High Lead Count LSI," *Proc. IEEE, Elec. Comp. Conf.*, pp. 221-226, 1983.

26. Angelucci, T., Sr., "Wafer Bumping," *Handbook of Tape Automated Bonding* (J. Lau, ed.), pp. 176-199, Van Nostrand, 1992.

27. Walker, J., "Wafer Fabrication and Design for Tape Automated Bonding," *Surface Mount Tech.*, pp. 45-48, January 1990.

28. Traut, J., J. Wright, and J. Williams, "Gold Plating Optimization for Tape Automated Bonding," *Plating and Surface Finishing*, pp. 49-53, September 1990.

29. Lo, H. Y., and E. Tjhia, "Backsputtering Etch Studies in Wafer Bumping Process," *Solid State Tech.*, pp. 91-94, June 1990.

30. Moskowitz, P. A., et al., "Aluminum Bump for Tape Automated Bonding," *Proc. 1992 Int. Elec. Packaging Conf.*, Austin, Texas, September 1992.

31. Van der Drift, A., W. G. Gelling, and A. Rademakers, "Integrated Circuits with Leads on Flexible Tape," *Philips Tech. Review*, Vol. 34, no. 4, pp. 85-95, 1974.

32. Loy, J., "Wire Bonding and TAB Technology," *Proc. ASM Elec. Mat. & Proc. Cong.*, pp. 133-136, September 1988.

33. Anderson, S. W., "Solder Attach Tape Technology (SATT) Inner Lead Bond Process Development," *Proc. 4th Int. TAB Symp.*, pp. 158-172, San Jose, CA, February 1992.

34. Unger, R. E., C. Burns, and J. Kanz, "Bumped Tape Automated Bonding (BTAB) Applications," *Proc. Int. Microelec Conf.*, pp. 71-77, Anaheim, CA, February 1979.
35. Kanz, J. W., G. W. Braun, and R. F. Unger, "Bumped Tape Automated Bonding (BTAB) Practical Applications Guidelines," *IEEE Trans. Comp. Hyb. & Manuf. Tech.*, Vol. CHMT-2, no. 3, pp. 301-308, 1979.
36. Lindberg, F. A., "Bumped Tape Processing and Application," Semiconductor Processing, ASTM STP 850, 1984, pp. 512-530.
37. Kuang, Y. X., and L. Liu, "Tape Bump Forming and Bonding in BTAB," *IEEE Proc. 40th ECTC*, Vol. 2, pp. 943-947, Las Vegas, NV, May 1990.
38. Eastman, D. E., et al., "Method for the Production of Self-Aligned Bump Tape for Automated Bonding," *IBM Tech. Disc. Bull.*, Vol. 30, no. 2, 1987, pp. 645-648.
39. Hatada, K., "TAB Confirms to Trends with Wider Formats," *JEE*, October 1990, pp. 96-100.
40. Taguchi, K., "Small Bump Development on Using TB-TAB for Fine Pitch Interconnect," *Proc. 4th Int. TAB Symp.*, San Jose, CA, pp. 236-241, February 1992.
41. Philofsky, E., *Solid State Elec.*, Vol. 13, p. 1391, 1970.
42. Gerling, W., *IEEE Elec. Comp. Conf.*, p. 13, 1984.
43. Clatterbaugh, G., et al., *IEEE Trans. Comp. Hybr. Manuf. Tech.*, Vol. 7, no. 4, p. 349, 1984.
44. Kang, S. K., "Gold-to-Aluminum Bonding for TAB Applications," *Proc. 42nd ECTC, CHMT Society of IEEE*, pp. 870-875, May 1992.
45. Khadpe, S., "Tape Automated Bonding–A Status Report," *Surface Mount International Proceedings*, San Jose, CA, 27 August 1991, pp. 1-6.
46. Moskowitz, P. A., H. R. Bickford, S. K. Kang, M. J. Palmer, T. C. Reilly, J. G. Ryan, and E. G. Walton, "Aluminum Bump for Tape Automated Bonding," *IEPC*, 30 September 1992, Austin, TX.
47. Sallo, J. S., "An Overview of TAB Applications and Techniques," *Proc. Nat. Electronic Packaging and Production Conf.*, West, February 1985, pp. 18-24.
48. Pitts, G. E., D. E. Boone, and D. M. Andrews, "Single Point Bonding Methods and Apparatus," U.S. Patent No. 4,766,509, October 1988.
49. Silverberg, G., *Microelectronic Manuf. and Test.*, February 1988, pp. 35-37.
50. Kelly, G., *Surface Mount Tech.*, August 1989, pp. 16-18.
51. Walshak, D., Jr., *Handbook of Tape Automated Bonding*, (ed. J. Lau), Van Nostrand, 1992, pp. 200-242.
52. Hammond, R., *Proc. ASM Int. 3rd Electronic Materials and Processing Congress*, San Francisco, CA, August 1990, pp. 1-9.
53. Spletter, P., and R. Crowley, *IEEE Proc. 40th Elec. Comp. & Tech. Conf.*, Vol. 1, 1990, pp. 757-761.
54. Emamjomeh, A., et al., *IEEE/CHMT 1991 IEMT Symposium Proc.*, pp. 21-26.
55. Zakel, E., G. Azdasht, and H. Reichl, *IEEE Trans.* CHMT, Vol. 14, December 1991, pp. 672-679.
56. Spletter, P., *Proc. 4th Int. TAB Symp.*, February 1992, San Jose, CA, pp. 58-71.

57. Kawanobe, T., K. Miyamoto, and M. Hirano, "Tape Automated Bonding Process for High Lead Count LSI," *33rd Electronic Packaging Conf. Proc.*, pp. 221-226, May 1983.
58. Zakel, E., R. Leutenbauer, and H. Reichl, "Investigations of the Cu-Sn and Cu-Au Tape Metallurgy and of the Bondability of TAB-Inner Contacts After Thermal Aging," *ITAB 1991 Proceedings*, p. 78.
59. Ramakrishna, K., D. W. Henderson, W. T. Chen, H. C. Lee, R. C. Eagle, and R. B. Hammer, "Thermal and Dynamic Analysis of TAB Inner Lead Bonding Process," *ASEM AMD* Volume 131, "Manufacturing Process and Materials Challenges in Microelectronic Packaging," pp. 73-86, 1991.
60. Chen, W. T., J. Z. Raski, J. R. Young, and D. Y. Jung, "A Fundamental Study of Tape Automated Bonding Process," *ASME* paper No. 90-WA/EEP-40, 1990.
61. Dawson, Paul, Yong-Shin Lee, W. T. Chen, and J. R. Young, "Simulation of TAB Upset Application," private communication, 1991.
62. Eagle, R. C., S. W. Anderson, and H. C. Wilson, "Six Sigma Implementation for TAB Inner Lead Bonding," *1992 International TAB Proceedings*, 16-19 February, 1992, pp. 140-156.
63. Brown, D. B., and M. G. Freedman, "Is There a Future for TAB?', *Solid State Technology*, September 1985, pp. 173-175.
64. Marshall, T. F., "New Applications of Tape Bonding for High Lead Count Devices," *Solid State Technology*, August 1984, pp. 175-179.
65. Speer Schneider, C. J., and J. M. Lee, "Solder Bump Reflow Tape Automated Bonding," *2nd ASM International Elec. Mat. and Proc. Cong.*, Philadelphia, PA, April 1989, pp. 7-12.
66. Speer Schneider, C. J., R. K. Spielberger, and P. G. Brusius, "Tape Automated Bonding with Controlled Collapse Chip Connections," International Electronic Packaging Society, San Diego, CA, September 11-13, 1989.
67. Spielberger, R., and J. Loy, "I/O Intensive Chip Packaging," VLSI and GaAs Packaging Workshop, Santa Clara, CA, 1988.
68. Silverberg, G., "Single Point TAB (SPT): A Versatile Tool for TAB Bonding," *Proc. ISHM Symposium*, September 1987, pp. 419-456.
69. Levine, L., and M. Sheaffer, "Optimizing the Single Point TAB Inner Lead Bonding Process," *Proc. 2nd International TAB Symposium*, February 1990, pp. 16-24.
70. Morris, J., "An Introduction to Tape Automated Bonding (TAB), The Next Generation in High Density Packaging," in *Proc. Int. Electronic Packaging Soc. Conf.*, October 1987, pp. 360-366.
71. Thompson, W. K., "Beam Tape Process Speeds Hearing Aid Production," *Electronic Packaging Proc.*, November 1982.
72. Cipolla, T. M., "Transfer Device," U.S. Patent 5102290, IBM, April 1992.
73. U.S. patent, Cipolla, T. M., P. W, Coteus, G. W. Johnson, P. Murphy, and C. W. Oden, "System and Method for Inspection and Alignment of Semiconductor Chip and Conductive Lead Frames," U.S. Patent, IBM, January 1990.
74. Cipolla, T. M., R. R. Horton, A. P. Lanzetta, M. J. Palmer, and M. B. Ritter, "Tape Automated Bonding Feeder," U.S. Patent 5052606, IBM, October 1991.

75. Palmer, M. J., R. R. Horton, H. R. Bickford, and I. C. Noyan, "HAT Tool for Fluxless OLB and TAB," *IEEE Proceedings*, 1991, pp. 507-510.
76. Attendorf, J., and R. Nicol, "Fine Pitch Case History: Mass Reflow Soldered, SMT Compatible TAB," *Surface Mount Technology*, June 1989.
77. Bredfeldt, K., J. Krol, and B. Wells, "Evaluation of Conductive Adhesives as a Medium to Attach Surface Mount Components and TAB parts to Printed Circuit Boards.'
78. Milewski, J., and C. Angulas, "TAB Design for Manufacturability and Performance," *ITAB '92 Proceedings*, p. 121.

5

Chip Level Interconnect: Solder Bumped Flip Chip

Lewis S. Goldmann and Paul A. Totta

5.1 INTRODUCTION

Although the concept of flip chip-on-board (COB) is in its infancy, flip chip on ceramic substrates is a mature and well developed technology, which is expected to provide, with at most minor modification, the basis for its COB extension. This chapter will describe the present state of the art of flip chips with solder bumps, and discuss the ramifications to the process when the chip is to be joined to an organic board. While some recent reports suggest that conductive polymer bumps[1-3] may have a role in future COB development, this chapter will be restricted to solder.

The solder bump technology was first introduced in the 1960s,[4] and the process then described by IBM for chip level interconnect[5] has remained remarkably unchanged.[6] The technology, referred to as the Controlled Collapse Chip Connection, or C4, was essentially unique to IBM in the 1970s, but gained wider utilization in the 1980s as the advantage of higher I/O density from its area array capabilities became apparent. Figure 5-1 shows a comparison of pad availability for area and peripheral terminal arrays. Announced products[6,7] have in excess of 500 C4s on a chip, with 100 micron pads on 225 micron pitch commonplace. A photo of a 29 × 29 762 pad area array is shown in Figure 5-2. Reductions to at least 80 on 200 microns are anticipated,[7] while development programs with 50 on 100 have been reported.[8] I/O counts on custom logic (ASIC) chips will certainly rise into the thousands. On an experimental basis, up to 16,000 C4s on a chip have been reported, with 25 micron bumps on 60 micron centers.[9]

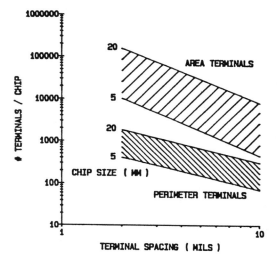

Figure 5-1 Peripheral vs area array pad availability.[8]

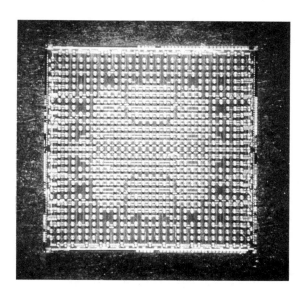

Figure 5-2 29 × 29 array chip with 762 solder bumps (IBM).[8]

In adopting the C4 process, many users have introduced material and process modifications to meet their particular process, yield and reliability requirements, and these have all contributed to making flip chip a particularly rich and vital field for development. Interest has been enhanced by the obvious cross-fertilization with Surface Mount Technology.

5.2 BASIC PROCESS DESCRIPTION

Figure 5-3 shows schematically a solder bump termination on a chip. First an insulating, or passivation, layer is coated over the terminal metal pad. A hole is then formed and the interconnecting metallization deposited. This is referred to in the literature as the Ball Limiting Metallurgy (BLM) or Under-Bump Metallurgy (UBM). This chapter uses the former designation. The solder bump is then added, and reflowed to take on a spherical shape. After testing and wafer dicing the chip is ready to be joined.

Each step in the process must balance cost, performance, yield and reliability considerations, and different manufacturers have arrived at different process solutions. The following sections will describe the most prevalent variations of each of these processes, and discuss their compatibility with COB processing.

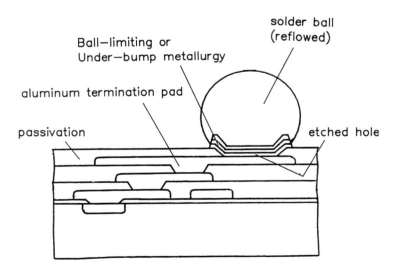

Figure 5-3   Schematic of solder bump for flip chip joining.[59]

5.3 CHIP PASSIVATION

A passivating layer is necessary to protect the aluminum metallization from the environment. In addition, if the chip has no wiring and insulation layers, the passivation also serves to insulate the devices on the silicon surface. Many types of films and processes have been described for interlevel dielectrics (an excellent survey may be found in Reference 10), and theoretically, any of these are applicable to top level passivation. Of these, two materials are predominant in flip chip processing: silicon dioxide[5,11,12] and polyimide.[13,14] Silicon nitride is also occasionally used.[15,16] All three are compatible with COB.

Desirable attributes of passivation coatings include:

1. resistance to penetration by moisture;
2. resistance to penetration by alkali ions;
3. etchability to a controlled via taper;
4. neutral or mildly compressive stress;
5. the ability to absorb or "cushion" thin film stress;
6. good adhesion to underlying metals and insulators; and
7. conformal step coverage.

5.3.1 Inorganics

Sputtered SiO_2 has been used on IBM flip chips for many years.[5] Sputtering is a physical process in which argon ions from a radio frequency (RF) glow discharge impact an SiO_2 "target", dislodging molecules which deposit amorphously on the surface. A bias is also applied to the wafer to enhance resputtering of the SiO_2, which redistributes material to cover the topography better and removes loosely adherent SiO_2. The resulting coating is smooth, adherent, pinhole-free, compressive, and does not significantly heat the wafer. Sputtered SiO_2 also has the lowest dielectric constant among the inorganic passivations, about 3.9.[10] The sputtered oxide is amorphous, although it is often incorrectly referred to as "quartz" (which is crystalline), alluding to the fused silica targets which were used in early stages in the development of the technology.

Silicon nitride shares many attractive features of SiO_2, but has a much higher dielectric constant,[10] which limits its use as a passivating layer. On the other hand, it exceeds SiO_2 in its impermeability to moisture and alkali ions, and can have high compressive stress. Like SiO_2, silicon nitride may be reactively sputtered,[12] but it is principally deposited by Plasma Enhanced Chemical Vapor Deposition (PECVD). In the latter process, gases are introduced into a vacuum chamber and ionized in an RF plasma. Chemical reactions – sometimes in the gas phase, but preferably on the surface – produce deposits on the wafer. The result is a high quality film deposited at a relatively low temperature.[10] As a passivator, PECVD silicon nitride is in compression.[15] Excessive thickness of the nitride has been known to cause high tensile stress in aluminum thin film lines and subsequent creep-induced voiding and failure.[18] Therefore nitride is kept thin, or modulated with an underlayer of SiO_2.

5.3.2 Polymers

Among polymeric materials, polyimides are the popular choice for chip passivation. They are stable at temperatures of 300 to 400°C, making them compatible with chip joining even with high lead solders. Film thickness and properties are readily controllable by the formulation and process parameters. A major advantage over inorganic passivations is the ability of polyimide to absorb thin film stresses imparted from the deposition of interconnection metallization and solder,[13] and from subsequent thermal exposures. This arises from its low elastic modulus and rapid stress relaxation, even at low temperatures.[19] Also, polyimides generally have a lower dielectric constant even than sputtered SiO_2. On the negative side, polyimides are more permeable to water and ionic contaminants, and may require adhesion promoters to insure adequate adhesion to the underlying organic or inorganic insulator.

Polyimide coatings are formed by spin coating a mixture of diamine and dianhydride monomers in a solvent, usually N-methyl-pyrrolidine 2 (NMP). For a fixed formulation, the thickness is directly related to the speed of rotation. The monomers combine at low temperatures to form polyamic acid, which, like its constituents, is soluble in NMP. The film is dried at about 100°C to remove most of the solvent, thence to 250-400°C to complete solvent removal and the conversion to polyimide. The commonly used pyromellitic dianhydride-oxydianiline (PMDA-ODA) family of polyimides are isotropic in behavior. New candidates, however, such as the long polymer chain biphenyldiaminine-phenyldiamine (BPDA-PDA) polyimides tend to be anisotropic in mechanical and electrical properties.[10]

5.4 TERMINAL METALS AND SOLDERS–MATERIALS

Following chip passivation, holes are etched with a photoresist mask, thus preparing the chip for terminal metals and solder. This section discusses the various material options available for the interconnection. Section 5.5 then describes the deposition processes currently in use.

5.4.1 Terminal Metals

The ball-limiting metallurgy (BLM) usually consists of three layers: (1) an adhesion layer such as Cr or Ti, capable of forming a strong bond with the passivation and with the terminating aluminum pad; (2) a solder wetting layer, such as Ni or Cu, which must remain at least partially intact through all the high temperature cycles-wafer reflow, card joining and possible reworks; and (3) a protective layer, Au or other noble metal, to retain wettability for the solderable layer when vacuum is broken. An additional, and vital, requirement of the adhesion and/or wetting layer is the formation of a barrier system to preclude solder penetration into the chip wiring or under the passivation.

The choice of BLM metals will likely depend on the choice of solder. For instance, gold is a poor wetting material with PbSn, in which it completely dissolves in seconds, but may be acceptable with indium in which it has a much lower solubility.[20] As a thin protective layer, however, gold is perfectly acceptable with PbSn, as its purpose has already been served by the time the solder is reflowed. Copper forms intermetallic compounds with the tin component in PbSn, and the amount of copper consumed depends upon the amount of tin available. Thus copper is acceptable for high lead solders, such as 95/5 PbSn, but not for high tin alloys, such as eutectic 37/63 PbSn.

The BLM adhesion layer is generally Cr,[5,8,13,21-25] Ti,[15,26] or TiW.[12,16,17,27] The latter, a codeposited alloy of about 10% Ti, is purported to enhance adhesion, corrosion resistance and barrier properties.[27] Aluminum has also been proposed,[28,29] although it presents potential corrosion concerns. To minimize stress on the passivation (especially from Cr), the adhesion layer is very thin, usually of the order of hundreds of ångstroms.

The second layer is invariably Cu[5,7,13-16,21-24,26] or Ni,[12,25,28-30] although other metals such as Mo[29] and Pt,[31] and composites,[32] have been proposed. Its thickness is typically 0.5 to 1.0 microns, a compromise between stress considerations and the need to withstand multiple reflow exposures. Stress imparted by the BLM to the passivation depends not only on the thin film stress generated by each layer, but on their thicknesses as well. While nickel develops a higher thin film stress than copper, it is much less reactive with tin, thus permitting thinner films. In those COB processes[33,34] in which melting adjacent to the BLM occurs only at wafer reflow, but not at card join or rework, a thinner Cu or Ni layer can be tolerated.

When Cr and Cu are used for the respective adhesion and wettable layers, it has been found[5,7,22] that these metals are immiscible and do not form a metallurgical bond, thus necessitating a codeposited or phased region of several hundred ångstroms to insure mechanical integrity. The SEMs of Figure 5-4 show how the mechanical interlocking typical of a well phased

  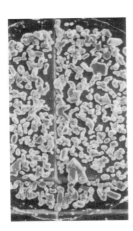

Figure 5-4 Phased Cr/Cu behavior. Cu_3Sn intermetallic morphology. Left, 1 reflow; center, 10 reflows; right, 25 reflows.

Cr-Cu codeposition results in retention of Cu_3Sn intermetallics through multiple reflows (Y. Tsang, IBM, private communication).

Gold is the dominant choice[5,6,12,13,21,22,24] for the final BLM layer, although Pd[7] has also been used. Cost considerations limit the thickness to a few hundred ångstroms, all that is needed for temporary protection. When electroplating solder, copper oxide is automatically removed in the plating bath, obviating the need for a protective layer.[15,16] Indeed, if subsequent BLM etching is required, gold may actually be detrimental as it is difficult to etch.

5.4.2 Solders

Lead-tin is by far the most common solder system used in flip chip bumps. Its phase diagram is shown in Figure 5-5. High lead (95-97% by weight), which melts in the 305-320°C range and requires a 360°C process temperature, has been traditionally used for joining to ceramics.[5,26,35] COB constraints on process temperature, however, suggest that a shift to the lower melting (183°C) eutectic may occur. This alloy, of course, is already widely used in a similar configuration, in the surface mounting of leadless chip carriers, and has also been reported for flip chips.[13,36] There have been processes described which could permit the continued use of high lead solders in a non-melting configuration. In one,[33] referred to as Surface Laminar Circuit Packaging, the board joining temperature is high enough only to melt a eutectic pad on the board. This "tacks" the solder bump without melting the entire joint. A completed joint is shown in Figure 5-6. In

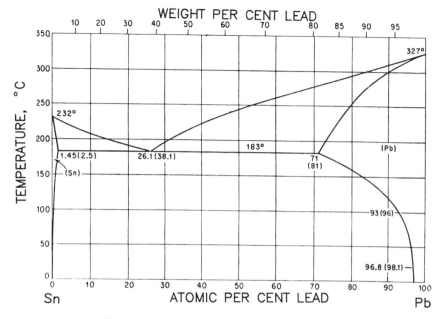

Figure 5-5   Equilibrium phase diagram for lead-tin.

Figure 5-6   Surface laminar circuit packaging (SLCP) solder joint.[33]

another,[34] the traditional tin-capped evaporated (or plated) bump bypasses the usual wafer reflow, and is joined to the board at a temperature just high enough to melt the tin cap (melting point 232°C).

Many other low melting point solders have been proposed for flip chip bumps, at least two of which may provide the additional benefit of improved thermal cycle reliability (as discussed in Section 5.4.3). These are the lead-indium system[37,38] and eutectic tin-silver.[39] Other low melt systems, such as binary tin-bismuth and tin-indium, and countless ternaries are worthy of study. Eutectic tin-bismuth (melting point 138°C), for instance, has recently provided promising results in a surface mount application[40] in which the board could not tolerate the usual joining temperature. An additional item of note is that future government anti-lead legislation could play a role in solder selection.[41]

5.4.3 Reliability Considerations

With the extension of flip chip technology to COB, an important concern must be thermal cycle reliability. The large thermal expansion mismatch between a silicon chip (2.5 ppm/°C) and an organic board (> 15 ppm/°C) produces a large displacement each on-off cycle, which must be borne almost completely by the solder joint, and which can lead to premature fatigue failure. Figure 5-7 illustrates the influence of board expansivity on joint lifetime. While most countermeasures involve strain-reducing organic encapsulants[33,42-44] or modification of board expansivity, conflicting constraints such as reworkability and cost could limit their use.

Within the bounds of the chip process, as described in this chapter, steps can be taken to improve mismatch reliability – not as powerful as encapsulants or board modification, but significant nonetheless.

Joint Geometry

Many studies,[13,23,39,45-49] have been made of the effect of geometry on reliability. The most accurate models are finite element representations which consider plastic flow properties of the solder, as well as, in the most sophisticated cases, time-dependent process, such as creep deformation and crack propagation. These, however, suffer from lack of parametric generality. Other models, which are analytic and parametric in nature,[23,45,46] are weakened by gross approximations in solder behavior and failure criteria. Despite the absence of a well-established parametric model or series of experiments to evaluate the impact of C4 geometry, several trends appear to be valid for unencapsulated joints.

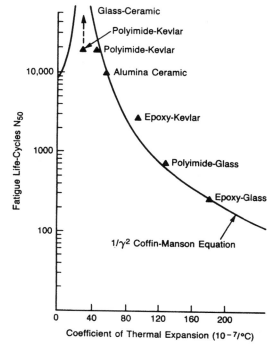

Figure 5-7   Influence of board expansivity on thermal cycle lifetime.[8]

There is an optimum ratio of bottom (board pad) to top (chip BLM) diameter which optimizes fatigue lifetime. This ratio, which must be determined empirically, is characterized by an approximately even division of fatigue cracks in the top and bottom interface regions of the solder.[45,50] The results of one experiment[51] are shown in Figure 5-8. Four cells, having nominally different interface ratios but the same volume, were thermally cycled to failure. Each part was subsequently microsectioned to obtain actual dimensions of the failed joints. Individual lifetime and dimensional data, and average cell failure mode indicate an optimum ratio of 1:1 for this case, and confirm the model prediction.

With no weight (beside the chip) for the solder joint to support, the joints assume the shape of a spherical segment. Within this constraint, the volume of the joint plays a role in thermal cycle reliability. Simplistically, more solder increases the joint height which reduces the thermal shear strain produced by the chip/board mismatch and thus increases lifetime. This is the rationale for processes such as double masking which increase the solder volume. The actual situation, however, is more complicated: as added volume increases height, it also increases girth, thus stiffening the center of

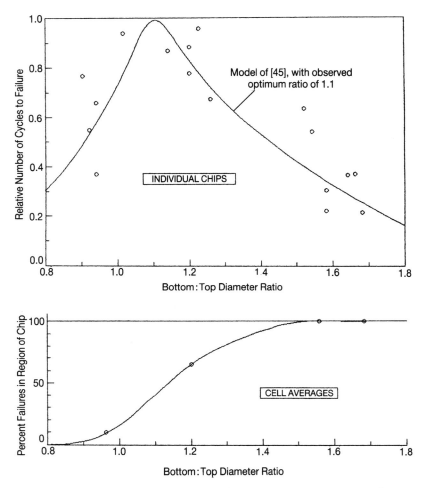

Figure 5-8   Influence of interface diameter ratio on thermal cycle lifetime.[45] 5/95 Sn/Pb, volume = 141 cubic mil-inches, 0-150°C, 3 cycles per hour.

the pad and concentrating the mismatch strain more and more at the interfaces. An optimum volume, above which the fatigue life turns around, has been hypothesized and expressed in a simple formula.[45] Both the trend and the formula have been validated,[23] but only in a mechanical simulation of thermal cycling. To date the existence of an optimum volume (or height) has not been verified in thermal cycling, although the beneficial effect of taller joints from increased volume is widely accepted.

Solder joints which are taller and more slender than would be dictated by the spherical segment geometry can substantially prolong fatigue life as

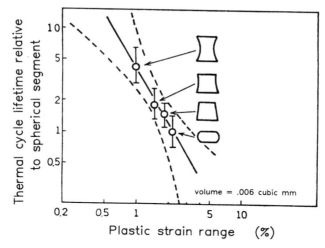

Figure 5-9 Departure from spherical height: effect on thermal cycle lifetime.[52] 5/95 Sn/Pb, −55 to 155°C, 1 cycle per hour.

shown in Figure 5-9. The techniques[33,34] to make high lead solders compatible with COB show this as a side benefit. Other methods have been devised to elongate joints,[52-55] but are of unproven utility for COB.

It bears repeating that suitably chosen epoxy encapsulants can greatly enhance fatigue lifetime,[42,43] and can overshadow the effects of joint geometry.

Material

As mentioned previously, alternative solders to lead-tin have shown fatigue life improvement in flip chip solder joints. Lead-indium has been studied extensively.[37,38] Figure 5-10 shows the increase in lifetime over 95/5 PbSn as a function of indium content. Among the low melt alternatives (indium > 50%), the higher the indium the longer the life. Indium, like tin, cannot be evaporated to significant thicknesses in a reasonable length of time,[37] and would thus require plating or other deposition technique. Caution has also been raised[37,56] about indium corrosion, particularly in a humid, chloride-contaminated environment; and about increased susceptibility to thermomigration.[57]

Recently, 96.5/3.5 SnAg, a 221°C eutectic, has been proposed[39] as a low melt fatigue enhancer. On the basis of fatigue striations in thermally cycled joints, a 2× lifetime improvement was estimated over eutectic PbSn.

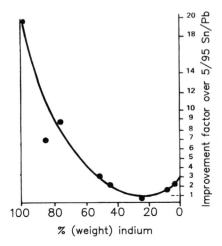

Figure 5-10   Thermal cycle lifetime of lead-indium solder joints.[37] 0-100°C, 3 cycles per hour.

5.5 TERMINAL METALS AND SOLDER–PROCESSES

5.5.1 Sputter Cleaning

Once *in vacuo*, and prior to the start of BLM deposition, wafers are usually given a cleaning step to remove passivation and photoresist residues, and aluminum oxide, at the bottom of the hole. This is done by RF sputter etching, utilizing argon gas ions to bombard the surface and physically dislodge the material.[58] This process, developed earlier for SiO_2 passivation, has also been shown to be the most effective preparation step for polyimide.[14] If a metal mask is used, the energy of the ions may be greatly amplified by applying a high dc bias to the mask,[57] thus using it rather than a remote surface as the cathode. If a lift-off polymer mask process (described below) is used, sputter cleaning is not possible, as argon ions impacting the wafer surface can heat it to the point where the photoresist will flow.[59]

5.5.2 Masking

Metal Masks

Until the 1980s, metal mask evaporation was used almost exclusively for BLM as well as for solder deposition. In this process,[5,9,23,60] the metals are deposited through a molybdenum mask which has had holes photolitho-

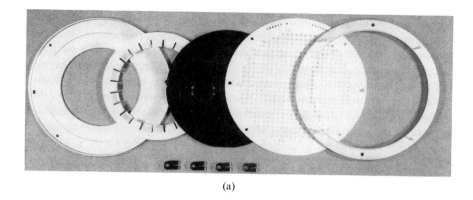

(a)

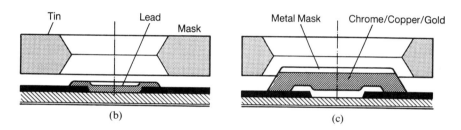

(b)

(c)

Figure 5-11 Metal mask technology.[8] (a) Tooling for alignment of mask to wafer. Left to right: backing plate; spring; wafer; metal mask; clamp ring. (b) Masking and evaporation of lead/tin. (c) Masking and alignment of chromium/copper/gold.

graphically etched from two sides. The fixturing is shown in Figure 5-11. Molybdenum is chosen for its excellent long-term dimensional stability at high temperature, and for its good thermal expansion match with silicon.

Solder may be deposited through the same mask ("unimask"), but in a separate evaporator; or through a new mask with larger holes. Unimask is less expensive, and makes it easier to detect blocked holes. Dual masking, on the other hand, allows for greater solder volume, which could benefit thermal fatigue, as discussed in Section 5.4.3.

Photolithographic Masks

In the last decade, many practitioners have replaced traditional metal masking with photolithographic processes. There are several reasons for this trend,[7] including (a) smaller interconnections, putting excessive demands on moly mask thickness and hole size; (b) larger wafers, exaggerating the detrimental effects of metal mask-wafer thermal mismatch; and (c) the requirement for tighter dimensional tolerances over larger chip surfaces.

Two basic photolithographic processes are used: sub-etch and lift-off. The former is illustrated schematically in Figure 5-12. Metal layers are blanket deposited on the wafer, and covered with a liquid (or dry film[12,61] if thicker material is required) photoresist. Areas to be cleared of metal are then exposed to UV light and the exposed photoresist removed or "developed" in an organic solvent. The exposed metal is then etched away (subtractively–or sub-etched), with the undeveloped photoresist serving as a mask. Finally the photoresist mask is stripped away, leaving the metal pad.

In the lift-off process, shown schematically in Figure 5-13, the resist is first applied and dried. The areas to be metallized are exposed and developed, leaving holes at the BLM sites. The metals are then blanket deposited, and the photoresist stripped off, lifting off with it the overlying metal. It is desirable that a lift-off stencil should have a knife-edge profile, similar to a

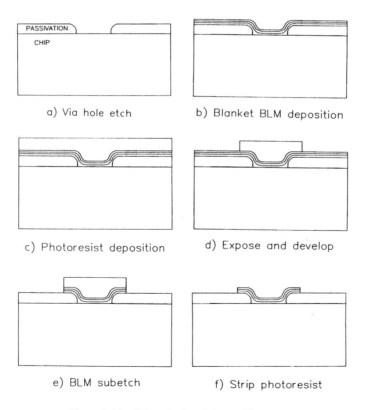

a) Via hole etch

b) Blanket BLM deposition

c) Photoresist deposition

d) Expose and develop

e) BLM subetch

f) Strip photoresist

Figure 5-12 Sub-etch photolithographic process.

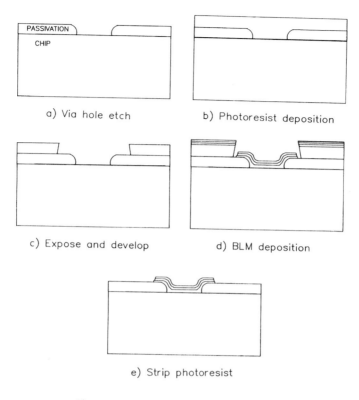

a) Via hole etch

b) Photoresist deposition

c) Expose and develop

d) BLM deposition

e) Strip photoresist

Figure 5-13   Lift-off photolithographic process.

metal mask, to insure the pattern is not contiguous with the removal layer, and that liquid can reach the base of the pad for efficient resist strip.

5.5.3 Deposition Processes

Evaporation

Until the mid 1980s, evaporation[5,12,13,26] was used almost exclusively for both BLM and solder deposition. In this process, "charges" are placed in individual "boats" which are sequentially resistance heated (electron beam activated evaporation has also become popular for bipolar devices).[7] The deposition rate depends on the power input and on the vapor pressure of the particular metal. Deposition of large thicknesses of tin, for instance, is impractical because of its low vapor pressure.

When evaporation is used with moly masking, the line-of-sight evaporant trajectory produces a tapered or conical segment pad. The tapered BLM produces less stress in the passivation than the straight sided or negatively tapered pad normally found after sub-etching. The tapered effect can be enhanced[62] by utilizing dome rotation and selective placement of the several boats.

Sputtering

As an alternative to evaporation, sputtering is frequently used[13-16,24,35] for blanket BLM deposition. Multiple materials may be deposited using individual targets which are alternately biased and shielded. Codeposition from a mosaic target has also been demonstrated.[24] Sputtering is more rapid than evaporation, and also has more process degrees of freedom with which to control film properties, thus for instance, to minimize film stress to the passivation. For example, both the adhesion of TiW on nitride passivation and the thin film stress of a Ti/Cu combination have been shown to be dependent on sputtering parameters.[15] While most sputtering applications have been with Ti based BLMs, Cr/Cu/Au deposition has also been reported.[13,24]

Plating

Electroplating in flip chip technology is widely described in the literature.[7,13-15,26,63,64] Solder components are plated atop a previously evaporated or sputtered Ni or Cu BLM which acts as a seed layer. A plated Cu[15,63] or Ni[26] BLM "extension" may precede the solder.

Plating, not subject to the vapor pressure constraints of evaporation, can produce thick films of all solders components, and in any order. Thus, thick films of tin are easily attainable, and may be placed on top of lead. This is an advantage when subsequent BLM etch removal is required, tin being less corrodible than lead in most common etchants.[26] Among potential disadvantages, electroplating is more susceptible to contamination (from the plating bath) than the relatively clean evaporation. Also thickness uniformity across the wafer may be an issue.[26]

Other Solder Deposition Methods

While evaporation and plating are the most widely practiced solder deposition methods, other techniques such as casting,[65] ultrasonic dipping[25,66] and placing solder balls in mask openings[30,67] have been demonstrated, but not, to this point, as manufacturing processes.

5.5.4 Representative Process Flows

As would be expected with such a broad selection of deposition and masking processes and materials, the technical and patent literature is rife with process menus, each designed to optimize some facet of joint integrity, performance, yield or reliability; or to minimize process time and cost.

The original C4 process,[5] described in previous sections, is still widely used in the industry. A Cr/CrCu(phased)/Cu/Au BLM is evaporated through a moly mask over sputtered SiO_2 or polyimide passivation. High lead PbSn is then evaporated through the same mask or one with larger holes. When used with "tack" bonding,[33] this process is compatible with COB.

The most commonly used photolithographic process sequence, also applicable to COB, is shown in Figure 5-14. Blanket BLM sputtering (usually TiW/Cu) is followed by photoresist application, exposure and development. Lead and tin, perhaps preceded by Cu or Ni, are then electroplated. The resist is then stripped and the BLM etched away, using the plated solder bumps as a mask.[7,14,26]

Other process sequences have been developed, incorporating such novel features as: utilizing both metal and polymer masks in the same sequence;[12,21,60] double lift-off masks for BLM and solder;[68] and reflowing the solder prior to photoresist strip.[69]

5.6 CONCLUSION

Solder bumped flip chip is a well established yet still evolving and expanding technology, which appears to be the most extendable of all bonding techniques for ceramic, glass, silicon and organic substrates. Each new conference or technical journal issue in the field of electronic packaging or hybrid microelectronics announces new processes or materials, or improvements in the existing ones.

In its present state, solder bumped flip chip is well situated for immediate entry into COB technology, the only significant caveats being the possible need to shift from the traditional high lead solders to lower melting alloys, and the potential gating factor of thermal cycle reliability. Coincidentally, but significantly, the surface mount industry continues to address both of these issues with vigor.

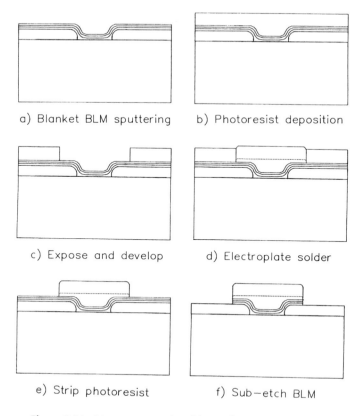

a) Blanket BLM sputtering   b) Photoresist deposition

c) Expose and develop   d) Electroplate solder

e) Strip photoresist   f) Sub—etch BLM

Figure 5-14   Most common photolithographic process sequence.

REFERENCES

1. Gilleo, K., "Direct Chip Interconnect Using Polymer Bonding," *IEEE Transactions on Components, Hybrids and Manufacturing Technology*, **13**(1), 1990, pp. 229-234.
2. Basavanhally, N. R., D. Chang, B. H. Cranston, and S. G. Seger, Jr., "Direct Chip Interconnection With Adhesive Connector Films," *Proc. 1992 Electronic Components and Technology Conference*, pp. 487-491.
3. Epo-Tek Polymer Flip Chip Technology (PFC), advertisements in trade journals, 1992.
4. Miller, L. F., "Controlled Collapse Reflow Chip Joining," *IBM Journal of Research and Development*, **13**(3), 1969, pp. 239-250.
5. Totta, P. A., and R. P. Sopher, "SLT Device Metallurgy and Its Monolithic Extension," *IBM Journal of Research and Development*, **13**(3), 1969, pp. 226-238.

6. Ray, S. K., K. Beckham, and R. Master, "Flip-Chip Interconnection Technology for Advanced Thermal Conduction Modules," *Proc. 1991 Electronic Components and Technology Conference*, pp. 772-778.
7. Yung, E. K., and I. Turlik, "Electroplated Solder Joints for Flip-Chip Applications," *IEEE Transactions on Components, Hybrids and Manufacturing Technology*, **14**(3), 1991, pp. 549-559.
8. Koopman, N. G., T. C. Reiley, and P. A. Totta, "Chip-to-Package Interconnections," *Microelectronics Packaging Handbook*, R. R. Tummala, and E. J. Rymaszewski (eds.), Van Nostrand Reinhold, New York, 1989, Chapter 6, pp. 361-453.
9. Weston, W., "High Density 128×128 Area Arrays of Vertical Electrical Interconnections," *Proc. 4th Annual Microelectronic Interconnection Conference*, July 1985.
10. Srikrishnan, K. V., and G. C. Schwartz, "Passivation and Dielectrics," *Handbook of Semiconductors*, S. Mahajan (ed.), Elsevier, Amsterdam, 1992.
11. Perri, J. A., "Glass Encapsulation," *Semiconductor Products and Solid State Technology*, **8**(5), 1965, pp. 19-23.
12. Imler, B., K. Scholz, M. Cobarruviaz, R. Haitz, V. K. Nagesh, and C. Chao, "Precision Flip-Chip Solder Bump Interconnects for Optical Packaging," *Proc. 1992 Electronic Components and Technology Conference*, pp. 508-512.
13. Heinen, K. G., W. H. Schroen, D. R. Edwards, A. M. Wilson, R. J. Stierman, and M. A. Lamson, "Multichip Assembly with Flipped Integrated Circuits," *IEEE Transactions on Components, Hybrids and Manufacturing Technology*, **12**(4), 1989, pp. 650-657.
14. Takada, N., Y. Moriyama, T. Ozawa, N. Fukuda, and N. Yoshioka, "Highly Integrated Hybrid IC's Using VLSI Flip Chip and Multilayer Thin Film Substrate Technologies," *NEC Res. & Dev.*, No. 97, 1990, pp. 18-24.
15. Warrior, M., "Reliability Improvements in Solder Bump Processing for Flip Chips," *Proc. 1990 Electronic Components and Technology Conference*, pp. 460-469.
16. Gibson, B., "Flip Chips: The Ultraminiature Surface Mount Solution," *Surface Mount Technology*, May, 1990, pp. 23-25.
17. Schwartz, G. C., "Interlevel Dielectrics and Passivating Films," *Electronic Materials and Processing*, P. Singh (ed.), ASM International, 1988, pp. 49-66.
18. Totta, P. A., "Stress-Induced Phenomena in Metallizations: U. S. Perspective," *Stress-Induced Phenomena in Metallization–First International Workshop*, C. Y. Li, P. Totta and P. Ho (eds.), American Vacuum Soc., Series 13 (1991), pp. 1-20.
19. Chen, S. T., C. H. Yang, F. Faupel, and P. S. Ho, "Stress Relaxation During Thermal Cycling in Metal/Polyimide Layered Films," *J. Appl. Phys.*, **64**, 1988, pp. 6690-6698.
20. Yost, F. G., "Aspects of Lead-Indium Solder Technology," *Proc. 1976 ISHM Conf.*, pp. 61-65.
21. Thompson, D. G., S. Y. Lien, and P. F. Hemler, "Computer Modelling Approach to Analyze HCC Solder Joints," *Proc. 1984 ISHM Conf.*, pp. 468-473.

22. Warner, D. J., K. L. Pickering, D. J. Pedder, B. J. Buck, and S. J. Pike, "Flip Chip-Bonded GaAs MMICs Compatible With Foundry Manufacture," *IEEE Proceedings-H*, **138**(1), 1991, pp. 74-78.

23. Kamei, T., M. Nakamura, H. Ariyoshi, and M. Doken, "Hybrid IC Structures Using Solder Reflow Technology," *Proc. 1978 Electronic Components Conf.*, pp. 172-182.

24. Taguma, Y., T. Uda, H. Ishida, T. Kobayashi, and K. Nakada, "Application of DC Magnetron Sputtered Cr-Cu-Au Thin Films for Flip-Chip Solder Terminal Contact," *Proc. 1991 International Electronic Packaging Conference*, pp. 619-623.

25. Myers, T. R., "Flip-Chip Microcircuit Bonding System," *Proc. 1969 Electronic Components Conference*, pp. 131-144.

26. Kawanobe, T., K. Miyamoto, Y. Inaba, and H. Okudaira, "Solder Bump Fabrication by Electrochemical Method for Flip Chip Interconnection," *Proc. 1981 Electronic Components Conference*, pp. 149-155.

27. Meyer, D., A. Kohli, H. Firth, and H. Reis, "Metallurgy of TiW/Au/Cu System for TAB Assembly," *Journal of Vacuum Science and Technology A*, **3**(3), 1985, pp. 772-776.

28. Van der Drift, A., et al., "Integrated Circuits With Leads on Flexible Tape," *Philips Tech. Rev.*, **34**(4), 1974, pp. 85-95.

29. "Semiconductor or Write Tri-Layered Metal Contact," U.S. Patent 3,409,809, 1968.

30. Tsijumoto, N., "Flip Chip and Micropackaged Chips," *Electronic Parts and Materials*, **19**(5), 1980, pp. 51-55.

31. "Flip-Chip Joining Semiconductor Chip to Circuit Board–Using Indium Solder, by Forming Chromium-, Platinum- and Gold-Layers in Joining Position of Chip and Board," Japanese Patent Application 2278743, November 1990.

32. "Stress Relief Substrate Metallization," U.S. Patent 4,773,523, 1988.

33. Tsukada, Y., S. Tsuchida, and Y. Mashimoto, "Surface Laminar Circuit Packaging," *Proc. 1992 Electronic Components and Technology Conference*, pp. 22-27.

34. "Low Temperature Controlled Collapse Chip Attach Process," U.S. Patent 5,075,965, 1991.

35. Takenaka, T., F. Kobayashi, T. Netsu, and H. Imada, "Reliability of Flip-Chip Interconnections," *Proc. 1984 ISHM Conf.*, pp. 419-423.

36. Kubota, T., K. Ichikawa, and M. Suzuki, "High Density Flip-Chip Bonding Technique for 400 DPI Thermal Print Head," *Proc. 1987 ISHM Conf.*, pp. 662-669.

37. Goldmann, L. S., R. J. Herdzik, N. G. Koopman, and V. C. Marcotte, "Lead-Indium for Controlled Collapse Chip Joining," *IEEE PHP-***13**(3), 1977, pp. 194-198.

38. Howard, R. T., "Optimization of Lead-Indium Alloys for Controlled Collapse Chip Connection Application," *IBM Journal of Research and Development*, **26**(3), 1982, pp. 372-389.

39. Takeda, K., M. Harada, T. Fujita, and T. Inoue, "LSI Packaging Technology for Mainframe Computers," *IEICE Transactions*, E **74**(8), 1991, pp. 2337-2343.

40. Gudul, H. E., and R. R. Carlson, "Low Temperature Lead-Free Wave Soldering for Complex PCB's," *Electronic Packaging and Production*, September 1992, pp. 29-32.

41. Derman, G., "Passive Activity: Getting the Lead Out, Part 1," *Electronic Engineering Times*, 21 September 1992, p. 58.

42. Nakano, F., T. Soga, and S. Amagi, "Resin-Insertion Effect on Thermal Cycle Resistivity of Flip-Chip Mounted LSI Devices," *Proc. 1987 ISHM Conf.*, pp. 536-541.

43. Suryanarayana, D., R. Hsiao, T. P. Gall, and J. M. McCreary, "Enhancement of Flip-Chip Fatigue Life by Encapsulation," *IEEE CHMT-14*(1), March, 1991, pp. 218-223.

44. Wang, D. W., and K. I. Papathomas, "Encapsulants for Fatigue Life Enhancement of C4 Connection," *Proc. 1993 Electronic Components and Technology Conference*, pp. 780-784.

45. Goldmann, L. S., "Geometric Optimization of Controlled Collapse Interconnections," *IBM Journal of Research and Development*, **13**(3), 1969, pp. 251-265.

46. Ohshima, M., A. Kenmotsu, and I. Ishi, "Optimization of Micro Solder Reflow Bonding for the LSI Flip Chip," *Proc. IEPC*, 1982, pp. 481-488.

47. Darveaux, R., and K. Banerji, "Fatigue Analysis of Flip Chip Assemblies Using Thermal Stress Simulations and a Coffin-Manson Relation," *Proc. 1991 Electronic Components and Technology Conference*, pp. 797-805.

48. Sherry, W. M., J. S. Erich, M. K. Bartschat, and F. B. Prinz, "Analytical and Experimental Analysis of LCCC Solder Joint Fatigue Life," *Proc. 1985 Electronic Components Conference*, pp. 81-90.

49. Wilson, E. A., and E. P. Anderson, "An Investigation into Geometric Influence on Integrated Circuit Bump Strain," *Proc. 1983 Electronic Components Conference*, pp. 320-327.

50. Goldmann, L. S., and P. A. Totta, "Area Array Solder Interconnections for VLSI," *Solid State Technology*, June 1983, pp. 91-97.

51. Goldmann, L. S., unpublished data.

52. Satoh, R., M. Ohshima, H. Komura, I. Ishi, and K. Serizawa, "Development of a New Micro-Solder Bonding Method for VLSIs," *Proc. 1983 IEPS*, pp. 455-461.

53. Matsui, N., S. Sasaki, and T. Ohsaki, "VLSI Chip Interconnect Technology Using Stacked Solder Bumps," *Proc. 1987 Electronic Components Conference*, pp. 573-578.

54. "Method for Joining Microminiature Components to a Carrying Structure," U.S. Patent 3,921,285, 1975.

55. "Flip Chip Module with Non-uniform Connector Joints," U.S. Patent 3,871,015, 1975.

56. Puttlitz, K. J., "Corrosion of Pb-50In Flip Chip Interconnections Exposed to Harsh Environment," *Proc. 1989 Electronic Components Conference*, pp. 438-444.

57. Roush, W., and J. Jaspal, "Thermomigration in Lead-Indium Solder," *Proc. 1982 Electronic Components Conference*, pp. 342-345.

58. Totta, P. A., "Flip Chip Solder Terminals," *Proc. 1971 Electronic Components Conference*, pp. 275-283.
59. "Method of Forming Metal Contact Pads and Terminals on Semiconductor Chips," European Patent Application 90480110.7 (Publ. no. 0 469 216 A1), 1990.
60. Brownell, D. L., and G. C. Waite, "Solder Bump Flip-Chip Fabrication Using Standard Chip and Wire Integrated Circuit Layout," *Proc. 1974 ISHM Conf.*, pp. 77-87.
61. "Solder Bonded Integrated Circuit Devices," U.K. Patent Application 8620287 (Publ. no. GB 2194386 A), 1988.
62. "Solder Mound Formation on Substrates," U.S. Patent 4,434,434, 1984.
63. Kimijima, S., T. Miyagi, T. Sudo, and O. Shimada, "High-Density Multichip Module by Chip-on-Wafer Technology," *Hybrid Circuits*, 21, 1990, pp. 33-35.
64. Rasmanis, E., "Fabrication of Semiconductor Devices With Solder Terminals," *Proc. 1973 ISHM Conf.*, pp. 3B-3/1-8.
65. Fisher, J. R., "Cast Leads for Surface Attachment," *Proc. 1984 Electronic Components Conf.*, pp. 219-226.
66. Inaba, M., K. Yamakawa, and N. Iwase, "Solder Bump Formation Using Electroless Plating and Ultrasonic Soldering," *IEEE CHMT*, 13, 1990, pp. 119-123.
67. Ariyoshi, H., M. Doken, K. Ohwada, and I. Miyata, "Hybrid IC Structure Made by Solder Reflow Technology," *Rev. Elect. Commun. Lab.*, 26(5,6), 1978, pp. 735-747.
68. "Method of Forming Solder Bump Terminals on Semiconductor Elements," U.S. Patent 4,273,859, 1981.
69. "Lift-Off Process for Terminal Metals," U.S. Patent 4,861,425, 1989.

6

Chip Attachment

Goran Matijasevic, Chin C. Lee, and Chen Y. Wang

6.1 INTRODUCTION

In the quest for the ever simpler microelectronic packaging, chip-on-board (COB) technology presents the ultimate savings in packaging. The chip package is eliminated altogether and the chip is attached (backside or flip-chip) directly onto the ceramic, glass, or printed-wiring board (PWB) substrate. Of the different types of microelectronic packaging, COB is the most cost-effective. Its use has therefore been primarily in high volume low-end applications. However, recent advances in materials and processes have provided the possibility for qualifying this packaging for harsh commercial environments as well as military applications.

With an ever-changing IC design, COB eliminates the concern of providing an appropriate package for each new chip. Beside the obvious advantage in process simplification, the major advantage of COB is a significant increase in the usage of the board area over through-hole or surface mount designs. The reduction in board land pattern allows the dice to be closer together resulting in shorter interconnections and faster speeds. Another advantage is the lower profile of the COB which allows for its use where other packaging technologies are not possible. Of the different kinds of multichip module (MCM) packaging, MCM-L (laminate) is basically an extension of COB technology and shares most of the characteristics. MCM-L modules operating at 40 MHz have been demonstrated and in the future could handle up to 2 GHz, which shows this to be a promising technology.[1]

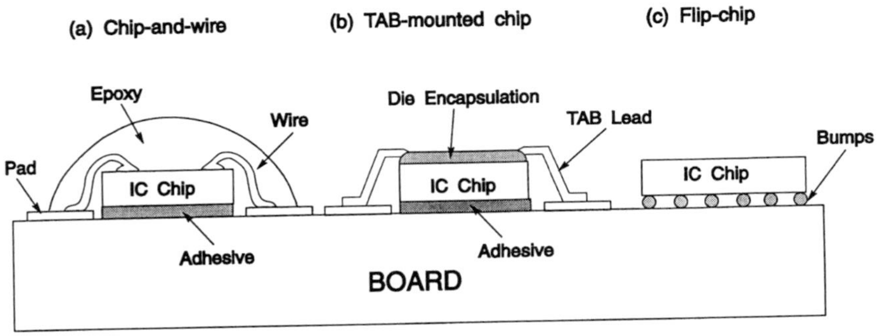

(a) Chip-and-wire   **(b) TAB-mounted chip**   **(c) Flip-chip**

Figure 6-1   The three major COB assembly techniques.

In this chapter, chip attachment methods will be reviewed, with an eye to both current technology and future trends. There are three basic methods for assembly of the bare die to the board. They are chip-and-wire, tape automated bonding (TAB), and flip-chip attachment. These are illustrated in Figure 6-1. If flip-chip attachment is used, the solder bumps that provide the interconnect also provide the mechanical attachment to the board, although sometimes an insulating resin is used as well. Discussion of this method is covered in Chapters 5 and 9 on Solder Bumped Flip Chips. In TAB (Chapters 4 and 8), after inner lead bonding is done, the chip is then attached to the board with an adhesive and the outer leads are then bonded. If the chip level interconnect is done by wire bonding (Chapters 3 and 7), the die is first attached with the backside to the board by a suitable adhesive and the chip interconnections are then made. For protection of the chips, a thin polymer or silicone overcoat ("glob top") is placed over the die to protect the die and the lead wires. For true hermeticity, glass, ceramic, or metal lids may be used for protection. Chip encapsulation is further discussed in Chapters 11 and 12.

6.2 PWB MATERIALS AND CHIP AND PAD PREPARATION

The choice of the printed circuit board (PCB), as the PWB is also known, in COB or MCM-L is driven by the choice of interconnect that will be used. For example, COB wire bonding technology primarily uses aluminum wire wedge bonding which can be done ultrasonically at room temperature. If gold wire thermocompression bonding is used, the board material must be able to withstand temperatures of 150°C or greater. Likewise, chip attach method will also depend on the type of board used. Most organic PWBs such as FR-4 boards, are unable to withstand temperatures higher than

125°C which is their glass transition temperature (T_g). These commonly used boards are suitable for controlled environments where aluminum wire bonding is suitable. If more strict operating conditions are demanded, gold wire bonding in combination with an FR-5, high-temperature epoxy/glass and polyimide/glass, or ceramic board is necessary.

As packaging becomes sophisticated, concurrent engineering becomes necessary, in which the materials and processes that will be used are developed at the same time as the package design is created. This will necessitate that the package designer work closely with the board fabricator. Table 6-1 presents some of the board choices. Besides the ones mentioned above, some special applications boards use high-performance laminates such as polyimide, bismaleimide-triazine (BT) resin, conductive polymers, teflon, and cyanate ester. Table 6-1 gives the maximum use temperature for the different materials, as well as the dielectric constant ε', the dissipation factor ε'', thermal coefficient of expansion (TCE), and thermal conductivity k.

Current boards have six to eight layers, with some up to 12 layers. The layers usually include two signal layers along with power, ground, and distribution layers. Use of very dense metal-on-organic dielectric materials such as copper-on-polyimide allows fabrication of uniform transmission line structures with low DC and AC resistive losses. Line patterns on the PCB can be 3 mils or narrower and are typically made of copper by print and etch or subtractive method. On the other hand, new transient liquid phase conductive inks can be used to directly print the traces on PCB substrates.[2] Good conductivity and additive processing makes them appealing substitutes for copper traces.

Table 6-1   Board Materials

Material	Max use temp (°C)	ε'	ε''	TCE (10^{-6}/°C)	k(W/m°C)
Epoxy-glass (FR-4)	120	4-5	0.05	15	2
Cyanate ester	230	3.8		10	
Epoxy (resin only)	180	3.7	0.02	55	2
Triazine	250	3.1	0.001	50	0.2
Bismaleimide-Triazine (BT) resin (laminate)	290	4.0	0.01	15	0.5
Polyimide (flexible)	400	3.6	0.01	50	0.07
Polytetrafluorethylene (PTFE)	320	2.2	0.004	80	0.3
Alumina (ceramic)	1600	9.5	0.003	6.5	30

To preserve the solderability of the PCB, various techniques are employed. One of the factors in solder joint defects is lead coplanarity. Hot-air solder leveling (HASL) provides a protective coating for the leads, but can produce differences in height from pad to pad. Solder electroplating on the PCB pads is used to provide a more closely controlled solder volume. Automated equipment can provide real-time plating thickness distributions. If solder mask is applied over bare copper (SMOBC), antitarnish coatings such as triazole and imidazole are used to preserve the bondability of copper.

For wire bonding, gold plating of the mounting pads is additionally required. The copper base metal layer is first coated with a nickel or palladium layer (4-6 μm) which serves as a barrier metal. For aluminum wire bonding, a "flash" gold plating of 0.2-0.4 μm is sufficient, while gold wire bonding requires 1-1.5 μm of gold. Both electroplating and electroless plating are employed.

The board wire bond pads and die pads also need to be placed appropriately to avoid bridging of the chip attachment material. This usually means that the wire bond pads should be at least 0.5 mm from the die pad. On the other hand, in consideration of encapsulant runout, this distance should not exceed 2.5 mm. Consideration to any SMT components placed on the board in close proximity must also be given. The COB process can be performed before or after SMT assembly depending on density of the boards and other factors. For rework consideration, COB should be performed first.

In some applications, it is required that the die be placed below the surface of the PCB, so that the TAB bonding is then done to the surface pads. Accurate milling of the PCB is then required to create this pocket or well for the die without disturbing the layers below or creating an uneven surface for subsequent die attach. In some cases, a through hole is created in the PCB, a thermal spreading metal is attached to the PCB from the backside and the die is then bonded to the metal plate. This thermal cut-out, in conjunction with a heat-spreading fin that can be attached to the metal plate, provides for good heat conduction and higher power dissipation.[3]

The board cleanliness prior to use is critical. It is necessary to make sure that any contaminants (esp. organic and ionic) are cleaned off. This may necessitate the baking of the board as many have a tendency to absorb some of the contaminants. During the baking, the contaminants as well as any residual moisture will evaporate leaving cleaner boards and pads.

Prior to chip attachment, the chip itself is prepared by proper hermetic passivation. A number of processes are available for this including silicon nitride and silicon dioxide as well as combinations of the two. A comparison of the different materials and deposition technologies done under various test conditions found that the best protection is obtained with 5000 Å of Si_3N_4 applied by room temperature reactive plasma deposition.[4]

6.3 CHIP ATTACH

6.3.1 Types of Chip Attach

If wire bonding is used for COB or MCM, chip attachment is the first process in die assembly. The chip is adhered to the die pad on the board, after which wire bonding is done using a wedge or ball bonder. Aluminum wedge bonding and gold ball bonding are the commonly used wire bonding methods. The disadvantage of using wire bonding in COB is that the die cannot be tested until it is fixed in place and wire bonded, and is thus not easily removable if it is found that it is faulty. This necessitates that the IC die is guaranteed on its electrical performance and reliability through adequate individual die pretesting and conditioning. When such Known Good Die (KGD) are used, use of wire bonded COB is not a problem.

Using TAB process for COB, inner leads are bonded to semiconductor chips by, for example, thermocompression to die metallization. Sometimes the chip has 250 μm-thick gold bump bonding pads (bumped-chip method), while another process uses bumped-tape where the tape leads have the bumps which facilitate the thermocompression bonding without interfering with device metallization. The die can now be tested and reworked in this tape state prior to attachment to the printed circuit board. Afterwards the die is encapsulated to prevent mechanical damage and then the outer-lead bonding is done. For lead pitches that are large (20 mils or more), screen-printed solder paste on the circuit board is commonly used. For finer pitches, solder is plated onto the pads of the board and after positioning the chip, a laser reflow system is used to carry out the attachment. If the board can withstand higher temperatures, thermocompression bonding is used just as it was in inner-lead bonding. In TAB bonding, the lead frame provides mechanical support for the chip. However, if high thermal or electrical conductivity is required, the die can be mounted with thermally or electrically conductive epoxy. For more details on TAB, please read Chapters 4 and 8.

Finally, if flipped-chip technique is used for COB, the IC die with controlled collapse chip connection (C4) bumps is attached face down forming at the same time mechanical, thermal, and electrical connections with the board. This can be achieved with solder reflow which will force proper alignment of the die. Other flip-chip attach uses eutectic, thermocompression, or just compression bonding. Adapting the C4 bonding schemes to COB bonding to FR-4 has been successfully done and is referred to as Flip-Chip Attach (FCA).[5] This is discussed in more detail in the chapter on Solder Bumped Flip Chip on Board.

One of the challenges in each of these COB package types is the bare chip placement. The need is to assemble complex devices at high speeds, with high consistency, and at low costs. Accurate placement of the bare die on the PWB with ever finer pitch and denser routing layouts is necessary. The most stringent placement requirements are for the compression type of flip-chip bonding where accuracy of 5 μm is needed. But even for wire bonded COB, inaccurate chip placement may cause insufficient adhesive coverage around the die and therefore imperfect die attach which can cause yield problems. There are a number of automatic die bonding machines on the market,[6,7] while the challenge of accurate chip placement on the PWB is being aggressively pursued by new ultraprecision die placement systems.[8]

The rest of this chapter will be devoted to chip attach done with the die backside to the board as well as some general bonding considerations. Bonding material requirements include high adhesion, high thermal conductivity, high electrical conductivity, and acceptable process temperature. The first three of these are relative in nature and depend on the application. In case of COB, the last requirement currently means a bonding temperature below 200°C, although with the progress in new materials, this could go higher. Furthermore, very important considerations in its application are avoidance of high stress on the IC die and, in case of solders, resistance to fatigue and creep rupture. Processing compatibility with regard to temperature, moisture and contaminants is also a factor in material choice. With increasing power requirements of the chip, quality void-free bonding becomes a special concern. For this, particular care must be given to clean die assembly to avoid solvent and residual hydrocarbon contamination. Finally, cost and flexibility of usage are likely to dictate the choice of a given material.

Bonding materials can be classified as polymers, which include the widely used epoxies and polyimides, glass, soft solders, and hard solders. Each of these has application advantages as well as disadvantages, as indicated in Table 6-2.

6.3.2 Polymer Bonding

Polymers such as polyimides, silicones, and epoxies filled with precious metals have widespread use in cost-driven microelectronic packaging. They are most commonly used in COB. Processing time on an automatic bonder

Table 6-2 Bonding Materials

Bonding materials	Advantages	Disadvantages
Organic adhesives	Low processing temperature	Poor thermal stability
Metal-filled epoxies	Stress relief	Low thermal conductivity
	Low cost	
Glass adhesives	Good thermal stability	High processing temperature
Silver filled glass	Low cost	Poor thermal conductivity
		No stress relief
Soft solders:	Stress relief	Thermal fatigue
Pb-Sn, Pb-In	Low cost	Creep movement
		Low strength
Hard solders:	No thermal fatigue	No stress relief
Au-Sn, Au-Si, Au-Ge	High strength	High cost

is typically one second. Epoxies usually offer the added advantage of single step, low temperature (150-200°C) cure processing,[9] while for polyimides T_g is higher. They do not induce high stress in the bonded materials as they deform inelastically to absorb the stress. They often come in form of pastes that can be dispensed at the bond location. On the other hand, film adhesives have advantages such as uniformity of film thickness and lower volatile component. Properties that are important in the evaluation of the die attach adhesives include viscosity, pot life, glass transition temperature, die shear, outgassing, electrical and thermal conductivity, as well as stability.[10]

Consistent viscosity is a necessity for adhesive dispensing, whether this is done by a syringe or by screen printing or stamping. Pot life determines the usefulness of a particular adhesive as there is an increase in viscosity over time. Definitions of this increase vary from 10 to 50% at which point there are problems in dispensing or screen printing of the epoxy and its subsequent wetting of the substrate.

Organic adhesives have poor thermal conductivity and metal fillers (usually silver) are added to increase both thermal and electrical conductivity. Volume fraction of filler to insulating polymer matrix at which the material becomes a conductor (V_c) depends on the geometry factor G (ratio of area of electrode contact/distance between electrodes) and size factor Γ (ratio of particle size to the smallest sample dimension).[11] In general, packing cannot be taken as random and the conductive pattern is far from isotropic. The conduction mechanism via the metal fillers can sometimes break down causing high resistivity after burn-in.[12] Other fillers include aluminum, and dielectric fillers such as BeO, AlN, Al_2O_3, and diamond. The last one, though not necessarily much better in thermal conductivity than the others as originally thought and quite expensive at the same time, can still be useful for high power devices if thin and void free bond lines are achieved, preferably using a flexible adhesive to reduce thermal stress.[13]

One of the biggest concerns with polymer usage is a type of degradation called reversion in which breakage of polymeric bonds occurs. The epoxies also have poor thermal stability and are therefore not used in high-reliability applications. Even 1 hour in 60% humidity greatly degrades anhyride epoxies.[14] They are also likely to release ionic contaminants due to outgassing, which is particularly detrimental to laser diode chips. Silver in the conductive adhesives was shown to increase outgassing. If oxygen is present, silver induced mass loss also increases.[15] Delamination can also occur with the use of die attach epoxies due to voids in the bonding, moisture absorption of the epoxy, poor adhesion, resin bleed, or CTE mismatch-induced stress.[16]

Polyimidesiloxanes (SIM) are film adhesives with good adhesion, low modulus, low moisture absorption, and low ionic impurities. Inclusion of siloxane into typical aromatic polyimide lowers T_g and gives good properties. Die attach carried out at 250°C with $7.5 \, g/mm^2$ force (10 psi) for 1-3 minutes gave good results.[17]

New thermoset adhesives can be cured at 200°C for less than 60 seconds with little weight loss and low moisture content.[18] The latter is of great importance for hermetic packages. Silicone die bond adhesive, owing to its low Young's modulus, has been shown to be virtually stress-free in thermally mismatched silicon on copper lead frame bonding.[19]

Direct chip interconnect (DCI) can be carried out electrically using anisotropic conductive adhesive films (ACAFs). The concept is to simultaneously attach and electrically interconnect IC dies to circuit traces on substrates.[20-22] They are called "Z-axis" thermally conductive adhesives as they conduct only in the z-direction, while providing resistance in the x- and y-directions. With millisecond curing speeds, ambient storage, low thermal and electrical resistances, and reworkability, they can be used in TAB and flip-chip applications. The resin formulation can also be made to match closely the TCE of adjoining materials. Conductive adhesives are thus moving to replace Pb-Sn solder in printed circuit board assembly.[23] They can achieve fine-pitch resolution at substantially lower viscosities than solder paste and can be screen-printed at thinner depositions.[24] However, since these materials contain metallic particles, their propensity for metal migration becomes a concern.[25] A suitable range of applications can be determined by evaluating the electric field around the particles as this field is the driving mechanism for the migration and corrosion.

6.3.3 Glass Adhesives

Metal-filled glass adhesives are designed to become inorganic after firing. This avoids outgassing and moisture problems associated with epoxy and polyimide adhesives.[26] Organics are removed by low temperature drying as the initial step during the firing cycle. Organic burn-out occurs at 130°C, causing shrinkage of the bond line and vertical collapse of the die. Final dwell at the bonding temperature is about 10 minutes.[27] Bonding temperatures, originally 400-450°C, are now below 400°C. If the glass, traditionally based on lead borate glass, is made from a melt of lead vanadium bismuth telluride with Ag content 75-90 wt%, die attach can be carried out at 260°C in 10 minutes.[28] This lower temperature makes glass adhesives usable on high temperature circuit boards. They can be used for bonding directly to silicon dies as the presence of a SiO_2 layer aids the bonding mechanism. However, glass has poor thermal conductivity and silver flakes have to be added similar to the epoxies. Furthermore, high stress is produced on the die backside due to high processing temperature and lack of plastic deformation of the glass adhesive. This can cause horizontal die cracking.[29] Use of solvents and binders for processing purposes also raises the concern of complete solvent removal.

Adhesion is achieved mainly by glass migrating to the die interface during the firing cycle and chemically bonding to the die surface.[27,30,31] An oxidizing ambient at high temperature is often needed for the glass die bonding and this may cause unwanted oxidation of other die or package metals. The oxidation is needed for metal coated surfaces so that they will be oxidized prior to joining. On the other hand, bonding is also achieved with gold plated substrates which do not oxidize. The bonding is produced by interdiffusion and formation of a metallurgical bond between silver and gold and subsequent mechanical interlocking with glass. As the metal-to-glass weight ratio goes above 4, metal becomes the matrix and glass the filler, decreasing the brittleness of the adhesive.[27]

6.3.4 Soft Solders

The most commonly used soft solders are lead-tin (Pb-Sn) alloys. Others include various low-melting indium, lead, and tin-based alloys. In flip-chip technology, terminal materials are most often soft solders which are combinations of the eutectic 37Pb-63Sn and the low tin 95Pb-5Sn alloys. In die attach applications, they are often used in paste forms which include fluxes which then need to be removed. Water soluble solder pastes (WSP) for surface mount attach applications are acceptable, but still inferior to standard RMA (rosin mildly activated) ones which require CFCs for flux removal.[32] Solder pastes are usually deposited through stencils onto the PCB. The stencils are made of brass or stainless steel, with the latter being compatible with increasingly used aqueous cleaning.[33] For flip-chip applications, in order to control the solder volume better, solders are often evaporated, but can also be electroplated.[34] Other techniques for solder deposition are being developed including solder jetting which uses inkjet technology for precision deposition of solder droplets.[35]

Soft solders are generally inexpensive, have acceptable thermal conductivity, but are mechanically weak. Though rigid at low temperatures, they become softened and lose tensile strength at temperatures above $0.75 T_m$. Materials bonded using soft solders do not experience high stress because the bonding layers deform plastically to absorb the stress developed.[36] When soft solder is used to bond the die, most of the stress would occur in the bonding layer because it is much softer than the die and leadframe or substrate.[33,37,38] Their yield strengths are low, usually less than 40 MPa. The capability of plastic deformation, however, makes soft solders subject to thermal fatigue[39] and creep rupture,[40] causing long-term reliability problems. As a result, during thermal cycling the bonding layer degrades because of thermal fatigue due to the constant plastic deformation produced by the dynamic stress. In eutectic soft solder joints, fatigue cracks were found due to grain boundary sliding.[41] Solder joint damage rate was studied under thermal cycling for various soft solders and it was found that 96.5Sn3.5Ag and 97Pb3Sn can absorb more strain before the onset of failure.[42] High Pb containing PbSn alloys are less strain rate sensitive than near-eutectic solder.[43] However, they require higher processing temperature. This can be lowered to 180-200°C by the use of an electroplated 37Pb63Sn solder volume on the board pad in combination with a 97Pb3Sn bump on the chip.[44] Microstructure was found to be of great importance in determining damage accumulation.[45]

Maximum stresses and strains occur at or near the interfaces.[46] If voids and cracks exist in the bonding layer, they will propagate during thermal cycling and thus cause the device to suffer an early failure. Another concern is the possible formation of intermetallics when the soft solders come into contact with conductor metallizations.[47] These stoichiometric binary compounds are generally very brittle and if there is a thick layer of this phase, it may cause cracking of the solder bond.[48,49] Therefore soft solders may not be appropriate if there are Au metallizations thicker than $0.25 \mu m$.[50]

Composite solders made of powdered intermetallics (10-40% Cu_6Sn_5) mixed into solder paste (60/40 Sn/Pb) have been prepared and characterized.[51] Mechanical properties were shown to be superior, but the mechanism of failure, that is, heterogeneous coarsening during thermocyclic fatigue, is the same as for 60/40 Sn/Pb.

Flux removal is a major concern in solder attachment as flux residues are reliability risks due to corrosion. Their removal using conventional solvents will no longer be possible due to environmental regulations, thus necessitating better water-soluble pastes. An alternative to this is the use of a fluxless soldering with an inert environment. One such process uses fluorine plasma,[52] while another uses evaporated multilayers which are joined and then heated to form the complete solder.[53] The latter technique will be described in greater detail in Section 6.3.6.

6.3.5 Hard Solders

Hard solders are low melting point gold alloys that have high strengths. The most commonly used hard solders include Au-Sn, Au-Ge and Au-Si eutectic alloys. Their properties, as well as those of soft solders, are summarized in Table 6-3.[38]

They all have high thermal conductivity and are free from thermal fatigue because of high strength which results in elastic rather than plastic deformation. They are not subject to fatigue or creep rupture during thermal cycling as are soft solders. Because of the lack of plastic deformation, dice bonded using hard solders have a certain amount of residual stress depending upon the melting temperature.[54] Hard solders transfer the high thermal mismatch to the device which can cause cracking of the die.

Hard solder bonding requires a gold pad on the substrate or board. In PCBs, this is done by first nickel plating the copper traces on the board

Table 6-3 Properties of Hard and Soft Solder Alloys

Alloy	Melting Point (°C)	Thermal Conductivity (W/m°C)	Coefficient of Thermal Expansion (10^{-6}/°C)	Yield Strength (MPa) at 23°C	100°C	150°C
Hard Solders						
Au-20Sn	280	57.3	15.9	275	217	165
Au-12Ge	356	44.4	13.3	185	177	170
Au-3Si	363	27.2	12.3	220	207	195
Soft Solders						
Pb-63Sn	183–188	50.6	24.7	35	15	4
Pb-5Sn	308–312	23	29.8	14	10	5

1 pascal (Pa) = 1 newton/m^2 = 0.145 × 10^{-3} psi.

followed by a gold plating. When using soft solders for bonding, the amount of gold has to be controlled so that the joint is not embrittled by gold intermetallics. However, with hard solders this is not a concern since they are gold alloys.

Of the three hard solders, the Au-Sn eutectic alloy has the lowest melting temperature and therefore has special use for devices sensitive to high processing temperatures such as GaAs dice.[55-58] Its use could also be appropriate in chip attachment to boards that can sustain the higher temperatures required for hard solders. The high thermal conductivity of the alloy makes it especially attractive in power amplifier packages which tend to run hot.[57] Successful attachment of large Si dice with Au-Sn alloy has also been demonstrated in spite of the large thermal expansion mismatch.[54,59] Furthermore, this eutectic alloy also has good mechanical properties, i.e. high strength and lowest Young's modulus of the hard solders. Aside from its use in die bonding, it is thus often used for TAB.[60-63]

Au-In alloy is another solder that could be used for bonding, but the melting temperature of the eutectic alloys is above 450°C which is substantially higher than the process temperature permitted for most semiconductor devices and circuit boards. Thermocompression bonding at a temperature of 150°C using low-melting In alloy on top of Au bumps has been used to bond liquid crystal display (LCD) panels.[64] Indium alone has been widely used as a soft solder for bonding delicate devices which need a low process temperature (157°C) and require plastic deformation of the bonding layer to absorb the induced stress.

6.3.6 Bonding Below the Eutectic Point of Solders

Solid state diffusion is a technique in which parts to be bonded are held together with sufficient pressure to allow close contact, but not cause macroscopic deformation.[65] Thermocompression bonding used in tape automated bonding is an example of diffusion bonding in which the parts are also heated to a temperature that assures faster interdiffusion. If the temperature is raised high enough for one of the components to melt, liquid phase bonding occurs. As diffusion in the liquid state is about three orders of magnitude faster than in the solid state, faster joining with less pressure applied is possible with liquid-phase bonding. Combining the two types, we have solid-liquid diffusion bonding. In this bonding mechanism, layers of alloy components are deposited on the two surfaces to be bonded. The use of pre-deposited multilayers eliminates the need for a preform and reduces oxidation. One of the alloy components has a melting temperature lower than the melting point of the final alloy desired. If the two surfaces are brought into contact and the temperature is raised above the melting point of one of the components (e.g., Sn or In), alloy formation will commence. If more of the second component material (e.g., Au or Cu) is present and an elevated temperature is maintained, solid-state diffusion will continue until the alloy is formed in the proportion of the layer compositions. This process is referred to as solid-liquid interdiffusion (SLID)[66] or transient liquid phase (TLP) bonding. This method permits the reversal of the conventional soldering hierarchy where processing temperatures of subsequent soldering steps become lower and lower. Bonding can be produced at a low temperature but can sustain high temperature afterwards.

The principle and results for the bonding technique which can be done at temperatures lower than the eutectic point are illustrated in the example of Au-In alloy. The phase diagram of this alloy is given in Figure 6.2.[67] The eutectic points of the phase diagram are at 456.5°C, 454.3°C and 495.4°C which corresponds to In weight percentages of 24, 28 and 42, respectively. The melting point of In is at 157°C, which is substantially lower than the eutectic points and this property of the phase diagram allows for the SLID (or TLP) bonding.

For Au-In multilayer bonding, Au and In layers are deposited on the die to be bonded and brought into contact with the gold-coated board. Gold,

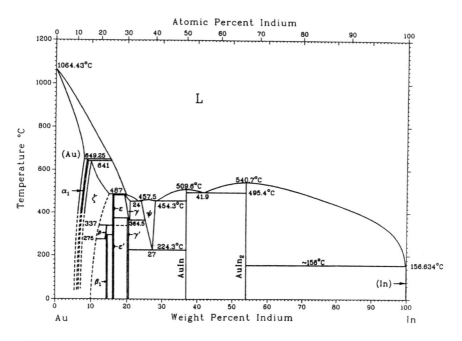

Figure 6-2   Gold-indium equilibrium phase diagram.[67] (Reprinted with permission of ASM International, © 1990.)

indium, and gold layers are deposited on the die in succession in vacuum to prevent indium oxidation at the Au-In interface. Interdiffusion between the two layers occurs and it was found that $AuIn_2$ forms almost immediately upon deposition because of the high diffusion rate,[68] as shown in Figure 6-3(a). The gold layer and the subsequently formed $AuIn_2$ prevent In oxidation which would otherwise impede the bonding with the substrate. Since the evaporated layers are made in a proportion that is lacking gold to form the Au-In eutectic, interdiffusion continues when this structure is joined with a layer of gold. Provided that enough pressure is applied to allow intimate contact between the surfaces and that the temperature is above the 157°C melting temperature of In, solid-liquid diffusion will occur as exhibited in Figure 6-3(b). When In melts, it wets the adjacent gold layers to form Au-In alloy, and liquid and solid components will interdiffuse. The mixture wets and dissolves the gold on the substrate to form more $AuIn_2$.

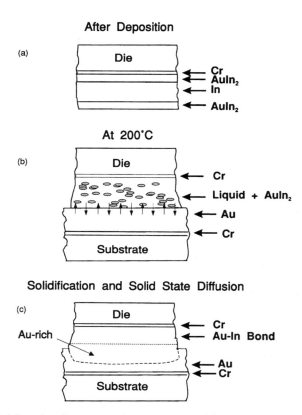

Figure 6-3 Multilayer bonding process description: (a) Multilayer Au-In composite on the die; (b) Melting of In and sold-liquid interdiffusion; (c) Solidification and solid-state diffusion.

This solid-liquid interdiffusion continues until the bond solidifies, followed by solid-state diffusion which makes the alloy close to the total composition of the layers joined as shown in Figure 6-3(c). The bond is thus complete and it has a melting temperature much higher than the temperature at which it is produced. Since here the bond is achieved at a temperature below the melting point of the final alloy, the residual stresses that form upon cooling of the bonded structure to room temperature will be smaller.

Au-In bonding was carried out with the Cr-Au-In-Au composite structure which alone would form an alloy of 86 wt% indium.[69] After the evaporation, energy dispersive X-ray (EDX) determined the composition of the top layers of the deposited structure at 53.5 wt% In which indicates that $AuIn_2$ forms on the structure immediately upon Au evaporation.

A GaAs die was bonded to 250-μm-thick alumina coated with chromium and 6.4 μm gold. If the gold layer on the substrate together with the Au-In composite on the GaAs die forms a uniform Au-In alloy, the alloy would then have 27 wt% indium. Bonding was carried out at 200°C for 10 minutes. Contact was made using 40 psi (0.276 MPa) of static pressure using a mechanical tool. Hot air thermode tool used for TAB bonding[70] can also be used.

Cross-sections were examined and the bonding layers were found to be uniform with a thickness of 12.6 μm. Au and In weight percentage across the bonding layer as determined by EDX shows gold composition ranging from 51 wt% right underneath the backside of the GaAs die to 100 wt% right above the alumina substrate. Since the original Au-In composite contained 14 wt% Au and 86 wt% In, according to the phase diagram, upon melting at 200°C, this mixture would dissolve the gold layer on the substrate and eventually become solidified at a gold content of 46 wt% ($AuIn_2$). However, from the EDX data, the lowest gold wt% over the entire bonding layer cross-section is 51% rather than 46%. Gold and indium atoms continue to interdiffuse by solid-state diffusion and form a mixture of AuIn and $AuIn_2$.

The quality of the bonding was examined using the scanning acoustic microscope (SAM). Figure 6-4(a) shows the SAM image of a specimen which has near perfect bonding. After 40 cycles of thermal shock, the specimens were examined by the SAM to identify bonding degradation and die cracking. It was found that well-bonded devices remained well-bonded and no die-cracking was detected after the test. Figure 6-4(b) displays the SAM image of the specimen shown in Figure 6-4(a) after the test. No bonding degradation is observed.

The unbonding temperature of the new bonding layer was determined to be 459°C (±5°C). Even though the bonding was done below 200°C, the bonding layer did not remelt until 459°C which is just above the melting temperature of the eutectic Au-In alloy between γ phase and ψ phase of 456.5°C.

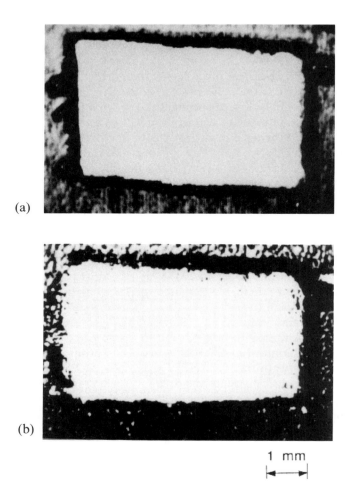

Figure 6-4   (a) SAM image of a GaAs die perfectly bonded to an alumina substrate using Au and In composite structure before thermal shock test. (b) SAM image of the same specimen as (a) after 40 cycles of thermal shock test between $-196°C$ and $160°C$.

This technique of transient liquid phase solding can also be applied to other alloy systems which have one component with a melting point lower than the alloy which is to be formed. It was demonstrated successfully with Au-Sn,[71] Pb-In-Au,[53] and Cu-Sn[72] alloy systems. Studies have been undertaken on the Ag-Sn and Ag-In systems.[73] It has also been successfully demonstrated with Pb-Sn alloys[74] and in conductive ink technology.[2]

6.4 ATTACHMENT QUALITY

6.4.1 Thermal Stress Considerations

If the thermal coefficient of expansion (TCE) of the board is greater than that of the die, there is a tensile stress produced on the top of the die. Since residual stress develops as the structure is cooled from the melting temperature, less stress will be generated by the chip attach material with a lower processing point.

Matching of thermal expansion coefficients is therefore of great concern in large power semiconductor devices. If we match all the materials to have the same thermal expansion coefficient, assuming uniform temperature distribution, there would be no stresses produced if no voids exist. If temperature distribution is not uniform, stress can be generated but it will not be significant and not be localized unless there are voids located underneath the die. If the two materials that are bonded are TCE matched but the bonding layer is not, it is advantageous to have a thinner bonding layer which will induce less stress on the die. Finally, when thermal expansion mismatch exists, vertical die cracking can be prevented if the stress field generated is uniform and below the tensile strength of the die, i.e., there are no voids to cause localized stress. Accordingly, the chips with no voids in the bonding medium would not crack during the bonding process or under thermal and power cycling after the bonding process.[75]

Sealing the die will be discussed in the chapter on Encapsulation of COB, but it is important to point out that TCE matching of the sealing material is of importance. It should be a compromise between the TCE of the substrate and that of the die in order to minimize thermal stresses that occur on the die, just as the die attach material had to fulfill a similar requirement. In addition to choosing a coating with a specific TCE, it is also possible to adjust the TCE of the epoxy-resin coating by adding an appropriate filler.

6.4.2 Attachment Quality

Voids are fairly common in the bonding layer and are either produced during bonding or introduced later during cooling, storage or use of the device. They can also be caused by various liquid or solid agents trapped during attachment or simply by uneven wetting of the surfaces to be joined. If large voids exist in the bonding layers near the die edges, they would induce high localized stress and consequently increase the probability of die cracking during thermal cycling or shock.[76,77] Voids in the solder bonds are formed due to material segregation as well as formation of oxides and carbon compounds on the molten solder.[78-81] These materials form a solid film on the surface of the solder solution, which prevents the alloy from producing a bond with the die and substrate. Edge voids can also occur due to incomplete spreading of the die attach material especially in the case of large dice. Metallurgical fatigue of the soft solders as well as creep and rupture may also cause disbonding. Formation of layers of brittle intermetallics at the interfaces can also cause failure. For hard solders, high TCE mismatch may introduce large mechanical stresses upon cooling which will cause die cracking or disbonding to occur.

Voids cause local stress concentration which is very much dependent on the location of the voids. Calculations showed that voids near edges of bonded materials are likely to cause high stress and possibly cracking.[77] Experimental observations of actual production samples as well as on samples with purposely formed voids confirmed this prediction.[29,81] Voids can also substantially degrade the thermal and electrical performance of the attached die by creating hot spots at the die location above the void.

Voids, especially built-in ones, can be detected nondestructively by examining the bond with X-ray radiography or with a SAM.[82] A new microfocus X-ray system uses a small focal spot to achieve improved magnification and resolution,[83] while cross-sectional X-ray laminography can be used to inspect solder joints.[84] However, X-rays will still not pick up delaminations which are less than 2% of total thickness. For this, as well as for detecting voids and cracks, the SAM is an invaluable tool.[85,86]

Since a major concern for die stress formation is voids and any unevenness in the bonding layer, die attach uniformity becomes crucial. Several methods have been employed to achieve voidless high-quality bonding. Doing the bonding in a very pure environment, without contamination of the die-attach materials, is of great importance. Application of uniform pressure to achieve good contact and therefore adhesion is also vital.

6.5 SUMMARY

In COB, chip attachment depends on the type of chip interconnect that is used. For chip-and-wire interconnect, chip bonding to the board has most often been achieved using polymer adhesives. However, as new boards are developed and requirements for high thermal conductivity of the bonding layer increase, other bonding mediums are necessary and are being developed. Whether a bonding material is a polymer, glass, hard or soft solder, requirements of high adhesion, good mechanical strength, high thermal conductivity, and acceptable process temperature are of importance. The lowering of the process temperature of the high strength and high conductivity attach mediums such as has been seen in glasses and hard solders allows the use of these materials in bare chip attach. New materials and processes that are being developed will further push the limits of the current technology.

REFERENCES

1. Tuck, J., "On board the MCM-L train," *Circuits Assembly*, vol. 4, pp. 20-21, March 1993.
2. Capote, M. A., M. Todd, P. Gandhi, C. Carr, W. Walters, and H. Viajar, "Multilayer printed circuits from revolutionary transient liquid phase inks," *NEPCON West Conference Proceedings*, pp. 1709-1715, 1993.
3. Yamada, K., A. Tanaka, H. Shinohara, M. Honda, T. Hatada, A. Yamagiwa, and Y. Shirai, "A CPU chip-on-board module," *Proceedings 43rd IEEE Electronic Components and Technology Conference*, pp. 8-11, 1993.
4. Gates, L. E., G. G. Bakhit, T. G. Ward, and R. M. Kubacki, "Hermetic passivation of chip-on-board circuits," *Proceedings of the 41st Electronic Components and Technology Conference*, pp. 813-819, 1991.
5. Powell, D. O., and A. K. Trivedi, "Flip-chip on FR-4 integrated circuit packaging," *Proceedings 43rd IEEE Electronic Components and Technology Conference*, pp. 182-186, 1993.
6. Iscoff, R., "Flexibility drives automatic die bonder market," *Semiconductor International*, vol. 13, pp. 80-88, September 1990.
7. "MMT Selection Guide: Die bonding," *Microelectronics Manufacturing Technology*, vol. 14, pp. 21-26, August 1991.
8. Chalsen, M. J., "Automatic chip placement: one solution, user-benefits, and future development," *Proceedings 41st IEEE Electronic Components and Technology Conference*, pp. 422-427, 1991.
9. Koopman, N. G., T. Reiley, and P. A. Totta, "Chip-to-package interconnections," in *Microelectronic Packaging Handbook*, pp. 361-453, ed. by R. R. Tummala and E. J. Rymaszewski, Van Nostrand Reinhold, New York, 1989.
10. Estes, R. H., "A practical approach to die attach adhesive selection," *Hybrid Circuit Technology*, pp. 44-47, June 1991.
11. Ruschau, G. R., S. Yoshikawa, and R. E. Newnham, "Percolation constraints in the use of conductor-filled polymers for interconnects," *Proceedings 42nd IEEE Electronic Components and Technology Conference*, pp. 481-486, 1992.
12. Opila, R. L., and J. D. Sinclair, "Electrical reliability of silver filled epoxies for die attach," *Proceedings 23rd International Reliability Physics Symposium*, pp. 164-172, 1985.
13. Bolger, J. C., "Prediction and measurement of thermal conductivity of diamond filled adhesives," *Proceedings 42nd IEEE Electronic Components and Technology Conference*, pp. 219-224, 1992.
14. Ameen, J. G., K. K. Nair, and D. W. Sissenstein, Jr., "Degradation effects of humidity and storage on anhydride cured epoxies," *Proceedings 42nd IEEE Electronic Components and Technology Conference*, pp. 939-944, 1992.

15. Phillips, T. E., N. deHaas, P. G. Goodwin, and R. C. Benson, "Silver-induced volatile species generation from conductive die attach adhesives," *Proceedings 42nd IEEE Electronic Components and Technology Conference*, pp. 225-233, 1992.

16. Comstock, B., "Die attach epoxy's role in thin package delamination," *Semiconductor International*, vol. 16, pp. 162-166, July 1993.

17. Sakamoto, Y., N. Takeda, T. Takeda, A. Tokoh, and D. Y. Tang, "New type film adhesives for microelectronics application," *Proceedings 42nd IEEE Electronic Components and Technology Conference*, pp. 215-218, 1992.

18. Nguygen, M. N., and M. B. Grosse, "Low moisture polymer adhesive for hermetic packages," *Proceedings 42nd IEEE Electronic Components and Technology Conference*, pp. 239-244, 1992.

19. Suzuki, K., T. Higashino, K. Tsubosaki, A. Kabashima, K. Mine, and K. Nakayoshi, "Silicone die bond adhesive and clean in-line cure for copper lead frame," *IEEE Transactions on Components, Hybrids and Manufacturing Technology*, vol. 13, pp. 883-887, 1990.

20. Gilleo, K., "Direct chip interconnect using polymer bonding," *IEEE Transactions on Components, Hybrids and Manufacturing Technology*, vol. 13, pp. 229-234, 1990.

21. Chung, K., G. Dreier, P. Fitzgerald, A. Boyle, M. Lin, and J. Sager, "Z-axis conductive adhesive for TAB and fine pitch interconnects," *Proceedings 41st IEEE Electronic Components and Technology Conference*, pp. 345-354, 1991.

22. Basavanhally, N. R., D. D. Chang, B. H. Cranston, and S. S. Seger, Jr., "Direct chip interconnect with adhesive-connector films," *Proceedings 42nd IEEE Electronic Components and Technology Conference*, pp. 487-491, 1992.

23. Bolger, J. C., J. M. Sylva, and J. F. McGovern, "Conductive epoxy adhesives to replace solder," *Surface Mount Technology*, vol. 6(2), pp. 66-70, 1992.

24. Nguyen, G. P., J. R. Williams, and F. W. Gibson, "Conductive adhesives," *Circuits Assembly*, vol. 4, pp. 36-41, January 1993.

25. Chang, D. D., J. A. Fulton, H. C. Ling, M. B. Schmidt, R. E. Sinitski, and C. P. Wong, "Accelerated life test of z-axis conductive adhesives," *Proceedings 43rd IEEE Electronic Components and Technology Conference*, pp. 211-217, 1993.

26. Kearney, K. M., "Trends in die bonding materials," *Semiconductor International*, vol. 11, pp. 84-88, 1988.

27. Dershem, S. M., and C. E. Hoge, "Effects of materials and processing on the reliability of silver-glass die attach pastes," *Proceedings of 42nd IEEE Electronic Components and Technology Conference*, pp. 208-214, 1992.

28. Smith, C., K. Le, T. Herrington, J. S. Jeng, and M. Akhtar, "Low-temperature die attach material," *Proceedings 42nd IEEE Electronic Components and Technology Conference*, pp. 235-238, 1992.

29. Kasem, Y. M., and L. G. Feinstein, "Horizontal die cracking as a yield and reliability problem in integrated circuit devices," *Proceedings 37th IEEE Electronic Components Conference*, pp. 237-244, 1987.

30. Davey, N. M., and F. W. Wiese, Jr., "Adhesion mechanisms in silver/glass die attachment of gold backed die," *Proceedings 1986 International Symposium on Microelectronics*, pp. 665-674.

31. Feinstein, L. G., "Die Attachment Methods," in *Electronic Materials Handbook, Vol. 1: Packaging*, pp. 213-223, ed. by M. L. Minges and C. A. Dostal, ASM International, Materials Park, Ohio, 1989.

32. Carpenter, B., K. Pearsall, and R. Raines, "Water soluble paste survey for fine pitch SMT attach," *Proceedings 42nd IEEE Electronic Components and Technology Conference*, pp. 49-56, 1992.

33. Crum, S., "Choosing stencils for solder paste deposition," *Electronic Packaging and Production*, vol. 31, pp. 66-68, October 1991.

34. Yung, E. K., and I. Turlik, "Electroplated solder joints for flip-chip applications," *IEEE Transactions on Components, Hybrids and Manufacturing Technology*, vol. 14, pp. 549-559, 1991.

35. Burkart, G. S., "The creation of a solder jetting consortium," *Circuits Assembly*, vol. 4, p. 42, April 1993.

36. Tribula, D., D. Grivas, and J. W. Morris, Jr., "Stress relaxation in 60Sn-40Pb solder joints," *Journal of Electronic Materials*, vol. 17, pp. 387-390, 1988.

37. Selvaduray, G. S., "Die bond material and bonding mechanisms in microelectronic packaging," *Thin Solid Films*, vol. 153, pp. 431-455, 1987.

38. Olsen, D. R., and H. M. Berg, "Properties of die bond alloys relating to thermal fatigue," *IEEE Transactions on Components, Hybrids and Manufacturing Technology*, pp. 257-263, 1979.

39. Frear, D., D. Grivas, and J. W. Morris, Jr., "A microstructural study of the thermal fatigue failures of 60Sn-40Pb solder joints," *Journal of Electronic Materials*, vol. 17, pp. 171-180, 1988.

40. Frost, H. J., R. T. Howard, P. R. Lavery, and S. D. Lutender, "Creep and tensile behavior of lead-rich lead-tin solder alloys," *IEEE Transactions on Components, Hybrids and Manufacturing Technology*, vol. CHMT-11, pp. 371-379, 1988.

41. Lee, S.-M., and D. S. Stone, "Grain boundary sliding in surface mount solders during thermal cycling," *IEEE Transactions on Components, Hybrids and Manufacturing Technology*, vol. 14, pp. 629-632, 1991.

42. Darveaux, R., and K. Banerji, "Constitutive relations for tin-based-solder joints," *Proceedings 42nd IEEE Electronic Components and Technology Conference*, pp. 538-551, 1992.

43. Solomon, H. D., "The creep and strain rate sensitivity of a high Pb content solder with comparisons to 60Sn/40Pb solder," *Journal of Electronic Materials*, vol. 19, pp. 929-936, 1990.

44. Boyko, C., F. Bucek, V. Markovich, and D. Mayo, "Film redistribution layer technology," *Proceedings 43rd IEEE Electronic Components and Technology Conference*, pp. 302-305, 1993.

45. Frost, H. J., and R. T. Howard, "Creep fatigue modeling for solder joint reliability predictions including the microstructural evolution of the solder," *IEEE Transactions on Components, Hybrids and Manufacturing Technology*, vol. 13, pp. 727-735, 1990.

46. Suhir, E., "Mechanical behavior and reliability of solder joint interconnections in thermally matched assemblies," *Proceedings 42nd IEEE Electronic Components and Technology Conference*, pp. 563-572, 1992.

47. Marshall, J. L., "Scanning electron microscopy and energy dispersive X-ray (SEM/EDX) characterization of solder–solderability and reliability," in *Solder Joint Reliability*, pp. 173-224, ed. by J. H. Lau, Van Nostrand Reinhold, New York, 1991.

48. Woychik, C. G., and R. C. Senger, "Joining materials and processes in electronic packaging," in *Principles of Electronic Packaging*, pp. 577-619, ed. by D. P. Seraphim, R. C. Lasky, and C.-Y. Li, McGraw-Hill, New York, 1989.

49. Chiou, B.-S., K. C. Liu, J.-G. Duh, and P. S. Palanisamy, "Intermetallic formation on the fracture of Sn/Pb solder and Pd/Ag conductor interfaces," *IEEE Transactions on Components, Hybrids and Manufacturing Technology*, vol. 13, pp. 267-274, 1990.

50. Hinch, S. W., *Handbook of Surface Mount Technology*, Longman Scientific & Technical, Harlow, England, 1988.

51. Marshall, J. L., J. Calderon, J. Sees, G. Lucey, and J. Hwang, "Composite solders," *IEEE Transactions on Components, Hybrids and Manufacturing Technology*, vol. 14, pp. 698-702, 1991.

52. Koopman, N., S. Bobbio, S. Nangalia, J. Bousaba, and B. Piekarski, "Fluxless soldering in air and nitrogen," *Proceedings 43rd IEEE Electronic Components and Technology Conference*, pp. 595-605, 1993.

53. Lee, C. C., C. Y. Wang, Y. C. Chen, and G. Matijasevic, "A joining technique using multilayer lead-indium-gold composite deposited in high vacuum," *Materials Research Society Spring Meeting, Joining and Adhesion of Advanced Inorganic Materials*, pp. 235-240, ed. by A. H. Karim, D. S. Schwartz, and R. S. Silberglitt, Materials Research Society, Pittsburgh, 1993.

54. Matijasevic, G., C. Y. Wang, and C. C. Lee, "Void-free bonding of large silicon dice using gold-tin alloy," *IEEE Transactions on Components, Hybrids and Manufacturing Technology*, vol. CHMT-13, pp. 1128-1134, 1990.

55. Pavio, J. S., "Successful alloy attachment of GaAs MMICs," *IEEE Transactions on Electron Devices*, vol. 34, pp. 2616-2620, 1987.

56. Matijasevic, G., and C. C. Lee, "Void-free Au-Sn eutectic bonding of GaAs dice and its characterization using scanning acoustic microscopy," *Journal of Electronic Materials*, vol. 18, pp. 327-337, 1989.

57. Humpston, G., and D. M. Jacobson, "Gold in gallium arsenide die-attach technology," *Gold Bulletin*, vol. 22, pp. 79-91, 1989.

58. Nishiguchi, M., N. Goto, and H. Nishizawa, "Highly reliable Au-Sn eutectic bonding with background GaAs LSI chips," *IEEE Transactions on Components, Hybrids and Manufacturing Technology*, vol. 14, pp. 523-528, 1991.

59. Uda, T., Y. Takeo, T. Sato, K. Sahara, S. Kuroda, and K. Otsuka, "Development of high reliability die-bonding technology in low temperature process," *Proceedings 37th IEEE Electronic Components and Technology Conference*, pp. 105-109, 1987.

60. Kawanobe, T., K. Miyamoto, and M. Hirano, "Tape automated bonding process for high lead count LSI," *Proceedings 33rd IEEE Electronic Components Conference*, pp. 221-226, 1983.

61. Zakel, E., and H. Reichl, "Au-Sn bonding metallurgy of TAB contacts and its influence on the Kirkendall effect in the ternary Cu-Au-Sn system," *Proceedings*

42nd IEEE Electronic Components and Technology Conference, pp. 360-371, 1992.

62. Atsumi, K., N. Kashima, Y. Maehara, T. Mitsuhashi, T. Komatsu, and N. Ochiai, "Inner lead bonding technique for 500-lead dies having a 90-mm lead pitch," *IEEE Transactions on Components, Hybrids and Manufacturing Technology*, vol. 13, pp. 222-228, 1990.

63. Walshak, D. B., Jr., "TAB inner lead bonding," in *Handbook of Tape Automated Bonding*, pp. 200-242, ed. by J. H. Lau, Van Nostrand Reinhold, New York, 1992.

64. Mori, M., M. Saito, A. Hongu, A. Niitsuma, and H. Ohdaira, "A new face down bonding technique using a low melting point metal," *IEEE Transactions on Components, Hybrids and Manufacturing Technology*, vol. 13, pp. 444-447, 1990.

65. Kang, S. K., and C. G. Woychik, "Diffusional processes in metallurgical interconnections," in *Materials Research Society Symposium Proceedings*, vol. 108, pp. 385-391, 1988.

66. Bernstein, L., "Semiconductor joining by the solid-liquid-interdiffusion (SLID) process," *Journal of Electrochemical Society*, vol. 113, pp. 1282-1288, 1966.

67. Okamoto, H., and T. B. Massalski, "Au-In (gold-indium)", in *Binary Alloy Phase Diagrams*, p. 382, ed. by T. B. Massalski (ASM International, Materials Park, Ohio, 1990).

68. Bjøntegaard, J., L. Buene, T. Finstad, O. Lønsjø, and T. Olsen, "Low temperature interdiffusion in Au/In thin film couples," *Thin Solid Films*, vol. 101, pp. 253-262, 1983.

69. Lee, C. C., C. Y. Wang, G. Matijasevic, and S. S. Chan, "A New Gold-Indium Eutectic Bonding Method," in *Materials Research Society Spring Meeting, Electronic Packaging Materials VI*, pp. 305-310, ed. by P. S. Ho, K. A. Jackson, C.-Y. Li, and G. F. Lipscomb (Materials Research Society, Pittsburgh, 1992).

70. Kang, S. K., 'Inner lead bonding mechanisms in TAB technology," *Materials Developments in Microelectronic Packaging Conference Proceedings*, pp. 187-192, 1991.

71. Lee, C. C., C. Y. Wang, and G. Matijasevic, "A new bonding technology using gold and tin multilayer composite structures," *IEEE Transactions on Components, Hybrids and Manufacturing Technology*, vol. 14, pp. 407-412, 1991.

72. Matijasevic, G., Y.-C. Chen, and C. C. Lee, "Copper-tin multilayer composite solder," *Proceedings of the 1993 International Electronics Packaging Conference*, pp. 264-273, 1993.

73. Jacobson, D. M., and G. Humpston, "High power devices: fabrication technology and developments," *Metals and Materials*, vol. 7, pp. 733-739, 1991.

74. Izuta, G., S. Abe, J. Hirota, O. Hayashi, and S. Hoshinouchi, "Development of transient liquid phase soldering process for LSI die-bonding," *Proceedings 43rd IEEE Electronic Components and Technology Conference*, pp. 1012-1016, 1993.

75. Shukla, R. K., and N. P. Mincinger, "A critical review of VLSI die attachment in high reliability application," *Solid State Technology*, vol. 28, pp. 67-74, 1985.

76. Van Kessel, C. G. M., S. A. Gee, and J. J. Murphy, "The quality of die attachment and its relationship to stress and vertical die-cracking," *Proceedings 33rd Electronic Components Conference*, pp. 237-244, 1983.

77. Chiang, S. S., and R. K. Shukla, "Failure mechanism of die cracking due to imperfect die attachment," *Proceedings 34th Electronics Components Conference*, pp. 195-202, 1984.

78. Pyle, R. E., and H. A. Stevens, "Characterization of die attach failure modes in leadless chip carrier packages by Auger Electron Spectroscopy," *Proceedings 22nd International Reliability Physics Symposium*, pp. 175-180, 1984.

79. Li, T. P. L., E. L. Zigler, and D. E. Hillyer, "AES/ESCA/SEM/EDX studies of die bond materials and interfaces," *Proceedings 22nd International Reliability Symposium*, pp. 169-174, 1984.

80. Kitchen, D. R., "Physics of die attach interfaces," *Proceedings 18th International Reliability Physics Symposium*, pp. 312-317, 1980.

81. Van Kessel, C. G. M., S. A. Gee, and J. J. Murphy, "The Quality of Die-Attachment and Its Relationship to Stresses and Vertical Die-Cracking," *IEEE Transactions on Components, Hybrids and Manufacturing Technology*, CHMT-6, pp. 414-420, 1984.

82. Markstein, H. M., "Inspecting assembled PCBs," *Electronic Packaging and Production*, vol. 33, pp. 70-74, September 1993.

83. Lehmann, D., "Electronic package inspection with microfocus X-ray," *Proceedings of the 1993 International Electronics Packaging Conference*, pp. 828-835, 1993.

84. Adams, J. A., "Non-destructive evaluation of advanced multichip modules using cross-sectional X-ray techniques," *Proceedings of the 1993 International Electronics Packaging Conference*, pp. 813-820, 1993.

85. Tsai, C. S., C. C. Lee, and J. K. Wang, "Detection and characterization of defects in thick multilayer microelectronic components using transmission acoustic microscope," *Proceedings 17th International Reliability Physics Symposium*, pp. 178-182, 1979.

86. Semmens, J. E., and L. W. Kessler, "Non-destructive inspection of multichip modules by acoustic micro imaging techniques," *Proceedings of the 1993 International Electronics Packaging Conference*, pp. 821-827, 1993.

7

Wire Bonding Chip on Board

Robert A. Christiansen

7.1 INTRODUCTION

With the need for automation and lower costs of manufacture for electronics leading to surface mount technology, several extensions of SMT have emerged for the direct incorporation of silicon chips to the circuit board, such as Chip-on-Board (COB) and Tape Automated Bonding (TAB). In both COB and TAB, board surface area requirements can be significantly reduced, and more compact, reliable packaging can be achieved at an economic cost. Indeed, the incorporation of modular grouping of multiple chips and passives together on a common substrate to reduce board-board I/O counts and distances has become the now popular concept of Multi-Chip-Modules (MCMs). Besides cost and physical package volume advantages, circuit speed, power requirements and EMC enhancements are achievable objectives. Both COB and TAB have zones of advantage and disadvantage, with gray areas of overlap. The packaging engineer should have a thorough understanding of each in his personal technology toolbox.

As with many lower cost packaging concepts, COB first appeared in low-cost products from the Far East, typically in popular game cartridges, calculators, watches, even CB radios (see Fig. 7-1). As the technology was developed, applications moved upward to digital thermometers (Fig. 7-2), high quality watch products, extended performance calculators (Fig. 7-3), memory storage (Fig. 7-4), and notebook computers. With the low-cost product volumes paving the way for the uplift of the technology as to better materials, process controls, and reliability, COB follows the concept of

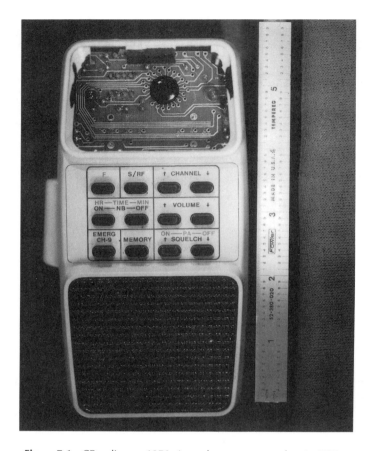

Figure 7-1   CB radio, *ca.* 1976. An early consumer product in COB.

"technology trickles *up*, not *down*". Today COB packaging is incorporated in volume production high quality office products, personal computers, and ancillary equipment such as disk drives where physical volumes are at a premium (Figs. 7-5, 7-6, 7-7). For the purposes of this chapter, chip-on-board (COB) description is as follows:

Silicon chip is directly attached to a circuited substrate without prior packaging.

Mechanical and electrical bonding and environmental passivation are made *in situ* on the circuit substrate.

Usually involves epoxy die attachment, wire bonding and encapsulation.

It is often used in a mix with other packaging technologies, such as SMT and/or PTH.

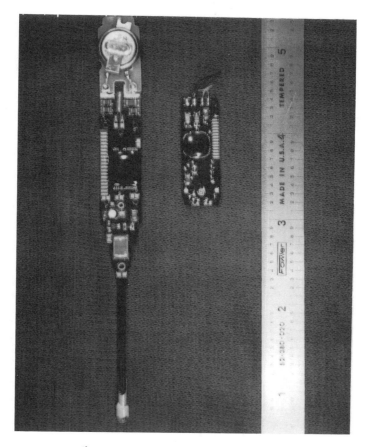

Figure 7-2 Digital thermometer, SMT/COB.

COB packaging offers significant packaging real estate reductions as well as improved electrical performance. With proper design and supplier and in-house quality controls in place, significant cost reductions are also achievable. A thorough understanding of product life, a stress-to-fail product development philosophy combined with strong failure analysis support, a close working relationship with chip and materials suppliers, and a manufacturing process that is well developed, automated and with process monitors and controls in place is essential. It is with the objective of defining key parameters in design, materials and processing that this chapter of this book is written.

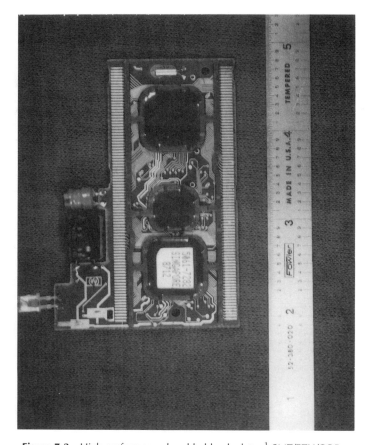

Figure 7-3   High performance hand-held calculator,[1] SMT/PTH/COB.

7.2 APPLICATIONS OF WIRE BONDED COB

Wire bonded COB is suitable for a wide range of I/O counts–certainly it is a competitive technology to 200 I/Os per chip with today's high-speed automated bonders. Typically the best design targets are:

High packaging cost chips
Supplier mark-ups
Long manufacturing "pipelines", often to Far East packaging house and
    return
Package ruggedization, reliability
Package volume, weight reduction

Figure 7-4   64K electronic memory storage, SMT/PTH/COB (top–3.6 mm/0.138 in. thick; bottom–7.7 mm/0.303 in. thick).

Typical "n", or number of chips per assembly is $1 < n < 7$, and dictated by the Shipped Product Quality Level (SPQL) of chips, process yields (enhanced in high-n designs by reliable repair processes), and design needs. The difficulty in producing high-n designs without a strong chip supplier SPQL and without development of reliable rework processes can be shown by a study of Figure 7-8, which plots percent card assembly yields as a function of chip SPQL and n (number of chips/card assembly).

Figure 7-5   Typewriter function card, incorporating mixed technologies: PTH, SMT, COB.

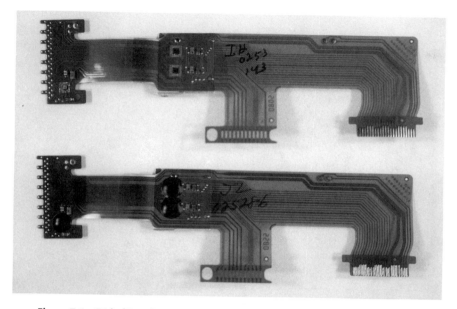

Figure 7-6   Disk-drive electronics, using chip-on-flex, a COB variation, with SMT.

When discussing I/O counts and packaging alternatives, the inevitable comparison to tape automated bonding (TAB) will occur. These technologies are *not* mutually exclusive, but complementary. Both should be in the packaging engineer's personal toolbox. Both technologies have zones of advantage, often with overlaps:

Low I/O (up to 150-200)   Wirebonding
Medium to high I/O (150+)   TAB
Very high I/O (400-600+)   Wedge-bonded (fine wire)

In either case, the objective is to reduce size and cost, and to improve performance. Product-specific needs and/or availability of manufacturing may shift zones also. A quick review of the two technologies is recapped in Table 7-1.

7.3 DEVELOPMENT OF THE COB DESIGN, MATERIALS, AND PROCESS SETS

Even within the boundaries of wire bonded COB there are many alternatives fitting wide design needs as to substrates, cleaning, wire bond process (i.e., gold vs. aluminum), type of encapsulation employed, etc. Before going into

Figure 7.7   Keyboard controller card (top), in PTH design, and (bottom) replacement designed in SMT/COB.

Table 7-1   COB/TAB Comparisons

COB	TAB
Low tooling cost	Tooling intensive
Quick EC changes	Change difficult, lead time
No co-planarity problems	Co-planarity problems with gang-bonding (not with single point)
Slower, serial bonding	Gang-bonding faster (SPT comparable to W/B)
Full passivation by glob encapsulation	Leads, chip edges exposed
Requires high SPQL chips in multi-chip applications	Bumping cost ($/chip)
Repair after encapsulation difficult	Chips testable on tape
Eliminate TAB decal cost	Remove/replace easier

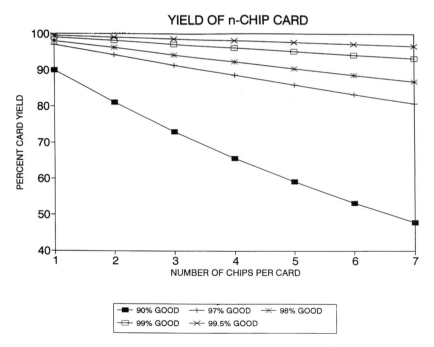

Figure 7-8   Assembly yields vs. chip SPQL.

Figure 7-9   Miscellaneous test sites (UTS).

detail with the various processes, the author would like to discuss a development tool we call the Universal Test Site, or UTS (see Fig. 7-9).

The UTS is a small, single chip test site that is pinned (soldered or compliant press-fit) to fit standard burn-in sockets and cards. Numbers of test specimens can be built up as compatible with development equipment (singles, paired, etc.) or full panels with subsequent post-build separation process for manufacturing. The prepared burn-in cards can then be subjected to various environmental stress and life tests conveniently, efficiently, in statistically significant sample sizes in a cost effective manner. They may be as simple as to fit P-DIP sockets or as high I/O (we use through 140) PGA sockets. The UTS makes it possible, especially for novice packaging engineers, to define many aspects of their COB design, as follows:

Define product life expectations/improvements
Board material suitability
Copper thicknesses, clearances, etch processes
Plating types, thicknesses, smoothness, purity
Cleaning processes, ionic cleanliness levels needed, ionic test
Die-attachment adhesives evaluation/qualification
Vendor chip–suitability for COB
Wire, wire bond process development
Passivation technique evaluation/qualification
Electrical test equipment
Rework processes
Compatibility with other processes (SMT, PTH)

Additionally, the UTS finds usage in production as a quality comparison, either to other lots used in COB, or to packaged device lots if chip is marketed in both packaged and bare die forms. Chip-related failure analysis is greatly simplified in the case of high-n chip designs. UTS-mounted die are commonly subjected to the same environmental stress levels in terms of temperature, humidity, cycling, etc., as are packaged die and commonly exceed packaged devices in reliability (no wire-wash, one step fewer interconnections). Fully populated, mixed technology (SMT and/or PTH, with COB) are normally subjected to lower stress levels similar to other post-assembly process stress testing (Example: assemblies at 50°C/80%RH versus UTS/COB or packaged chips at 85°C/85%RH).

In conclusion, the use of the UTS has proven to be a valuable tool in the development of and defining interactions between the various materials, processes, chips and environmental testing for COB designs.

7.4 SUBSTRATES

A number of options for substrates exist, including standard FR4, Bis-Maleide triazine (BT) resin, polyimide (both rigid and flex), and thermal-cored composites. The choice of substrate and type of wire bonding, i.e. aluminum wedge vs. gold thermosonic, must be made together. The choice will be influenced by trading off substrate costs per unit area against wire bond density per area. For very small package physical size (for example, digital LCD thermometer), BT resin with its 170°C glass transition may be affordable because of high wire bond count/area. On a full, mixed technology (PTH/SMT/COB) card, where only a small area is occupied by the COB sites, the additional cost incurred in utilization of a premium substrate for the whole circuit board would quickly negate COB packaging savings. As an example, note the areas occupied by the six COB-mounted chips in Figure 7-5 compared to the SMT/PTH devices. This card is an FR4" + " material having a glass transition of 130°C, entirely compatible with aluminum wedge-bonding, as well as die-attach and encapsulation processes. Conversion to gold thermosonic wire bonding with its 170°C (approx.) substrate heating would require selection of a premium more thermally resistant material, such as BT Resin or polyimide at approximately 2-3× cost factor (based on quotations for that card in a two-sided, no-inner-plane design). On the other hand, the digital thermometer (Fig. 7-2) and extended performance calculator (Fig. 7-3) are high chips and wire bond count per unit area designs, making gold thermosonic bonding, with its faster bond speed (than aluminum wedge bonds) the economic selection. Even higher wire bond count per substrate area will be found in the higher performance MCMs that are becoming the prevalent computer packaging concept. For thermal dissipation reasons ceramic, silicon, or metal-cored substrates may be required.

An alternative to FR4 that is more commonly seen in Far East products is G10 laminate, which lacks the flame retardant chemistry and/or additives of FR4. No free bromine was extractable, and there were no biased T/H result differences in a multi-manufacturer product sampling of FR4 and G10 materials prior to release of the typewriter function card (Fig. 7-5).

Another substrate which finds utilization in COB packaging is the polyimide flex cable. This option finds usage in designs for example when a flexible interconnect cable is required already, but a separate circuit board, connector, etc., can be eliminated. An example of this design is Figure 7-6, used in a 3.5 in. direct access storage disk (DASD) hard file, in personal computers. The disk drive is 320 megabytes, formatted, or 371 unformatted in a package envelope of 1.63 × 4.00 × 5.75 in. (4.1 × 10.2 × 14.6 cm). MTBF specification is 150,000 hours for the drive assembly. This design incorporates metal heat spreaders (chips bond directly to the metal through windows in the flex circuit) that also serve as attachment points to the drive assembly. This design moves the chips from the prior design separate card to the drive head end of the flex circuit, lowering power requirements to drive the circuit as well as enhancing speed. Many unique differences exist from hard-card (FR4, etc.) designs. More (custom, increased cost) fixturing is required to clamp the flex circuit in the chip/bonding area to prevent movement during those processes. Usage of stiffeners at the chip site is desirable, as one must prevent bending stress on the chip and avoid peel loading of the epoxy encapsulant to avoid moisture intrusion or total delamination. For automated, machine vision diebonding and wire bonding, fiducial marks must be nearer the chip sites as well as more in number than on the equivalent size FR4 card. The reason for this is dimensional accuracies. Variations in FR4 boards are commonly held to within 0.001 in./in. (mm/mm) from datum point. For flex circuitry, 0.002 in./in. (mm/mm) from true position is common, and as much as 0.005 in./in. (mm/mm) has been observed in localized areas. One primary cause is the relief of stresses in the film with the subtraction of copper in the etch process. The result of this is that to maintain accuracy, additional referencing cycles are required. When machine vision is in search mode, no wire bonding is occurring, hence throughput will be somewhat affected on flex circuiting as opposed to FR4. Other process modifications that may be needed in working on flex circuitry as a substrate are:

Cleaning process adjustment, due to tendency for organic solvents to solubilize, "dragout" the commonly used acrylic adhesives, re-depositing as organics onto wire bond pads.

Working with flex circuit supplier to eliminate adhesive voids in the wire bond pad area, as these both reduce bond clamp pressure and affect the ultrasonic energy input.

Chip-on-flex represents a packaging opportunity for the engineer, but one must be aware of its unique properties as compared to the rigid substrates.

7.5 CIRCUITIZATION

Copper is the near universal metallization. It is with the subsequent surface plating sequences that variations in product environmental stress resistance need discussion. Thicknesses used in the COB industry today range from 0.5 to 2 ounce copper (0.0007-0.0028 in., or 17.8-71 micron). Copper circuitry from most experienced circuit board houses has been satisfactory with one quality exception: smoothness. The problems lie not with the (gradual) weave replications, but in abrasive cleaning process scratches, either pumice or pad. Chemical brightening for solder mask adhesion promotion is preferred. For 1 to 1.25 mil wire (0.001-0.00125 in., or 25-32 micron), scratches in the bond area of 100 microin. (0.0001 in., or 2.5 micron) width/depth range can reduce bond area and quality significantly and randomly. Wire bond pull strength will be affected adversely. An opinion offered here is that, if one can hold all other variables (cleaning, plating/purity, bond process, etc.) constant, then smoother is better.

Below the 100 microin. (2.5 micron) point, additional bond strengths can be achieved, but one must be careful to understand and isolate other variables. Below 40 microin. (1 micron), two things occur: very little increase in detectable bond strengths, and a significant increase in measurement cost to the product. A more utilitarian, cost-effective approach is to consign a manual wire bonder to the circuit board vendor, with a post-plating/pre-ship wire bond pull test at a mutually agreeable value. This offers the advantage of detecting variances in plating hardness, impurity, contaminants (organics, etc.). The pull-testing is post-plating of the circuit. Studies on the effect of surface roughness on wire bondability have been published by Seshan and Ray[2] and specifically for aluminum wedge bonding by DiGirolamo.[3]

7.6 LAYOUTS

This topic will be controversial as many different product targets and motivations to design in COB will be found, from the low I/O calculator or clock to the large, high I/O ASICs and processors used in personal computers. A compact, totally COB packaged design will have different constraints than a mixed technology full function card in terms of affordable etched and plated circuit boards. Below are listed some considerations in choosing board sizes, pitch, and layout orientations, as well as some references on design capabilities.

Overall board size   Production equipment today is readily available to do sizes up to 6 × 12 in. (152 × 305 cm). Design, if smaller, can be panelized up to this size. This allows panels to be also run directly onto 12 in. SMT and PTH lines.

Line spacing   Typically from 2 mil lines/2 mil spaces (4 mil pitch) to as much as 8 mil/8 mil (16 mil pitch). In metric units, from 0.1 mm pitch to 0.4 mm pitch.

Influences
  • line pitch needed to accommodate chip I/O
  • cost/unit of card area vs. line pitch
  • individual cards vs. 100% good cards in panel requirement

- bonding wire diameter affects maximum wire span, chip to pad
- design current capacities (though usually exceed bondwire capacity)
- cleanliness, passivation to keep ionics low (current leakage, shorts, migration, corrosion)
- environmental exposure severity

Wire bond pad length Must accommodate original bond length (see Fig. 7-21), plus accuracy tolerances. It is recommended, if design space permits, that pad include length sufficient to make one or two rework bonds on fresh metallization, as well as an area for board tester contact (at back, or solder mask end of pad).

Wire bond pad orientation The most common encountered is orthogonal layout, where lead tips are oriented normal to the die edge. Shortcomings to this layout appear quickly in higher I/O device plications. Approaching corners, wirespan increases. Wire line of flight comes at an angle to the bond pad, perhaps crossing over adjacent pads creating shorts possibilities. Particularly with the elongated wedge bond, space for bond length may be inadequate. The principal advantage is ease of layout coordinates by most board design systems.

Alternative pad layouts are available which eliminate many of these problems, but are more complex in design. These are variable center/variable radius, variable center/fixed radius, and die-centered fixed radius. While directed at quad flat pack lead frames, an excellent treatment to pad layout optimization is given by Jahsman.[4]

Fine pitch developments Oki in the development of a 400 dot per inch (~16 dot per mm) LED printhead[5] has demonstrated production wire bond pitch of 60 micron, or about 2.4 mils for a 4000+ wire assembly. This required reducing wire size from 1 mil (25 micron) to 0.7 mil (18 micron) and utilization of wedge rather than ball bonding. A ceramic substrate is used.

Oki has subsequently developed 40 micron (1.6 mil) pitch wire bonding for 600 DPI LED printhead using 10 micron (0.4 mil) gold wire.[6]

A number of other articles of interest to those interested in wire bond densification have been published.[7,8,9]

Wire bond span As a guideline, an aspect ratio of 100D maximum is easily remembered and is a good design point, where D is wire diameter. For example, with a 1.25 mil (0.00125 in. or 32 micron) wire, a normal maximum span is 125 mil, or 0.125 in. (3.2 mm). For 1 mil wire, 100 mil (2.5 mm), for 0.7 mil wire, 70 mil (1.8 mm), etc. This is a good design point for manufacturing yield. Trade-off in yield may need to be made when for example, an alternative must be selected between finer pitch/higher cost substrate vs. extending a few corner wires to 140 mil (3.6 mm) run with 1.25 mil (32 micron) wire. This might be on only one large high I/O chip on a multi-chip assembly. The decision here would most likely be to make the 140 mil (3.6 mm) runs, but assure minimum stress in handling through encapsulation. Again, the 100D is a design guideline, not a specification. An element of wire bond span is the separating of diebond pad from the wire bond pads. As a minimum it should equal or exceed line-line clearances. Die-attach epoxy, if of a bleeding nature, may dictate a separation of as much as 30 mil (0.030 in. or 0.75 mm). See diebond pad size notes also.

Diebond pad size Epoxy die-attach materials have a high affinity for cleanly etched FR4 epoxy surface, leading to a tendency to bleed. This is reduced if die-attach filleting area is provided on the gold-plated pad. This also contributes to wire bond span as does diebond pad to wire bond pad distances. The following guides have been successfully utilized:

	Suggested perimeter	
Relative die size	Longest die side	Adder per side
Small	<0.125 in. (3mm)	0.015 in. (0.4 mm)
Medium	<0.125–0.250 in. (6 mm)	0.022 in. (0.55 mm)
Large	>0.250 in. (6 mm)	0.030 in. (0.75 mm)

It is also suggested that use of non-bleed or limited bleed die-attach compositions should be investigated when bond density becomes a span issue.

Vias Open vias should not be used in the encapsulation area of the COB design due to capillary wicking and leakage. Bottom side pre-tenting of vias will result in air entrapment and subsequent expansion and voiding in the encapsulation process.

Fiducial marks For accurate, automated assembly, fiducial marks, or reference points must be used. Design should be coordinated with the machine vision system manufacturer. It is recommended that plating match the diebond and wire bond pads and not be over-coated with solder masking (or overlay film in chip-on-flex designs).

Miscellaneous As with packaged devices, chip sites may need stationing, or placement to enhance thermal performance on the card. Excellent thermal bonding to the card is an attribute of chip-on-board. Additional area gain for dissipation can be obtained by the design of thermal paths from the diebond pad corners to unetched copper areas (see Fig. 7-10).

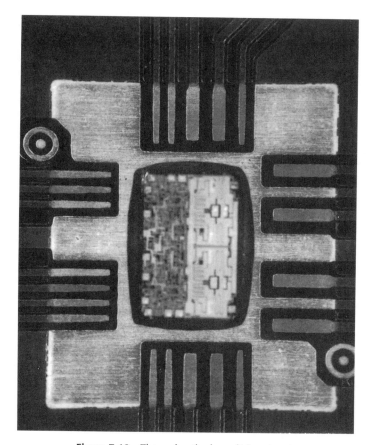

Figure 7-10   Thermal paths from diebond pad.

7.7 SUBSTRATE PLATING

The most common COB metallurgy is probably additive electroplated gold over nickel over the copper circuit, with other technologies such as semi-additive gold/nickel, immersion gold/electroless nickel, and nickel and its alloys. As with most engineering choices, selection of plating hierarchy, quantity and process will be influenced by the design needs.

7.7.1 Electroplated Gold/Nickel (Full Additive)

A full additive plating process results in the purest gold deposit. This gold should be a minimum of 99.9% pure gold, with no thallium or hardeners.

Thickness is optional, but most high quality COB designs will fall into a grouping bounded by 20 to 40 micron. (0.5-1.0 micron) gold, over a minimum of 80 micron. (2.0 micron) nickel. Nickel maximum is usually in the 200-320 micron. range (5-8 microns) on FR4, although 500 micron. (12.5 micron) products have been observed. On flex circuitry, nickel will be kept at minimum due to hardness and crack sensitivity of circuit tracings. The full additive process gives full conformal coating of the copper/nickel circuit and thus provides maximum environmental resistance. Numerous designs have successfully exceeded qualification requirements of 2000 hours at 85°C/81%RH with bias as well as other tests. The drawbacks to this process are usually highest cost, and the challenge in circuit layout to get all plating shorts out to card edges (or plating bus) in multi-chip, high I/O designs. Various novel ways have been incorporated in COB designs to solve this problem. One product was observed to have shorted all wire bond pads to their diebond pad, bringing out only one trace per die, followed by mechanical operation to sever shorts. All exposed copper was covered by the subsequent encapsulation. Cost reduction can be achieved by selective plating.

7.7.2 Electroplated Gold/Nickel (Semi-Additive)

In this method, for a quick review, a thin strike of copper is plated universally over substrate surfaces. After artwork/resist exposure, circuitization copper is electroplated up to desired thickness, followed by the nickel and gold, likewise electroplated. Following removal of resist, the thin strike of copper between circuits is etched away, commonly using the gold plating as the etch resist. The resultant product is very similar to the above full additive process, with one exception: sides of the circuit traces have bare nickel and copper exposure, and no conformal coverage of gold exists. As one would expect, a greater vulnerability to environmental exposure exists. In a number of design evaluations using this technology, the following observations are offered:

2000 hour 85°C/81%RH stress test results indicate some design/card supplier combinations would pass, some design/supplier combinations would fail.

Failures were not in the encapsulated COB area of card.

Greater attention to monitoring general board cleanliness, solder mask porosity and encapsulated COB area fully meeting solder mask edges aids success in meeting 85°C/81%RH.

In no case did card assembly testing of 1000 hours at 50°C/80%RH yield fails.

While the ability to effect economies by selective gold-plating is not feasible with this technology, it does offer a solution to plating access of all diebond and wire bond pads without layout constraints. Plating thicknesses used are consistent with the full additive process. In both of the above processes care must be also taken in monitoring and specification control of maximum gold thickness (as well as minimum for wire bonding) on mixed technology card designs due to gold dissolution into solder joints (see Fig. 7-11) or, in case of wave soldering gold contamination of the pot, either case potentially

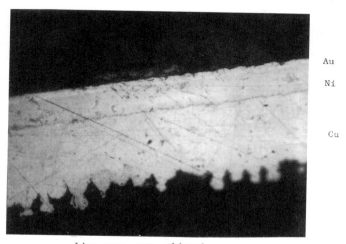

Line area, not soldered

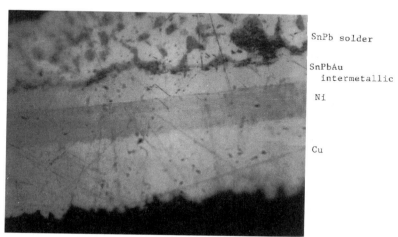

Joint area, soldered

Figure 7-11   Gold dissolution into SMT solder paste reflow.

Figure 7-12 AuSn$_4$ "needles" appearing in SMT solder paste reflowed joint on gold plated (55 microin.) COB card.

resulting in solder embrittlement. The author has observed onset of "needles" as in Figure 7-12, with SMDs at a gold thickness of 55 microin. (1.4 micron).

7.7.3 Electroless Nickel/Immersion Gold

This technique offers the lowest cost method of obtaining a full (albeit thin) conformal nickel/gold coating of the copper lines, without plating shorts. A layer of phosphorus nickel 100-400 microin. thick (2.5-10 micron) is immersion gold-plated 2 to 7 microin. (0.05 to 0.18 micron). As the gold thickness is very thin and solder mask utilized as selective plating resist, the quantity, and cost, of gold is minimized while providing the desirable full conformal coverage of copper. The caveat is that much greater controls on the bath and resultant platings must be maintained. As the gold is so very thin, the hardness and smoothness is critical to the wire bonding process in bondability and yields. Likewise the phosphorus content in the nickel influences bondability. As the process is immersion plating (a surface nickel atom replaced by gold), nickel, a gold hardener, must be controlled in the bath. Although operational process bandwidths may be narrower than with electroplated gold, with careful process quality monitors in place an economic extension to multi-chip, intermixed technologies (SMT, PTH) can be economically obtained, with much less risk of gold contamination and/or embrittlement of solder joints. The typewriter function card (Fig. 7-5) is an example of this technology.

7.7.4 Nickel and Alloys

Probably the lowest cost in terms of wire bondable (certainly with the scrub of the ultrasonic aluminum bonder) platings is the electroless nickel. While bath-to-bonder times need definition and control for most nickel plates, one plated alloy, nickel-boron, has excellent shelf storage capability (one year minimum, factory environment).

Standard nickel-phosphorus electroless platings are found in very low cost consumer goods (dashboard clocks, musical greeting cards, some disposable watches, etc.). Almost universally aluminum wedge wire bonding is incorporated with these substrate surface platings. High volume, in-line minifactories assure controlled bath-to-bonder times. Solderability is quite good, unaged, and little tendency to be scavenged by wave soldering is observed, minimizing solder pot contamination.

The advantages of Ni-B alloy plating for wire bonding were recognized early though primarily for thin film hybrids. G. J. Estep, of Bendix Electrodynamics reported[10] on the successes using aluminum ultrasonic wedge, although unsuccessful with gold (thermocompression, no thermo-sonic), as well as the excellent solderability of Ni-B platings. In a later publication[11] D. J. Baudrand of Allied-Kelite acknowledged Estep's work and confirmed the following:

Ni-B platings maintain better solderability after steam aging than Ni-P platings.

Ni-B yielded excellent aluminum ultrasonic wire bonding results (thermosonic, in practice by the 1983 article also yielded excellent results).

Baudrand cites header-pin bonding with elimination of gold plating on substrate pads and header pins, replacement of gold wire by aluminum wire, and elimination of defects (Kirkendall voiding).

In the late 1980s, T. H. Chiles of IBM Office Products Division in Lexington, Kentucky (now known as Lexmark International, Inc.) initiated a technology study to determine if Estep and Baudrand's published findings could be applied to electronic circuit boards for COB (and intermixed PTH, SMT) designs on FR4 substrates. The Baudrand and Estep findings were maintained, with shelf storage studies producing high quality wire bonds after one year in factory storage, and excellent environmental reliability. Steam-aged solderability also remained excellent for Ni-B applied to organic substrates as in the published ceramic hybrids. While much of this effort has been held proprietary, some studies are now being published.[12,13]

While technical performance has been excellent for Ni-B, and the electroless nature allows easy plating of all bond pads, with excellent solderability for boards designed in COB mixed with SMT or PTH technologies, production acceptance has been slow, probably for the following reasons:

Ni-B is a hard, relatively brittle plating and should not be used alone ≥75 microin. (1.9 micron) thickness, which by itself would be marginal for copper protection, thus still requiring a two-bath, dual plating, most commonly with a Ni-P as underplate.

Ni-B is not a common circuit board supplier plating bath. Most applications in production are wear or corrosion protection of mechanical parts. These suppliers do not normally have circuit board drill/rout/etch/plating facilities. In either case, a significant capital investment and operating costs are involved. These start-up costs are reflected in cost quotations. The alternative is the split operations of a two-tiered supply set-up.

7.7.5 Platings–Miscellaneous

The importance of supplier pre-shipment wire bond pull testing has already been discussed. Additionally the COB user will want to specify and monitor plating thicknesses, purity and visual requirements. For any of the nickel-phosphorus electroplates, the phosphorus content of the nickel plate deposit should be specified and controlled. For plating thicknesses the high quality electronic, X-ray based plating thickness monitors, when used with a calibrated standard in materials and thickness ranges being measured have proven to be an accurate, highly productive method of testing plating thicknesses. In the very thin soft platings, accuracy is believed better than the time-honored sectioning/polishing principally because of the gold smear tendency. Additional discussion of nickel and gold platings for wire bonding has been published by Lovie[14] and Endicott and Nobel.[15]

7.8 CHIP METALLURGY

In the interests of applicability and brevity, only the chip bonding pad top metallurgy will be reviewed here for its effect on COB design and manufacture.

Aluminum pad metallization is the industry norm, and given reasonable protection from oxidation, organic contaminants, mis-masked passivations, etc., aluminum is an excellent material to wire bond to. As amperage densities increased, concerns for electromigration of aluminum arose, leading to the addition of copper to the metallization deposit. The amount of copper addition has been a controversial subject, ranging from 0.1% to 4 wt% Cu by design, and as high as 8 + % has been observed in problem tracking analyses. Thomas and Berg of Motorola reported[16] on the formation of Al_2Cu intermetallics and the subsequent dissimilar metal micro corrosion and pit formation. Further, in aqueous exposure, the $Al(OH)_3$ product can be solubilized and forms a glassy halo deposit around the pit about 10 microns (0.4 mil) in diameter. This will definitely adversely affect wire bondability ("no-sticks") and reliability in thermal cycling of product built with these metallurgies, if moisture exposure of the pads has formed these micro corrosion sites. Lexmark in 1989 experienced similar problems with a chip supplier who was firm on a 4 wt% Cu addition to the aluminum pads. Identical pitting, glassine haloes were observed, with a serious no-stick production problem. Even more troubling from a processing stability viewpoint was the extensive development of Al-Cu-Si intermetallics of varying degrees dependent on wafer furnace zone temperature variances. Wire bonders tuned for optimal wire upsets (typically 1.8 wire diameters) for 99% Al/1% Si wire would move to poor chip bonding (lifts), then when re-adjusted for the "hard, rough sites" (see Fig. 7-13) would approach bond overworking to near heel-cracking. This would occur within single wafer lots. With extra manufacturing costs and controls, this assembly completed production run, and the chip supplier has since reduced copper content to 1-2%.

Figure 7-13 Al-Cu-Si intermetallics development, 2000×.

Similar intermetallic formations experience to that of Motorola and Lexmark is reported by Mayumi, et al.[17] The starfish-like intermetallic formation in Figure 7-13 at Al-Si-Cu (0.4%), Al-Si-Cu (0.8%), and Al-Si-Cu (1.6%) clearly shows the increasing formation of these intermetallics. Mayumi et al. state that "The 0.1% Cu addition substantially suppresses the failures in spite of little appearance of the precipitates. Further Cu addition from 0.1% gives no more effect on suppressing the failure."

It is the opinion of this author that Cu additions should be minimized at a level sufficient to resist electro- and stress migration, but beyond 0.5-1.0 wt% Cu, detrimental effects on a stable, consistent automated wire bonding process will occur and should be disallowed.

Other applicable reports on Al-Cu chip metallizations have been published by Charles Whitman (now with Lexmark) et al.[18] and Puttlitz, Ryan and Sullivan[19] of IBM/Burlington, VT.

The addition of small amounts of titanium (typ. 0.15%) in substitution for Cu is becoming more prevalent, especially for stress-migration resistance.[20,21] Hosoda, Yagi and Tsuchikawa of Fujitsu found that aside from greatly improved EM lifetimes with the 0.15 wt% Ti, addition of 0.1 wt% Cu to aluminum substantially suppressed the stress migration open failures.[20]

In summary to this section on chip metallurgy: If wire bond lift problems suddenly occur in a developed process it is imperative to determine and correct the cause. Too often in a manufacturing environment with operator control of bonding parameters the tendency in "no-stick" situations is to increase power/time variable with resultant microcracking under pad metallizations (incipient thermal cycle fails) or, (in case of aluminum wedge bonding only) drive toward heel cracks (not to be confused with tool marks). Plasma cleaning of chip surfaces is done in many shops to enhance bonding. Plasma cleaning of chips to improve bondability after diebonding to an organic substrate should be approached knowledgeably and cautiously in view of potential for halocarbon dissociation and free halogen embedment into organic substrates, later to be encapsulated with polymeric (non-hermetic) compounds. Corrosion can occur as moisture eventually permeates the polymer packaging. Off gas control of die attachment compounds will be discussed in the appropriate section.

7.9 SUBSTRATE CLEANLINESS

The objectives of the substrate cleaning process are principally to remove ionic contaminants, remove organics from bond pads, and remove particulates.

A major key to extending the field service of any plastic encapsulated (whether COB or conventional) packaging is the reduction of ionic contamination. Halogens are especially to be avoided (Cl^-, Br^-, F^-) for their ability to corrode metallizations when combined with water permeation of the COB encapsulant, just as it occurs in molded plastic packages. Mobile ions such as sodium (Na^+) and potassium (K^+) are also to be avoided, due to silicon interactions. Fortunately, as more devices are available with wafer level passivation systems suitable for plastic molded devices, COB has benefited. If there is a question as to suitability for packaging a given chip in COB (new wafer passivations, no plastic package experience, etc.), then an ionic susceptibility evaluation can be performed. Such a test may include normally processed (cleaned substrates, die- and wire bonded) test sites. After electrical parametric testing is performed, sites are doped with an addition with known concentrations of NaCl solution (e.g., 0, 5, 50, 500 parts per million), heat dried *in situ*, then encapsulated and environmentally stressed, preferably with bias. Appropriate control samples should be included. A failure curve family should develop, with the COB user matching results to his product needs. For quality office products, Lexmark has found a maximum ionics equivalent of 1 microgram NaCl per square centimeter to be suitable. Ionometers capable of accurately monitoring such ionic levels are readily available. This cleanliness level must be maintained through the encapsulation process. With boards cleaned to these levels, and with usage of low mobile ionic (as per Mil-Std-883/Method 5011) content die-attach and encapsulant materials, COB assemblies are currently manufactured which are capable of withstanding stress testing at 85°C/81% relative humidity for in excess of 2000 hours, with bias.

Another aspect of cleaning besides ionic effects is the removal of organic deposits from bonding surfaces that could adversely affect wire bond yields (including "no-sticks") and reliability (expecially thermal cycle life) or with diebonding shear strength. Sources for these organics have been identified as solder mask bleed or development residuals, flex circuit thermoplastic adhesive dissolution/re-deposit in cleaning operations, diebond adhesive off gas products during cure cycle as well as human origin. Materials manufacturers now offer solutions to many of these historical problems, but it is up to the COB user to specify, qualify and have quality control monitors in place to preclude these problems. Mil-Std-883/Test Method 5011 has provided a commendable uplift in the quality of die-attach materials in terms of ionic contents, allowable off gas limits, as well as more traditional mechanical properties.

Organic solvents, typically halocarbon-alcohol blends have been heavily used in board cleaning processes and are capable of providing normal organic and ionic level contaminants (to meet the $< 1\,\mu g/cm^2$ requirement).

Usage of these materials is rapidly being eliminated due to the Montreal Protocol. The newer, non-ozone affecting (claimed by manufacturers) HCFC solvents offer an alternative that for Lexmark met the $1\,\mu g/cm^2$ requirement on boards with minimal existing manufacturing equipment impact. A change of expansion valve and increased freeboard (wall height above vapor zone) along with expected robotics travel adjustments was found to meet the needs. Further discussion on cleaning alternatives has been published.[22]

Oxygen-based plasma cleaning to eliminate organic contamination from bond pads has been used at various companies. One report, appropriate to COB addressed the effective removal of solder mask residual.[23] Trade-offs are costs of elimination of contaimination at the source by material and process control versus relatively expensive Auger and SEM analyses (destructive) and the throughput limitations of (necessarily) batch processing of plasma cleaning. Chlorofluorocarbon plasmas should be used with caution with organic (FR4, flex) substrates due to concerns of dissociated halogens impinged into the organic material, later to combine with moisture permeation to provide a potentially corrosive environment.

Particulate contamination presents issues beyond ionics, and is especially troublesome to maintenance of high wire bonding yields and reliability. Present chip pads are typically 4 mil by 4 mil (0.1 mm × 0.1 mm) with trends to 2 mil by 4 mil (0.05 mm × 0.1 mm). Wire is typically 0.7 to 1.25 mil (18-32 micron) in majority of applications. With human hair being typically 2-3 mil (50-75 micron) diameter, the problems are easy to perceive, the board fiducials are read, the chip fiducials are read, any offsets from the embedded program are computed, and wire bonding begins in the correct dimensional coordinates. Particulates in the programmed bonding site will result in faulty bonds, or even machine shutdown. Levels of cleanroom classes used in COB assembly vary from little or no control to class 100 or better, with classes 1000 to 10,000 frequently used. The objective should be to properly protect the hardware, not to make workers uncomfortable. A class 10,000 cleanroom with all work in progress (WIP) covered or in nitrogen dry-boxes, cleanroom coats, gloves, shoe cleaning, "salad bar" guards at operator stations and careful attention to elimination of moving parts (lead screw bellows, plastic robotic drive chains, etc.) above the COB assembly station can produce high quality hardware with good operator morale. Again, each COB application will have unique reliability vs. cost decisions.

Reliability problems from particulate contamination can be from shorts (metallics), moisture intrusion path formation, etc.

7.10 DIE ATTACHMENT

Many die-attach systems are available in the electronics industry, including filled epoxy, silver-glass, gold-tin and gold-silicon eutectics, tin-lead euetectic solder, etc. Discussion will be concentrated here on systems that are compatible with low-cost COB substrates, i.e. filled epoxy systems.

The die attach material must, besides reliable attachment of die to the substrate:

- Provide thermal path to substrate.
- Absorb differential thermal expansion between the FR4/copper and the silicon chip.
- Process at low to moderate temperature (below 160°C preferred).
- Provide electrical conductivity (or isolation if design requires) for chip to substrate.
- Not introduce detrimental effects (ionics, organic off gassing) during process.

7.10.1 Materials

The die attach formulation will consist of at least three basic materials: the epoxy resin, curing agent, and filler. The resin should be ionically clean, that is, of low mobile ionic content. Various molecular weights are available that will give trade-offs in surface wetting and scrub versus tendency to bleed into unwanted areas. Fine pitch packaging will require a non-bleed type resin carrier, at often a very slight loss in absolute shear strength for most applications.

The curing agent (typically tertiary amine, benzimidazole or acid anhydride) controls process cure time and temperature, the elevated temperature properties of the bond, the work life and the allowable maximum "open time" (moisture susceptibility of the curing agent). The trend of the industry has been to meet the needs of lead frame-chip bonding for the packaged device industry which have more tolerance for high speed, high temperature curing than typical COB substrates, and so require cure temperatures in the 150-175°C range for reasonable flow-processing times. High purity, low off gassing die-attach adhesives in use on COB will typically be required to cure at or below 150°C for minimum effect on the FR4. The filler likewise must be low in mobile ionics and of uniform small particle size (smaller than desired bondline thickness!). Chemically-precipitated (for high purity) flat-flake aspect ratio silver is commonly used where a *conductive* chip to substrate bond is desired. Loading levels are commonly at or near 65% by weight, at which point low electrical resistances (0.001-0.0001 ohm/cm) are achieved. Additional silver loading (80% by weight) can improve thermal conductivity, but contributes only limited electrical improvement. Of insulating fillers, silica (SiO_2) is most commonly used due to higher purity, followed by alumina (Al_2O_3) and calcium carbonate ($CaCO_3$). A premium material found in high performance applications with ability to absorb the cost impact is diamond filler, at present approximately a 5× cost factor, gaining 20% approximately in heat transfer capability. These are used in applications requiring the bottom of the silicon chip to be maintained at a differing voltage than diebond pad site.

Seeing limited use in special applications is polyimide resin/silver which offers high purity resin with low ionics and high temperature strength but at a premium price in terms of material cost, substrate material upgrade and lengthy high temperature cures. Another attachment that has been used in COB with FR4 substrates is eutectic Sn-Pb solder attachment. While offering simplified attachment in a mixed SMT assembly, special attention is needed in both cleaning, to remove flux, ionic and organic residuals (directly contacting the chip metallurgies) and in bondline thickness control. This is usually thicker than epoxy die-attach material in order to absorb the silicon/FR4-copper thermal mismatch without cracking the silicon chip. As stated earlier, the die attach must accommodate this stress. The modulus of the joining material, the thickness of that material, the differential between joining (solidification) temperature, subsequent operational temperature and stress relaxation tendency will affect joint reliability. Higher power dissipation and thermal performance can be achieved with eutectic Sn-Pb attachment, but some gains are offset by the thicker bondlines (0.003 in./0.08 mm minimum, recommended) compared to epoxy (0.0005-0.003 in./0.012-0.08 mm typical). Some properties are given in Table 7-2.

Most die-attach materials are one part premixed, but two part pastes and even pre-cut "B-staged" (partially cured) films are available and used. Many suppliers today offer materials meeting the stringent requirements of

Table 7-2   Mechanical Properties of Die Attach Materials

Property	Silver–Epoxy	Eutectic Sn–Pb solder
Mod. of elasticity	1.2–1.8×10^6 psi (8.3–12.4 kPa)	4.35×10^6 psi (30 kPa)
Thermal conductivity	(decreases with increasing temp.) 1–4 BTU/hr-°F-ft^2/ft	(nearly constant until melt point) 20–30 BTU/hr-°F-ft^2/ft

Mil-Std-883/Method 5011 at little or no cost premium. Familiarization with these requirements is recommended.

The evaluation testing leading to the selection of a die-attach adhesive should include not only the parameters described in Mil-Std-883/Method 5011, as bonded, but should be environmentally stress tested sufficiently to define aging effect of the product usage on the diebond functionality. Lexmark uses a dieshear test equivalent to Mil-Std-883/5011, with sampling as follows for office environment COB applications:

As bonded (time-0)
Dieshears @ $-30°C/+23°C/+85°C$ (extended up to 150°C if product warrants)
10 thermal cycles, $-40°C/+60°C$
Dieshears @ 23°C
2000 thermal cycles, $0°C/+125°C$
Dieshears @ 23°C
4000 (cumulative) thermal cycles, $0°C/+125°C$
Dieshears @ 23°C
6000 (cumulative) thermal cycles, $0°C/+125°C$
Dieshears @ $-30°C/+23°C/+85°C$ (extended up to 150°C if product warrants)

The COB user should select his die-attach material considering not only assembly reliability requirements (shear strength with temperature and fatigue endurance, low ionics, low cure off gassing, etc.) but also manufacturability such as flow control, balanced worklife and cure requirements, and rework capability.

7.10.2 Process

The processes of dispensing die-attachment materials, placement of the silicon chip, and subsequent cure are widely varied in degree of automation from hand operation through machine vision-equipped automatic die bonders.

There are three basic methods of dispense in use today (excluding film preforms): syringe, or "dot-dispense", screen printing, and stamping, or transfer printing.

Most commonly used is the dot-dispense, using equipment and process similar to the SMT bottom-side attachments. A pressure/pulse-time controller (the operator in a hand-operation) dispenses a specific amount of adhesive. The chip is pick-and-placed into the wet adhesive and seated to desired bondline thickness. While much of the yearly output is assembled with manual operations, automatic diebonders provide much more precise control over:

Dispense quantity (milligrams) and placement
Chip orientation, placement accuracy (x,y)
Down pressure to seat chip in adhesive (ergo bondline thickness)

Dot-dispense of die-attach materials offers fast, automated processing of the adhesive dispensing operation, with excellent protection of the unused adhesive from moisture, contamination, etc., and yields relatively void-free diebonds. Further, it lends itself well to mixed technology boards that already have devices present and are no longer planar. There are some inherent limitations in the use of this process as chip area increases. The circular conical dispense undergoes deformation to a flat disc as the chip is seated. Sufficient adhesive to flow to the furthest corners is required to prevent "cliff-hangers", or unsupported, cantilevered wire bond pad sites. Simultaneous adhesive squeeze-out begins to accumulate at the centers of die-sides as the corners are filled. This can progress to "bridges", where die-attach material bridges the span between the substrate diebond pad and the top surface of the chip, resulting in electrical shorts or at the least, epoxy contamination of the chip wire bond pads, or even a wire bonder head crash if die-attach material exceeds the plane of this top surface. The chip size at which "cliff-hangers versus bridges" becomes a yield problem will shift with adhesive rheology, but author experience has indicated that this occurs in the 4-5 mm (approx. 0.16-0.2 in.) die size range.

Some companies offer multiple-dot dispense heads for their equipment to somewhat pre-pattern the dispensed material for large chips. While providing a solution to the cliff-hanger/bridge problem, the chip placement process captures air, entrapped at the interstitial areas of the multidot dispense. The subsequent oven cure (typically 150°C) causes expansion of this entrapped air, and has resulted in significant voiding, chip lofting and/or tilting.

In summary, for smaller die, i.e. less than 4-5 mm (0.160-0.200 in.) sides, dot dispense offers a high productivity, flexible manufacturing process.

Screen printing, with roots in the ceramic hybrid industry offers a fast method of placement of very controlled, pre-patterned adhesive. Process equipment again can be modeled after SMT equipment. The drawback to this process for COB is that the substrate must be highly planar and this precludes mixing of process sequences if using SMT or PTH in conjunction with COB, or if using solder masks. Further, much "working" of the adhesive over a large area by squeegee and screen offers the opportunity for contamination of the adhesive (moisture, from screen, airborne particulates, etc.). As to void formation, this process lacks the raised center, sweep-air-to-sides action of either stamping (below) or dot-dispense (above).

Stamping, or transfer printing offers a flexible adhesive placement process that is compatible with non-planar, mixed technology substrates, that overcomes the die-size limitations of dot-dispense. In this process a flat circular dish is filled with adhesive (see Fig. 7-14). Prior to the immersion of the stamp tool into the adhesive a rotation of the dish under a pre-set doctor blade provides a consistent material height. The stamp tool embeds a controlled depth into the wet adhesive, withdraws, and places the quantity of adhesive on the substrate. Design details of stamping tools are commonly proprietary to the user, but with good design a consistent, uniform pyramid of adhesive (see Fig. 7-15) can be placed that allows near-void-free bonding on chips as large as 10 mm (approx. 0.4 in.) per side, and to a bondline consistently within ±0.25 mil (0.00025 in. or 6.2 micron) with close monitoring of adhesive viscosity. Exposure to contamination (moisture, particulate) with covered dish approximates dot dispense and is much less than in screenprinting. Because of the individual travel-times, dish to substrate, productivity is less than for screenprinting and probably less than dot-dispense. However, referring to Figure 7-8, "n" is usually ≤7 chips/

Figure 7-14 Stamping type semi-automatic diebonder. Note toroidal dish with adjustable doctor blade and the turret-style stamping/chip placement tool which facilitates handling of different sized die.

Figure 7-15 Silver-epoxy die-attach stamping prior to chip placement.

substrate in COB applications, and one diebonder usually handles the throughput capability of several wire bonders.

In summary, stamping offers a flexible, productive process which gives accurate, high-yield, low void diebonds for large area chips but is somewhat more sensitive as to adhesive viscosity control.

Dispense quantity of any of the above processes should be monitored and controlled if consistent diebonding is to be achieved. A simple method utilized at Lexmark is to program a high site density dispense or stamp onto a pre-weighed substrate with re-weigh and calculation of milligrams/dispense.

Table 7-3 Die Attach Material and Process Controls

Raw Material	Process
Viscosity	Viscosity
Ionics	Shelf life/storage
Filler/%	Dish time elapsed
	Tool design
IR spectra	Doctor blade height
Shelf-life	Tool immersion
Therm stab (off-gassing)	Board contact time/pressure
	Deposition quantity
	Open-time (before chip-place)
	Chip placement force (grams × sec)
	Open-time (before oven)
	Cure time/temp
	Visual inspect
	Diebond thickness
	Dieshear strength
	% Voids in bondline

7.10.3 Quality Controls

The control measures should be appropriately chosen to represent the process utilized. Table 7-3 reflects the material and process parameters monitored and controlled in a Lexmark business equipment application, where the stamping process is used. Dot dispense and screen print processes would modify the list to reflect those processes.

7.11 WIRE BONDING

7.11.1 General

This section will deal with COB wire bonding options, wire materials, bonding processes, and necessary controls. For an in-depth study of wire bonding, testing and problem analyses, it is strongly urged that the COB engineer obtain and keep at hand as a reference the text *Reliability and Yield Problems of Wire Bonding in Microelectronics*, by George Harman.[24]

Surface topography and metallurgies for both chip and substrate have already been addressed, but will see some review in this section as they apply to COB applications.

7.11.2 Wire Bonding Options

For COB the choices are basically two: aluminum wedge ultrasonic or gold thermosonic ball and stitch. The choice is influenced, as previously discussed under "Substrates" by the amount of board space required for the COB needs versus the integrated (if any) design SMT/PTH needs. It also is influenced by non-technical choices such as capital equipment processes and experience base that may already be developed and in place for a hybrid packaging house adding COB to its packaging capabilities. Some of the characteristics of gold and aluminum are given in Table 7-4.

Table 7-4   Wire Bond Options for COB

Ultrasonic Aluminum Wedge	Gold Thermosonic
Slower—must rotate bond head to direct line between bond pads	Faster—preheat to 170°C, ball bond can go any direction from 1st bond
Cold process—can use lower cost substrates (FR4, Flex)	Requires premium substrate materials (typically 3X$)
Cold process—less opportunity for development of Au/Al intermetallics	Bonding at Au_xAl_n intermetallic growth temperatures
Finer pitch possible due to elongated bond	Less chance for wire or chip pad corrosion due to ionics
Longer span between pads	
Lower, more uniform loop height	
More tolerant to oxides, contamination	

Aluminum Wedge/Ultrasonic

A majority of COB designs will incorporate other devices, either SMT or PTH, or both. FR4 epoxy glass laminate and related materials (G10, etc.) are the near-universal substrate and offer relatively low costing per unit area, as discussed in the Substrates section. Properly plated FR4 offers a reliable, cost-effective COB substrate for aluminum wedge wire bonding that is compatible with these other packaging processes. Ultrasonic wedge bonding produces an oval, or elongated bond at both first and second bonds of the wire as opposed to thermosonic gold bonding (see Figure 7-16). This facilitates effective fine pitch bonding.

Because of this elongated bond, it is necessary to align the line of flight of the wire between the chip and substrate pads so as to not impart a sidewise tearing motion on the deformed (flattened, work-hardened) aluminum wire. This requires a rotation of either the bonder head or the workstation. Wirebonders are marketed with each type, and both have been used satisfactorily for COB assembly. When larger panel assemblies are processed on a rotating workstation wire bonder, the workstation, hold-down tooling, board assembly, etc., can result in significant mass to accelerate and stop, acting both to increase bearing wear and to increase flight times between bonds slowing throughputs. It is this rotational sequencing that adds time to the ultrasonic wedge bonding process over the gold thermosonic ball process. On the other hand, the narrower elongated wedge bond allows finer pad pitches than does the (round) ball bond, giving more potential for denser I/O counts such as the 0.0016 in. (0.04 mm) pitch cited in Layouts, Fine Pitch Developments, in Section 7.6. The low, flattened bond leaves the bond pad surface at a lower angle allowing a low profile and more uniform loop height. This facilitates lower encapsulated height for thin-card applications ("smart cards", etc.). The strengthened 99% Al/1% Si alloy wire allows longer pad-pad fanout spans, enhancing higher I/O counts (or to coarser pitch cards, thus decreasing card costs).

The narrow, elongated wedge bond does not completely "encapsulate" the chip pad metallurgy with noble metal as does the gold ball bond decreasing the ionic corrosion resistance of the interconnection (as well as the aluminum wire). Fortunately the availability of ionically clean encapsulant and die-attachment materials on the market since about 1986 has negated much of this difference. The narrower wedge bond does facilitate (if required) rework on fresh chip pad metallurgy. This is not possible with gold ball bonding.

This ultrasonic action tends to break up and displace contaminants, especially brittle oxides making the aluminum wedge process more tolerant in manufacturing processes.

A final "plus" for the aluminum wedge bonding process is that no heating is required. Krzanowski et al. found the bond to be free of melt zone.[25] The temperatures that create the often-cited purple plaque of Au-Al inter-metallics in the 300°C thermal compression bonding of years ago are not present in the ultrasonic aluminum wedge process. Selikson and Longo found that even after baking aluminum wire bonds for 1000 hours at 300°C, no bond degradation was noted, while rapid degradation occurred in the gold wire bonded units.[26] Few problems were found to exist in the aluminum rich phases. $AuAl_2$ is metallic in nature and exhibits low

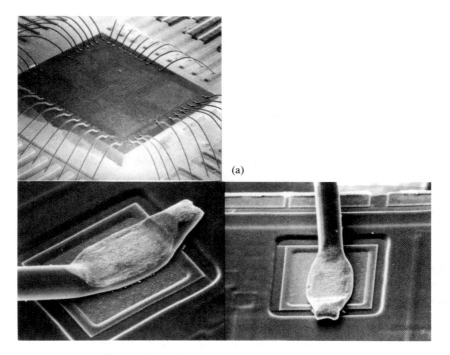

(a)

Figure 7-16a   Aluminum ultrasonic wedge wire bonds.

resistance (Selikson and Longo cite 8×10^{-6} ohm/cm). Similar results (7.9×10^{-6} ohm/cm) were observed by Haag in 1991.[27] In the gold-rich intermetallics (gold wire on aluminum pads) Haag cites 25.5×10^{-6} ohm/cm for Au_5Al_2 and 37.5×10^{-6} ohm/cm for Au_4Al, which correlates with findings by Kashibara and Hattori.[28] Selikson and Longo discuss the question whether the aluminum-gold intermetallic problems have been merely shifted to the (in their case) aluminum wire, gold plated substrate bond and find that not to be the case. Production of over a million COB assemblies at Lexmark without evidence of aluminum wire-gold substrate intermetallics failure would tend to support Selikson and Longo's statement. Again, the amount of gold (2-7 microin., even 20-40 microin.) present in the bond area is small compared to the aluminum available in the wire. While 300°C exposure of thermocompression bonding exceeds the normal 160-175°C processes of thermosonic gold ball bonding, Au-Al intermetallic formations do occur below 175°C.[29] There will be proponents of thermosonic bonding at lower temperatures to decrease adverse effect on

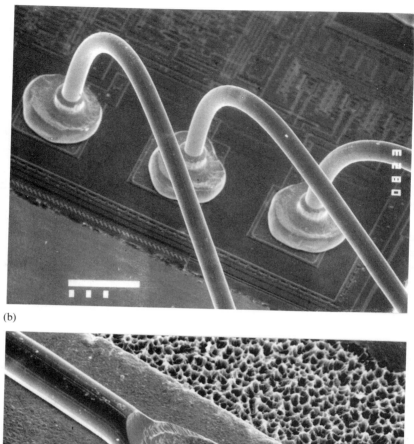

(b)

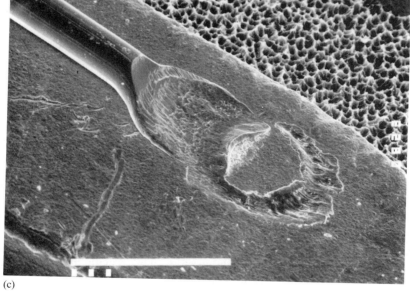

(c)

Figure 7-16b, c   (b), gold thermosonic ball, (c) stitch.

circuit boards used in COB, but bonding below 160°C gave pull values of less than 4.0 gram for 0.9 mil (23 micron) gold wire in a study by Hu et al.[30] For comparison using 1.25 mil (32 micron), 99% Al/1% Si ultrasonic wedge bonding at Lexmark, the bond process is oriented toward consistency and reproducibility rather than maximum pull strengths, yet yields average pull strength of 12 gram with standard deviation of 1-2 gram.[31]

Gold Thermosonic Ball

The vertical exit of the bondwire from the deformed ball bond (see Fig. 7-16) allows the traverse to the second, or stitch, bond in any angular direction. This eliminates the motion requirements of acceleration/deceleration of the work table or bond head as required by the wedge bonder. Current best-of-breed gold thermosonic bonders are capable of production speeds of 6.5 wires/second compared to aluminum wedge bonders of 5 wires/second. If substrates costs and intermetallic formations could be ignored, gold thermosonic wire bonders could become as prevalent in the COB assembly as lead frames in packaged devices. Gold thermosonic bonding applications in COB become significantly increased in importance as the wire bond count increases relative to total board area. A good example of this is observed in the calculator board of Figure 7-3, where a significant portion of the board area is COB relative to PTH and SMT, as opposed to the typewriter function card of Figure 7-5 where six COB sites share board space with over 200 other components. There will be a crossover point economically where to reduce time on the bonder will pay for the premium substrate cost per unit area. There are two precautions when comparing bonder speeds: (1) fiducial reading times are comparable between bonder types, thus decay the burst speed rate advantage, and (2) burst speed is exactly that, a burst, over a series of in-line bonds and does not consider lost traverse times between (multiple) chip sites. Overall, application oriented assessments should be given in reviewing COB wire bonding options rather than simple burst-rate comparisons.

The passivation advantages of "encapsulating" the chip pad with the deformed gold ball have been discussed, as well as the potential for intermetallics.

A recent, not often considered advantage of ball bonding is the ability to gold bump chips for TAB bonding, at least in early development and prototype stages or to convert available wire bond technology chips to (currently unavailable) TAB-bondable chips. This is with the caveats that the wire bonder software accepts a change to first-bond (ball) bonding only, without sequencing through the stitch bond, and that appropriate gold wire is utilized.

7.11.3 Wire Metallurgy/Controls

The near-universal aluminum bonding wire is the 99% Al/1% Si alloy. Pure aluminum is not only very soft, but has a low strain-hardening rate making it difficult to draw small diameters (< 0.003 in. or 77 micron). The alloying of 1% silicon strengthens the wire without losing much of its electrical conductivity or its ultrasonic bondability. At a temperature of 525°C the silicon is in solid solution.[32] With dropping temperature the amount of silicon that is held in solid solution decreases and silicon precipitation begins and continues until almost all the silicon exists as a precipitate at room temperature. For optimum wire bonding, 99% Al/1% Si wire should be used within a shelf life of six months from date of manufacture. Wire sized above 0.003 in. is commonly 99.9% pure aluminum as it demonstrates adequate stiffness and strength at this cross-section with primary applications those of higher power dissipation and current carrying. Aluminum-magnesium (0.5%, 1.0%) wire is also readily available. ASTM specifications are available for all the above aluminum and alloy wires. The specification for an aluminum bonding wire should include an elemental analysis for aluminum and alloy element, and should also limit the amounts of impurities, especially sodium, lithium and potassium elements. The specification should define tensile and elongation properties. Suggested values for 99% Al/1% Si wire are given in Table 7-5 below.

Due to the silicon precipitation, silicon particle size should be limited to 0.5 micron, maximum. Also controlled will be the visual inspection requirements, packaging, etc. It is recommended for instance, if multiple wire (diameters, alloys) are in simultaneous production, that spools be ordered color-coded from the vendor for ease in tracking. Stocked wire should be stored with its axis in the horizontal to prevent loop-over and subsequent wire feed problems in the bonder. Before the preparation of the COB user's wire specification it is suggested that the publication "Bonding Wire: A Practical Guide for the Packaging Engineer" should be studied.[33] This report also delineates allowable dopants in gold bonding wire. Although, gold wire is normally 4-9's (99.99 wt%) purity gold, some

Table 7-5 Typical Specification Values for Al/1% Si Wire

Diameter, in. (micron)	Tensile Strength grams	Elongation %
0.001 (25)	15–18	1–3.5
0.00125 (32)	19–22	1–4
0.002 (51)	38–44	2–6

dopants may vary for manual bonder usage versus high speed auto-bonders, and the bond wire vendor should be involved in this specification item. Dr P. Douglas has published additional information on gold wire properties.[34] Gold bonding wire does not have shelf storage limitations beyond environmental housekeeping (and spooling as mentioned above). Other inspection, horizontal stocking, color coding, etc., apply also to gold wire.

The application of gold bumping of chips for TAB applications using a thermosonic ball bonder will utilize a somewhat harder wire material to provide an intentional break in the work-hardened neck, just above the flattened ball with the post-bond lift-off of the bonder.

7.11.4 Thermal

The linear thermal expansion coefficient of the wire materials is useful from the standpoint of thermal stress calculations and for matching expansion with the encapsulant to be used (see Table 7-6).

Table 7-6   Linear TCEs of Bonding Wire

Material	Range	CTE, in./in./°C	Source of Data
99.99% Au	20–100°C	14.2×10^{-6}	AFW REPORT #7
99.99% Al	20–100°C	23.6×10^{-6}	AFW REPORT #7
99.99% Au	20–300°C	15.3×10^{-6}	AFW REPORT #7
99.99% Al	20–300°C	25.3×10^{-6}	AFW REPORT #7
99% Al/1% Si	20–300°C	25×10^{-6}	AFW REPORT #7
99% Al/1% Si	23–300°C	25.4×10^{-6}	Lexmark Mat'1 Lab

7.11.5 Electrical Current Capabilities

Two considerations must apply here. First, the COB encapsulant, depending on its thermal conductivity will provide heat-sinking capability and therefore current-carrying capacity beyond that of the normally reported circumstance of flying wires (unencapsulated). Because of variance in wire materials, diameters, design currents and duty cycles, encapsulants and heat sinks, little published data is available. Second, flying-wire data is available. The COB designer should consider this data, if not for direct applicability to the end-use encapsulated device, then for the post-wire bond, pre-encapsulation electrical test. Due to the difficulty in economic and reliable repair of encapsulated COB assemblies, a rigorous pre-encapsulation electrical test is normally applied. Thus, flying-wire data is of interest.

A variety of flying-wire continuous and 25% duty cycle data is reported in American Fine Wire Technical Report #7.[33] This reports burn-out current values for various wires and applications, then suggests:

$$\text{Design maximum current} = \frac{\text{Burn-out value}}{\sqrt{10}}$$

Transient current capacities in the form of mathematical formulae are reported by Coxon, Kershner and McEligot.[35]

Reported by R. A. Munroe and J. O. Honeycutt,[36] the allowable current for any flying wire can be defined by solving the equation developed by Jacob.[37]

$$I = \sqrt{\frac{K}{2P}}\left(\frac{\Pi D^2}{L}\right)\sqrt{\Delta T}$$

Where:

I = current in amperes
K = thermal conductivity in watts/cm/°C
P = resistivity in ohm/cm
D = diameter of wire in cm
L = length of wire in cm
ΔT = temperature rise in °C

If assumptions of $T = 25°C$, $L = 0.150$ in. (0.38 cm), span $= 0.1$ in. (0.25 cm) were made, Munroe and Honeycutt[36] prepared Table 7-7.

7.11.6 Wire Bond Process

Equipment   There are numerous wire bond equipment suppliers with lines of bonders from the manual, one bond at a time, through full machine vision and materials-handling automation. Each has strengths and weaknesses for COB assembly. The following comments apply to both gold thermosonics and aluminum ultrasonic wedge bonders.

Wire bonding equipment is normally expensive capital equipment with long lead times (units rarely commissioned on speculative basis), which get increasingly expensive in terms of cost and lead time as custom

Table 7-7 Current Carrying Capacities of Bond Wire

Wire Diameter, in. (mm)	Iau amps	Ial amps
0.0005 (0.013)	0.055	0.035
0.0007 (0.018)	0.11	0.075
0.0010 (0.025)	0.22	0.15
0.00125 (0.032)	0.34	0.23
0.0015 (0.038)	0.50	0.35
0.002 (0.051)	0.88	0.60
0.003 (0.076)	2.00	1.35
0.004 (0.102)	3.50	2.40
0.005 (0.127)	5.50	3.75

modifications are required (table size modifications, custom materials handling, externally driven by computer, management information handling, etc.). Because of the wire counts versus chip counts, a single diebonder can handle the throughput of several wire bonders. An error in sizing wire bonder suitability and throughput thus becomes a multiple loss. The best advice that can be given is to define your needs carefully, start early, be thorough in machine evaluations, and have a strong delivery/ performance contract in place with the vendor.

Some of the items that should be in addition to normal technical and throughput items are:

- True multichip bond cycle time. This is in reference to the previously discussed "burst speed" which is commonly reported by machine vendors and is usually taken for a line or row of wire bonds of short span length. Real designs have two and four sided I/O chips, scattered over a COB assembly. Real cycle time will consider chip and board fiducial-read times, wire spans of actual length used in design, machine table accuracy (versus re-reading fiducials at subsequent chip site-again, a period when no wire bonds are in process), any extra X, Y, or Z direction traverse (to clear any obstacles onboard), etc.
- Machine "up" time. Contact other customers using this machine for their experiences and press the vendor for an up-time guarantee statement in the purchase agreement.
- Maintenance. Define warranty coverage, maintenance-response times in purchase agreement. Decide degree of maintenance that will be performed by user personnel, and further, if by owner/operator

concept or by separate maintenance team. Specify training extent required, and whether on-site or at vendor's site.

- Wire count capability in memory. With common COB packaging or microprocessor and ASIC (Application Specific Integrated Circuit), along with other memory devices, wire counts can accumulate, greatly exceeding those of packaged devices whose wire bonder programs "reset" as each device proceeds under the head. The on-board memory in the wire bonder should be capable of storing locations of all wires to be placed on a given assembly, at a minimum that of a single COB board assembly of a multiboard panel, allowing step and repeats once new fiducials are acquired.
- Work table size. Today, bonders are available which can bond over an area of 6×12 in. (15×30 cm) and more, whereas a few years ago they were limited to 2×2 or 4×4 in. (5 or 10 cm) working area. This opens the opportunity to intermix other assembly processes such as SMT and/or PTH with COB, in a panelized format.
- Wire tear-off feature. This is applicable to wedge bonding. Two types of tear-off are used, with mixed opinion. Clamp tear has been utilized for years. While the ultrasonic wedge is still in downwards contact with the completed second bond, the clamps exert a pull on the wire through the wedge, causing an elongation and break in the wire between the clamp and the bond. The clamps then loosen and additional wire is fed forward to begin the next bond. This method provides a downward holding force on the completed wire bond nugget while the tear force is exerted. Some new bonders use "table-tear", where, as the bonder tool lifts, the table is laterally moved to break the wire from the previous bond. Some force is exerted on the completed wire bond in table-tear. The amount of force in a vertical, peel (weakest) direction is affected by "home" height setting of a bonder wedge above the plane of the board assembly. One aspect of table-tear is that substrate "no-sticks" are immediately obvious without proceeding to the electrical tester, and the bonder will probably shut down in a logjam of untorn wires. A positive aspect is that some force is applied to each bond in a pull test of each substrate bond, unless "back-bonding" or "reverse bonding" where first bond is at substrate, second bond at chip. On the negative side, improperly adjusted table-tear bonders have been observed to allow no "escapes" to the pull tester of less than 8 grams for 1.25 mil (32 micron) 99% Al/1% Si wire bonds in an environment of Mil-Std-883 requirements of 3 grams minimum, Lexmark 5 grams minimum (engineering specification) and 6 grams minimum (manufacturing guard-banding of process). The clamp-tear machines avoid this pull on the bond, but have the potential to allow low-pull escapes unless either 100% pull test, usually uneconomic for COB, or the statistical quality

control sampling plan provides coverage. COB users should be aware of the choices.

• Automation level. Most wire bonder manufacturers will offer a range from manually operated through full machine vision and material handling. Manual bonders are entirely satisfactory for operator training, development prototype building, incoming raw material (chips, boards) bondability testing, etc. Semi-automatic bonders will perform full chip wire bonding after program is stored and board/chip fiducials are "punched in." Fully automated machines will be able to locate fiducials on both the board and chip (within manufacturer's target specification) using machine vision, then complete the assembly bonding. Materials handling capability at the table is fairly universal. However, if magazine transport, unload, work, reload output becomes a requirement, more difficulties in lead time requirements and dramatic cost increases usually occur. Each bonder manufacturer offers materials handling systems of their own design or procurement to handle hardware such as selvaged lead-frame "sticks" and other packages for devices. The ability to transport flat circuit boards (at least 6×12 in. or 15×30 cm) laterally across the bonder table is readily available today, with work-in-waiting and work-completed stations. To meet the diverse COB users' needs for panel sizes, magazine handling, etc., may require the wire bonder manufacturers to call in internal or external engineering specialty assistance if that is outside their core experience area. Likewise, if management information input/output data is required in software. Such features may inflate price and delivery times by easily 30-60%, as well as impact up-times of the composite equipment.

• Machine vision. This must be compatible with the product. For Lexmark, an automated bonder that worked well in fiducial recognition on gold-plated copper fiducials on FR4 was having difficulties with chip-on-flex. Some angular copper circuits near the fiducial marks were covered with a polyimide film overlay. The machine vision system was purchased by the wire bonder manufacturer from a subcontract supplier. The bonder manufacturer could offer only limited assistance. Lexmark optical engineers were able to solve the problem by changes of wavelength of the work lighting system. The COB user must work with the bonder manufacturer to determine the design of mutually acceptable fiducials prior to release of board artwork for prototyping and/or machine acceptance testing. This may require compromise design by the COB user if several bonder companies are competing to supply the bonders.

• Vibration compatibility. The room area in which the bonders are to be located ideally would be on an isolated concrete slab ground floor to minimize transmission of vibration (air handling units, fork trucks, doors, etc.). In bonding of 1-mil (25-micron) wire, micro-movements of 0.1 mil (2.5 micron) between substrate and bondhead can cause broken wires, missed wires, cracked, nicked or otherwise damaged wires, and no-stick incidences. In general, wire bonders are supplied with moderately good isolation systems. On occasion this can be overcome either by severity of vibration or inadvertent modifications that bypass the isolation. An example known was where a manufacturing wire bond area was located in a ground floor area that had a basement. Along one wall of the cleanroom was a major interplant fork truck artery. Behind the cleanroom was the fork truck overnight park/recharge facility. At the close of the workshift, high traffic was reinforced with the dropping of forks (often with gusto at close of shift) to the concrete floor. Bond tool motions of 1.6 mil (40 micron) peaks were recorded. In another example, a manufacturing engineer attached materials handling magazine elevators to either end of a well isolated bonder table, then bolted the elevator bottom mounts to the concrete floor. Needless to say, vibrations created at elevator transport/stops greatly decreased bonder yields. Both of these incidents range outside the technical content normally within a handbook, but were real occurrences and demonstrate the need for the COB user to consider the proposed location of his bonders before placing the purchase order, and perhaps even to record the vibrations experienced in this proposed area as part of the procurement specification to the bonder manufacturer, as well as alert his industrial and manufacturing engineers as to the sensitivity of this equipment.

• Operator friendliness. Ease of threading, wire spool replacement, wedge replacement, etc., should be reviewed as well as control clarity and placement and programming ease.

• Support fixturing. While absolutely critical in the proper hold-down of flex circuitry for wire bonding, adequate clamping of even simple FR4 boards is essential to eliminate warpages and/or motion during bonding. Design will be largely custom, and up to the COB user, but certainly interaction with the wire bonder manufacturer is needed, especially as to attach-points and to mass (weight) added to the bonder table, particularly with rotating workstation wedge bonders (acceleration/deceleration elapsed times, bearing wear, etc.).

7.11.7 Wire Bond Process Set-Up

When performing bonder set-up operations, the printed circuit boards and chips should be of the same manufacturer's lots as the parts to be wire bonded. Chips on the set-up PCBs may be electrical rejects, but all pads should be in place and of correct metallization. Likewise, PCBs used in set-up may have defects rendering them unsuitable for deliverable product such as bridges, shorts, solder mask bleed (not on pads), etc., but gold/nickel/copper metallization and surface smoothness, etc., should be to specification requirements.

In the set-up operation a bonding matrix should be performed including time, energy level, wedge (or capillary) down force and for gold thermosonic bonding, heater platen temperature. Down force for aluminum wedge should give a deformation in the range of 1.2-1.3 D (wire diameter) in resting on the 99% Al/1% Si wire, without ultrasonic pulse applied. Post-bond inspection should be made for bond deformation (~1.8D target), and a destructive wire bond pull test made for $\bar{X}$ and standard deviation. In addition, parts should be inspected for bond placement on pads, wire clearances and damage (heel cracks, etc.) die clearance and tail lengths. Adjustments should be made and the set-up process re-cycled until meeting all criteria.

It is further important in set-up (and in process quality control) destructive pull testing that a record be kept of the mode of failures. Optimized bonding will include a mix of bond fail codes. Fail codes used in the industry are relatively standard:

Code 1 - Bond lift at primary (chip) bond
Code 2 - Wire bond break at heel* of primary (chip) bond
Code 3 - Wire break in non-upset area
Code 4 - Wire bond break at heel of secondary (substrate) bond
Code 5 - Bond lift at secondary (substrate) bond

* For aluminum wedge. For goldball, the neck.

The COB user will find a number of references useful in understanding the bonding processes.[38-40]

7.11.8 In-Process Monitoring

Because of the diverse usage of COB in product lines, no one set of accept/reject levels is universally applicable. In general, wire bonding is a process that lends itself well to statistical process control (SPC). At Lexmark, for each wire bond machine, each operator-shift, destructive pull testing is performed and monitored for minimum pull strength of 5 grams specification for 1.25 mil (32 micron) 99% Al/1% Si wire. Further, manufacturing provides guard banding on critical processes, and in wire bonding, stops and re-cycles the bonder set-up procedure at a 6 gram minimum pull. Additionally $\bar{X}$, (mean), σ (standard deviation) and $\bar{X} - 3\sigma$ are monitored. Typical values observed are 12 grams $\bar{X}$ with σ of 1-2 grams.

Should a problem such as "no-sticks" develop, a reference bonder with pre-tested "golden cards" is run to sort out whether a board or machine problem exists.

In addition to the destructive pull testing described above, non-destructive pull testing can be used from a sampling program through 100% on critical assemblies, usually at a reduced requirement (Lexmark requires 4 gram on the 1.25 mil (32 micron) 99% Al/1% Si wire. For gold thermosonic bonding, ball shear has become increasingly popular. Descriptions of these techniques are given by Peter Singer,[41] and, in more detail in ASTM F 459[42] and in Mil-Std-883D.[43] Additional references are given in the bibliography.[24,44-49]

7.11.9 Visual Inspection of Bonds

Visual inspection of wire bonds has traditionally been at $30\times$ magnification with optical microscope. Currently a variety of methods are adaptable. For example, Lexmark finds useful an automated x-y positioning table with digital readouts and a video camera complete with VCR to tape results for quality control record-keeping.

It is suggested that COB users obtain original quality copies of both Mil-Std-883D and NASA Reference Publication 1122 to review and keep in the inspection area. Some general visual criteria (especially for aluminum wedge bonding, commonly used on COB) are given below.

Bond dimensions. The completed wire bond upsets are usually reported as ratio to original wire diameters. An example is shown in Figure 7-17, including 1.25 mil (32 micron) Al wire. For wedge bonding the optimum

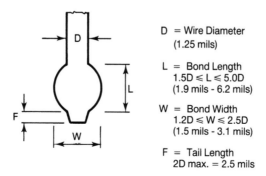

Figure 7-17   Bond dimensions.

upset for operation seems to be at or near 1.8× the original bond wire diameter. This should produce (given clean, bondable chip and board surfaces) the highest pull strengths with minimum overworking of metal and/or wire damage.

Bond width. Another bond parameter that is guard-banded by Lexmark manufacturing. While specification allows W to be from 1.2 to 2.5 D, guard bands are set to intervene outside a window of 1.5 to 2.2 D.

Tail lengths. Should be a maximum of two wire diameters to preclude shorts to adjacent chip metallurgy or to adjacent substrate pads.

Bond placement (primary, or chip bonds). Many conditions exist and are addressed in the NASA publication and Mil-Std-883D but intent is for 75% of the bond to be on the pad metallization. No portion of the bond foot is to be on top of the metallization line leaving the pad or a metallization line near the pad. Rework of bonds on top of the line leaving the pad may cause an open if damage to the passivation occurred during the initial bonding. Shorts may occur to metal lines near the pad if the passivation over them is damaged by the bond foot.

Wire-die clearance. At a point 5 mils (0.13 mm) from the heel of the bond, the wire clearance over the die must be a minimum of 1 mil (0.25 mm) (see Fig. 7-18) to assure electrical clearance through encapsulation process.

Bond placement (secondary or substrate bonds). Substrate pads (at Lexmark) are made typically 25-30 mils (0.6-0.8 mm) in length. Intent is for original bonds to be made on the forward (chip) 30-50% end of the finger metallization, with reworks, if required, to be made on fresh metallization rearward, without danger of solder mark contamination or bonding in the area previously probed (dented?) in raw board electrical testing.

Wire clearances. Wires, unless at same potential, should be spaced no closer than two diameters from another wire, unglassivated metallization or unpassivated die area.

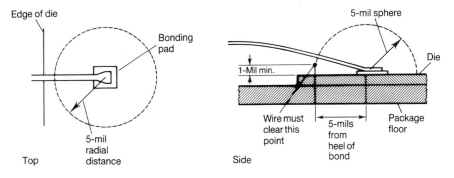

Figure 7-18   Wire/die clearance requirements.

Wire damage. Wires should not exhibit nicks, bends, cuts, crimps, scoring or neckdown that reduces wire diameter by more than 25%. Torn wire bonds are unacceptable (see Fig. 55 of NASA Ref. Pub. 1122, part of which is reproduced in Figure 7-19.

Of special interest in inspection of ultrasonic wedge bonds of aluminum are *heel cracks* (see Fig. 7-20) and overworked bonds. All wedge bonds will show a tool mark at the upper radius of the heel as it leaves the primary (chip) bond pad especially at the transition of "worked" metallization to the unworked wire diameter. Some scuffing will occur in the ultrasonic process. This is not to be confused with heel cracking, a reliability problem, which has origins in overworking of the ultrasonic bond and/or forward movement of the bonding wedge and wire as it first lefts from the bond, then traverses rearward to the second (substrate) bond. Damaged wire bonds can ultimately fail in brittle fracture to vibration or shock (unencapsulated) or in thermal cycling in product life or testing, especially if encapsulant thermal expansion significantly differs from the aluminum wire. Heel cracking of wedge bonds is addressed by Harman.[24]

Additional information on heel cracking of Al-Si wire is provided by Fitzsimmons and Miller.[50] Overworking of the aluminum wire bonding is visually identifiable by the halo-effect (see Fig. 7-21) and, usually, excessive upset (width > 2.5 D) which can contribute to heel cracking. Overworking may be evidence of other bonding problems such as change in hardness of chip metallurgy due to intermetallics formation (for example, excess copper and/or silicon doping of the aluminum metallurgy of the pad) or of

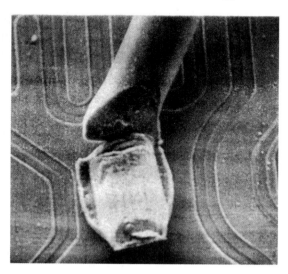

Figure 7-19 Torn bond.

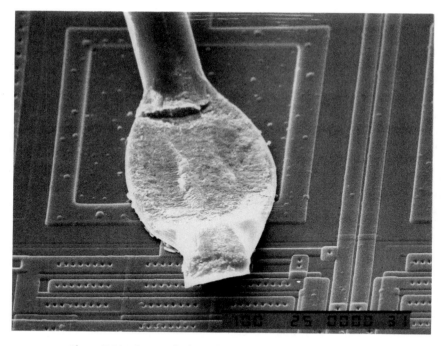

Figure 7-20   Overworked wire bond, exhibiting cracking at heel.

contamination. Besides the potential for wire breakage problems, the hazard of initiating micro cracking in the device below the pad metallurgy, ultimately leading to thermal cycling failure exists. If a shift in bonding parameters occurs in a developed, in-production COB product, the cause should be identified and corrected. Unfortunately, almost all wire bonding manufacturing organizations go through a first experience of dialing up more power/time/down force energy when encountering "no-sticks" early in their experience curve.

Overworked bonds, heel cracks and chip pad microcracking are the tracks left behind of such tendencies.

7.11.10 Rework

Wirebond rework is a normal expectation of microelectronic manufacturing. Proper rework can yield bond quality equal to first-time bonding if proper guidelines are followed. There also should be identified test cells in the product qualification cycle that include production rework samples. With diebond rework, thorough removal of organics without damage to the substrate metallurgy is key. With wire bonding, the substrate can readily in most cases be designed to accommodate wire bonding on fresh metallurgy. More care must be exercised in re-bonding to chip pad metallization to assure reliability. Reliable rework at chip pad is enhanced by attention to such detail as when setting bond placements initially, designing in a skew to the bond pad corner farthest from the entry line, thus assuring no break in the entry line in the rework.

Lexmark allows rebonding to the chip pad only under certain conditions. If there are no bond nuggets or residual segments left on the pad, the pad may be rebonded only if enough pad metal remains to bond at least 75% of the rebond to pad metal.

If a bond nugget or residual segment is present on the pad, rebonding may be performed on top of the nugget as long as the following requirements are met:

- At least 50% of the bond nugget must be flat and bonded to the pad metal.
- No portion of the bond nugget must protrude in the z-direction more than two wire diameters.
- At least 75% of the rebond must be on the bond nugget and/or pad metal.
- The rebonding process must have its own unique set optimized bonding parameters. Rebonding requires more power than bonding to pad metallization.

Note: NASA Reference Publication 1122 states that rebonding on top of a bond nugget is not allowed. Reliability testing has shown that this practice is acceptable as long as the above requirements are met.

Reworked wire bonds are tested in accordance with Mil-Std-883D, Method 2023.4 (non-destructive) except that maximum pull requirement on the bond is 4.0 grams (Mil-Std-883 requires only 3 grams for 1.25 mil (32 micron) 99% Al/1% Si wire bonds). Vogel, at Lexmark has found reworked wire bond reliability to be indistinguishable from first-time bonds if correctly implemented.[51] This is further substantiated by an Air Force study[52] to define acceptable rework conditions and methods, reported by Malloy and Kondounaris of Hughes and Farrell of Rome Air Development Center, who found that reworked epoxy diebonds and wire bonds reworked on epoxy-diebonded devices were as reliable as first-processed parts for both aluminum wedge and gold thermosonic ball bonding. On the other hand, reworks to eutectically attached die were inferior to the original product.

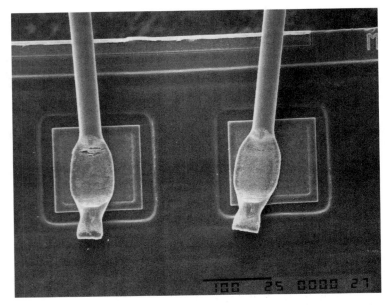

(a)

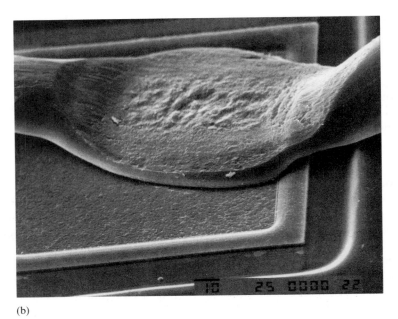

(b)

Figure 7-21 Overworking of aluminum wedge bond. (a) Normal wire bond aluminum at wire upset = 2D, some scuffing at heel, (b) slight overworking, (c) severe overworking of bond, (d) severely overworked, heel cracking exhibited.

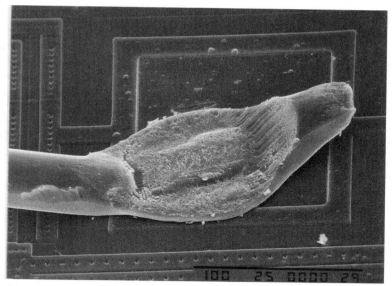

(c)

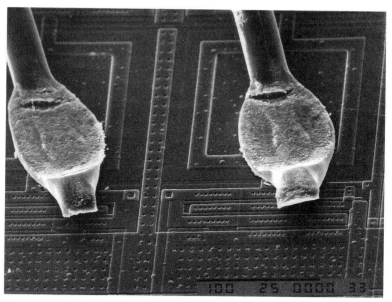

(d)

Figure 7-21 (continued)

7.12 ELECTRICAL TESTING

In the normal flow of packaged die, electrical testing is performed at wafer level, module level and at functional board assembly level. With COB assembly, the chip goes from wafer level directly to the board assembly. Module testing is eliminated. The questions then are how to assure chip performance in the product and to avoid scrap due to test inabilities? The answers lie in extending testing across the missing module level testing from either end, i.e. increased wafer testing as far as economically feasible, and to push the manufacturing assembly test both early in assembly where the COB assembly is reworkable, and by test coverage beyond the normal manufacturing test, that is typically: is the correct part present, correctly oriented, and all hooked up?

Other testing decisions will be the trade-offs between test coverage versus packaging density for both wafer test and assembly test.

7.12.1 Wafer Testing

Wafer testers provide a means of early capture and the avoidance of the costs of packaging die which are obviously bad electrically. There are shortcomings in its capability to replace module testing:

- Large capital expense (ergo, $/operating hour).
- Small, tungsten probes often cannot provide sufficient current for full power-up test.
- Vacuum chuck to position wafer may not provide sufficient heat sinking for power testing.
- Chip real estate dedicated solely to increasing test coverage may cause chip size increase, and so give fewer chips/wafer, and cost increases per "good" chip.

The chip manufacturer will therefore optimize the economics of his operational balance between die-sizing and test coverage versus scrap costs for modules. The COB designer must work with the chip supplier to move the SPQL as high as possible until tester time and die area costs meet or exceed assembly rework and/or scrap costs. On developed technology microprocessor and ASIC die today, SPQLs of 99-99.5% are marketed. Power device testability (probe resistivity, cooling) is improving and die SPQLs of 99% are available from a limited number of suppliers.

7.12.2 Raw Card Testing

This will be only briefly discussed as the economic drivers are analogous to other packaging technologies, i.e., SMT, PTH, TAB, etc., therefore it doesn't pay to put good devices on bad boards. Diebonders and wire bonders can be bought with recognition/no-populate capability for bad cards within an otherwise good panel. Early generation SMT and PTH equipment are likely not to have this capability thus driving the mixed technology assembly to require 100% good boards in a panel.

7.12.3 Board Assembly Testing

To extend beyond the standard manufacturing test philosophy of testing for correct part presence, oriented correctly and hooked up will require some test equipment enhancement. To do a "bare die" test the passive devices normally found in the full assembly will be dummied into the tester. The first test cycle will follow die- and wire bonding, and will be an opens and shorts test expanded to check for:

- Punch-through with mild stress on die (example: 7V on 5V die)
- Some functional testing within the die.
- Now that chips are wire bonded, diebonded to board, the user can drive power to maximum loads (with the caveat of not exceeding tester limits) to eliminate weak chips not detected at wafer test.
- Microprocessors are stepped through subsets of instructions while verifying each input could detect high and low signals.
- ASICs, as one knowledgeable test equipment engineer put it bluntly, are "ugly" to test because of their custom architecture.

Other steps may be necessary to cover the absence of module testing, such as clock speed. This is normally a function of chip design, layout and build technology, but is testable at module level if necessary. In this case, the COB chip could be tested on a sample basis as a module using the universal test site (UTS) concept earlier mentioned.

The post-wire bond electrical test is repeated at Lexmark following COB encapsulation, and again after SMT and PTH processes (of course, with the dummied-in passive device loads replaced by the real devices). It is of high economic importance in a multichip, multipackaging technology assembly to have in place a strong pre-encapsulation electrical test to catch wafer test escapes and in-process fallout with reliable, qualified and economic chip removal and replacement and wire bond rework processes. Finally, to introduce the next section, the encapsulation material process and equipment must be optimized to minimize in-process damage incidence.

7.13 ENCAPSULATION

While all the prior steps are important in determining the manufacturing yields and costs, certainly in terms of product service and reliability, one of the most crucial elements in making COB assembly an economic success is the choice of an encapsulation material and careful attention during processing to not damage "good" assemblies. To make multi-chip COB manufacture reliable and economic requires the COB user not only, to iterate, to drive chip vendors to high testability and SPQLs on their product, followed by building testability into both the COB assembly and manufacturing test equipment, but also, finally, not to lose the parts either by selection of an inadequate encapsulation material and process or by mis-handling in the process. Rework of traditional, glob-topped COB assemblies is rarely successful in sense of both high reliability and economics.

7.13.1 Materials

The solution and development of the COB encapsulation system requires the understanding of failure mechanisms related to these materials. Major frequency factors are wire bond breaks or lifts due to thermal expansion mismatching, and to ionic corrosion.

7.13.2 Silicone

Early glob-top materials commonly used (especially in ceramic hybrid microcircuits) were of high purity silicone, either of a crosslinking low durometer elastomeric consistency, or of a tacky, gel-like material. These materials offered low alpha particle emissions, high purity (much higher than epoxies of that period) and low modulus of elasticity. High thermal expansion coefficients were somewhat offset as to stress on wires due to ability to creep and decay stresses with time. The high water vapor permeability allowed moisture transmission readily, but actual water content of the polymer at saturation was low and with low ionics levels, corrosion was muted. On the negative side, the low modulus provided little mechanical protection, and the gel-type remained tacky, holding contaminants. Neither system is compatible with solvent-based cleaning processes if the COB design is intermixed with PTH or SMT using rosin base fluxes. A mechanical capping, sometimes hermetic, was used to provide mechanical and environmental protection but at a severe economic penalty for consumer product applications. Lexmark evaluated both silicone and filled epoxy encapsulant systems and found that in small (1.0-1.25 mil or 25-32 micron) aluminum wire bonds greater incidence of thermal cycle fatigue damage occurred than in the higher modulus/lower thermal expansion filled epoxy systems on test.

7.13.3 Epoxies

Epoxy systems, as in die attach materials, are typically formulated with a variety of base epoxy resins, curing agents (amine, acid anhydride, or benzimidazole) and mineral fillers such as silica (SiO_2), alumina (Al_2O_3) or calcium carbonate ($CaCO_3$). The mineral fillers (as the epoxy/curing agent system) must be of low mobile ionic content (Na^+, K^+, Cl^-) and low thermal expansion coefficient. Generally silica offers higher purity, and is the filler most often utilized in COB glob-top applications. Some of the requirements of an epoxy glob-top system are:

1. Expansion coefficient close to that of wire material
2. Glass transition (Tg) higher than circuit application
3. Low cure shrinkage
4. Void-free fill over wires, chip
5. Acceptable flow control
6. Low ionic content (<20 ppm Na^+, K^+, Cl^-).

Other discussions on selection criteria have been published by Burkhart[53] and Wong.[54] The latter reference is oriented to materials for packaged devices rather than COB, but contains useful materials property information.

The matching coefficient of expansion to the embedded wire is to prevent thermal cycling breakage or lifts. The addition of the low expansion, no-shrinkage material can reduce the TCE of a thermoset epoxy from 80-100 $\times$ 10^{-6} in./in./°C to as low as 14-15 $\times$ 10^{-6} in./in./°C (mm/mm/°C). Simultaneously, cure shrinkage may be reduced from 1% linear ($\approx$ 3% volumetric) to typically 1% volumetric, likewise reducing wire stress. Actual pictures of an unfilled epoxy encapsulant breakage of 1.0 mil (25 micron) gold wire are shown in Figure 7-22. These wires failed in less than 100 cycles of 0°C to 125°C. Conversely, with 99% Al/1% Si, 1.25 mil (32 micron) wire with TCE of 25.4 $\times$ 10^{-6} inch/inch/°C, Lexmark has gone through the

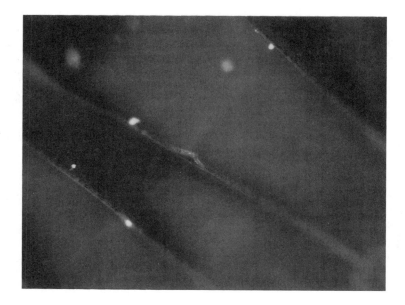

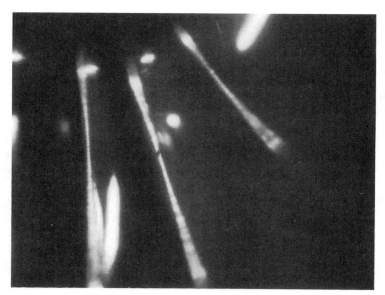

Figure 7-22   Breakage of 1.0 mil (25 micron) gold wire by high thermal expansion coefficient epoxy encapsulant (<100 cycles, 0°C to +125°C).

qualification cycle on three hardware generations without a single wire bond fail attributable to the encapsulant. The qualification cycle includes:

10 cycles, $-40°C/+60°C$ (ship/handle/store test)
6000 cycles $0°C/+125°C$ (life stress test)
2000 hours $85°C/81\%$ RH with bias
2000 hours $125°C$/bias

Silica filled epoxy glob tops are today available with as much as 75% filler loading and TCE as low as $14\text{-}17 \times 10^{-6}$ in./in./°C which accommodate the gold wire (14.4×10^{-6} in./in./°C). In the qualification testing of various candidate encapsulant systems for Lexmark products we found for 99% Al/ 1% Si, 1.25 mil (32 micron) wire exposed to 0-125°C thermal cycling:
The objective of a high glass transition temperature (Tg) is to minimize

Epoxy TCE, 10^{-6} in./in./°C	Result
25–29	No wire breakage
30–33	Very isolated breakage $\sim$1500 cycles
34–38	Low percent breakage initiated $\approx$1000 cycles

the thermal expansion stresses. Filled epoxy systems evaluated saw TCEs above the Tg temperatures typically triple the rate of below-Tg expansion. This assumes more importance in mixed technology designs which will after COB assembly be subjected to wave, vapor phase, or IR-reflow solder operations. Epoxy encapsulants are available today with Tg of 175°C, near the temperature of eutectic solder melt.

As mentioned earlier in the cleaning section (Section 7.9), in any plastic package, be it P-DIP, SOIC, PLCC, PQFP, or COB, moisture will eventually permeate the package, accelerating corrosion if free ionics (especially chlorides or other halogens) are present. To extend the product life, sources of ionic contamination should be reduced or eliminated wherever possible, whether surface cleanliness or within materials contacting chip or wire. In the early 1980s, glob-top materials of controlled (100 ppm Cl^-, 20 ppm Na^+, K^+) ionics were offered by at least one major US electronic materials company. By 1986-87 compounds were available at less than 20 ppm Cl^-, Na^+, K^+. Has corrosion of aluminum chip pads and/ or wire bonds been experienced using LMI (low mobile ion) encapsulants?

In the production of three hardware generations and over a million multichip COB assemblies, the following Failure Analysis Reports are filed:

1. One microprocessor chip, FR4 universal test site, 85°C/81% RH test. Three adjacent wires, all on one chip corner failed due to corrosion. Cl^- presence detected. Total cumulative wire length involved less than 0.4 in. Source never identified, believed to be fingerprint on wire. Ionics content of encapsulant 43 ppm (vs. spec. of 100 ppm maximum). This was only fail of hundreds of pieces in test cell.
2. Numerous fails of microprocessor chip lot on universal test site FR4 substrate qualification test, due to corrosion. Fails occurred rapidly in first few hundred hours of 85°C/81% RH. No fails were occurring in:
   - Alternative vendor microprocessor on same substrates, same build line, same encapsulant lot (8 ppm Cl), same test chamber.
   - Other chips (motor drivers, ASICs) simultaneous process.

Analysis yielded sulfur, potassium, calcium contaminants. These were ultimately traced to ESD packaging materials used by the chip vendor's subcontract wafer dicing house, and further, to the cosmetic make-up used by an employee. The above cited examples, while ultimately exonerating the encapsulant (the topic of this section) illustrate both how subtlely contamination can enter the COB process, and how the capability to analytically resolve such sources is essential. In both cases above, the encapsulant (non-personal, non-procedural, inaminate) was initially indicted by manufacturing only to be exonerated by subsequent thorough analyses. Encapsulant materials available today for COB assembly are adequate with a matching expansion rate and ionic cleanliness levels equivalent to or better than transfer molded packaging materials, and certainly provide less encapsulation stresses (wire wash, etc.).

7.13.4 Other COB Encapsulation Systems

At least one Far East material supplier offers "dry" preforms for COB encapsulation in two types. The first is a pressed "B-staged" (partially cured) pellet, shape and thickness of the customer's choosing. The pellet, with area suitable for covering outer wire bond pads is placed on a pre-heated COB assembly (140-160°C) and reflowed to encapsulate the chip and wire. After epoxy, the assemblies proceed to curing ovens as regular glob top encapsulants. The primary advantages are simple, vacuum pick-up tool application, ease of automation, predetermined encapsulation material volume and shape factor (although ultimate cured footprint approximates the liquid dispersed globs). Disadvantages are cost, lead time, and length of cure of the resin-curing agent used.

A second type of preform, available from the same Far East company is based on a slit or die-cut preform cut from a sheet of epoxy and filler that has been coated onto a thin non-woven carrier fabric. Amount of encapsulant desired can be effected by thickness of coating on the fabric. There are some advantages over the above pellet system in that preforms are more easily prepared from bulk material at the end (COB) user's site but the major advantage is controlled footprint in any pre-ordained shape. Fillets will extend beyond the cut preform configuration no more than 0.1 in. (0.25 mm) (thin preforms) to 0.02 in. (0.5 mm) (thick preforms). This can free up board space in densely packaged assemblies. Disadvantages are cost (as above, 5 to 10 times bulk dispensed material cost) and that the film is directional and must be placed resin-coated side down to chip and wire. Application is by pick and place, reflow, as above.

In the quest for a repairable passivation, thin organic coatings have been evaluated for COB applications both of the poly-paraxylylene vapor polymerization type and of the solvent dispersed high purity liquid systems. The former was neither technically nor economically successful at Lexmark due to: (1) poor adhesion of coating to unprimered substrate/wire/chip assembly, (2) poor stability of coupling agent and incompatibility with chip metallurgy when adhesion promotion evaluated, and (3) cost/assembly, and throughput for the batch operation.

The latter, thin coating of the COB assembly with a thin, high purity, solvent-borne coating was successful technically through 1000 hour stress testing (in environments listed in Table 7-8 for COB) with one caveat: The

Table 7-8 COB—Environmental Stress

Stress Cond.	Tech Dev't.	Design Appl'n.	Mfg. Qual'n.
Temp Shock −40°C/+60°C	10 cycles	10 cycles	10 cycles
Biased T/H 85°C/81% RH	2000 hr	2000 hr	1000 hr[a]
Temp Cyc 0°C–125°C	6000 cyc (2000 hr)	6000 cyc (2000 hr)	3000 cyc (1000 hr)
High Temp w/Bias 125°C	2000 hr	2000 hr	1000 hr
Corrosive Gas	500 hr	not reqd	not reqd

[a] Test sites only, Mixed Tech Assembly given 1000 hr of 50°C/80% RH.

individual wires must not bridge, or web with the coating (unfilled, flexible, TCE $\approx 200 \times 10^{-6}$ in./in.) or wire failure in thermal cycling would occur. Very strict process controls were required (for consistent, non-webbed coating) as to temperature and relative humidity in area, coating viscosity, dry-down rate (solvent system, percent solids), and surface tensions of coating and wires. The reward was a solvent-strippable, reworkable coating (reworks met environmental requirements as new-builds) with a penalty of costing as much as glob encapsulation to apply but without the mechanical protection of glob encapsulation. A molded cap was press-fitted into drill-holes in the FR4 to provide mechanical protection with ease of removal. Design was not appropriate for subsequent intermixing with SMT or PTH process cleaning.

One note on safety: many past (and some present) epoxide curing agents have been identified as carcinogenic or suspect-carcinogenic materials. Alternative, non-carcinogenic materials systems are available to meet COB user needs. Vendor Material Safety Data Sheets (MSDSs) should be obtained and reviewed for all candidate materials. Lexmark Electronics Packaging Technology will not consider usage of carcinogens in manufacturing.

7.13.5 Process

Encapsulation materials are normally available in either the two-part mix, stored until use at room temperature, or as premixed, de-aired frozen compound, either by material vendor or by the end-user. Assuming the proper material is specified and incoming analysis meets specification (such as ionics, filler, etc.) it is now the process engineer's task to provide void-free fill over wires and chip with simultaneous flow control not to contaminate areas unwanted, and very importantly, not to cause any wire displacement or other damage to the completed, tested (and after this operation, non-recoverable) assembly.

The heavy filler-level loading of the epoxy resin in order to lower compound TCE to that of either aluminum or gold wires will certainly impart a very high viscosity at room temperatures, and "tenting" (bridging over wires, from chip to board, without encapsulant under wires) can occur. The result will be air expansion and blow holes during the oven-cure process. Heating of the material dispense container can cause rapidly changing rheology and thus dispense characteristics and quantities due to epoxy crosslinking in the dispenser. For high yield of reliable encapsulation of COB assemblies, the following controls are recommended:

1. Tight viscosity control of uncured encapsulant

2. Storage temperature/time
3. Temperature, humidity, cleanliness of work area
4. Assembly preheat from dispense (aid flow through wires to chip base) through gelation of glob. This fixes glob height, depth over wires, precluding subsequent "wet" slumping, wire exposure, etc.
5. Open time (from dispense in room environment until entering curing oven). Acid anhydride curing agents are commonly used in COB glob top formulations. These are subject to reaction with environmental moisture and if over-exposed (Lexmark controls to 4 hours at 50% RH), will result in materials of lowered glass transition temperature and high thermal expansion (increase from $25\text{-}26 \times 10^{-6}$ in./in./°C to $32\text{-}33 \times 10^{-6}$ in./in./°C has been observed). Problems can also occur in subsequent wave or vapor solder operation (if used).
6. Cure time "clock" starts when the COB assemblies reach stated temperature, not when oven door closes (thermal lag due to fixturing mass). Actual thermal profiles of the oven, fixturing and assemblies should be done as a part of process development and manufacturing implementation (and for each process change).

COB process engineers responsible for the encapsulation process should obtain and read "The Seven Sins of Globbing."[54]

7.13.6 Equipment

COB encapsulation dispense equipment is available from the simple air pressure/timed pulse syringe to full machine-vision, materials handling automation with pre-heat/post-heat stations. Features the COB user should be able to discuss with the equipment supplier besides throughputs are:

- "Air over" syringe (easier clean-up) vs. positive displacement (highest accuracy)
- Stationary or on-the-fly dispense (preferred, less "tenting")
- Pre-heat/post-heat requirements
- Board sizes (flexibility without load, unload of workboard holders)
- X, Y accuracy
- Shot size ranges (Lexmark shot sizes range from 0.7 to 500 milligram)
- Workstation safety (fumes, access to hot surfaces, resins)
- No contamination sources (overhead friction sources, plastic chain, etc.)

Additional information on dispensers has been published.[56]

7.14 ENVIRONMENTAL STRESS TESTING

The environmental stress sequencing used by Lexmark for the qualification of COB for high quality office equipment (typewriters, printers, peripherals) is given in Table 7-8.

The temperature shock test is a low cycle count exposure to simulate environments that office products can be exposed to in shipping, handling and/or storage *en route* to end user (trains, planes, trucks, etc.).

The biased temperature humidity is an accelerated stress test normally given to packaged devices for usage in this product line. For this test, the single chip, "universal test sites" are used for comparison to the same chip in vendor-packaged devices and 85°C/81% RH environment is used. Full, mixed-technology (COB/PTH/SMT) assemblies are commonly given a reduced exposure of 50°C/80% RH because of the lesser reliability seen with packaged components (in particular SMT, with its reduced passivation over PTH devices) after solder/cleaning operations. Some incidence also of copper migration has been observed in this biased T/H environment with the semi-additive (bare Cu-Ni-Au sidewalls) type metallization in this test at 85°C/81% RH. Other than the two isolated cases of corrosion fails (Section 7.13) previously mentioned, no production program COB chip has failed qualification to this environment at Lexmark.

7.15 MANUFACTURING PROCESS FLOW SEQUENCE

A typical manufacturing process flow for assembly of circuit boards incorporating COB mixed with SMT and PTH packaging technologies, and as incorporated at Lexmark, is given in Figure 7-23.

7.16 SUMMARY

- Wire bonded COB has proven to be a viable, cost-effective, reliable electronic packaging technology.
- Wire bonded COB follows the axiom of "Technology trickles up, not down." What began as a low cost games, calculators, and watch electronic packaging technique can be (and has been) uplifted with engineered and controlled materials and processes to a packaging technology suitable for high reliability business equipment.
- Wire bonded COB and TAB are not mutually exclusive packaging technologies. Both offer zones of advantage, and only in some applications overlap one another.

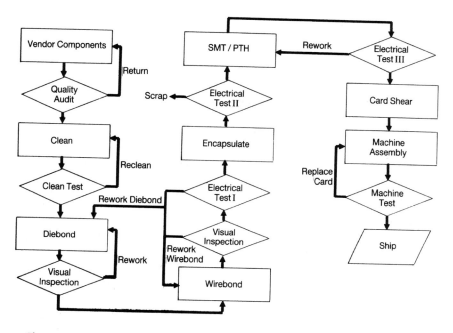

Figure 7-23 Manufacturing flow, COB assembly, intermixed with other technologies.

- Wire bonding is not a stagnant technology. Through use of automation, machine vision, and other enhancements to improve productivity, high-end viable wire I/O counts are increasing.
- Wire bonded COB reliability is, as with packaged devices, dependent on ionic cleanliness, both for board and chip surfaces, as well as that of materials used in assembly (die attach adhesive, encapsulant, etc.).
- Economic success of wire bonded COB will be heavily affected by numbers of chips per assembly, the SPQLs available for the chip set, and the testability and reworkability of the assembly prior to encapsulation. Special attention should be given the encapsulation process so as not to cause failures of assemblies going through this process.

ACKNOWLEDGMENTS

The author would like to express appreciation to the support technicians who worked to gather the data from which COB decisions could be made, to the engineer co-workers, to the management who made this possible (L. R. Steward and J. D.

Zbrozek) and to the word processing organization for their cheerful help in publishing the information.

REFERENCES

1. Gauge, Donley et al., "A Multi-Chip Printed Circuit Board Hybrid for the HP18C and 28C Handheld Calculators," *ISHM Proceedings 1987*.
2. Seshan and Ray, "Effects of Surface Roughness on Wire Bondability," IBM/ East Fishkill, IEEE/CHMT Doc. No. CH 2295-0/86/000-0109, 1986.
3. DiGirolamo, J., "Surface Roughness Sensitivity of Aluminum Wirebonding for Chip on Board Applications," Lexmark, *Electronic Packaging and Production*, March 1991, pp. 65-68.
4. Jahsman, "Lead Frame and Wire Length Limitations to Bond Densification," *ASME Journal of Electronic Packaging*, December 1989.
5. Watanabe, Arao, Tokura, and Shibato, *ISHM Proceedings 1988*.
6. Yamashita, et al., *IEICE Transactions*, vol. E74, No. 8, August 1991.
7. Shu, "Fine Pitch Wire Bonding Development using a New, Multi-Purpose, Multi-Pad Pitch Test Die," *IEEE Transactions on CHMT*, vol. 14, no. 4, December 1991.
8. Shah, Levine, and Patel, "Advances in Wire Bonding Technology for High Lead Count, High Density Devices," *IEEE Transactions on CHMT*, vol. 11, no. 3, September 1988.
9. Otsuka, et al., "Development of Wire Bonding Technology for Fine Pitch," Hitachi, IEPS Conference Proceedings 1992.
10. Estep. G. J., "Electronic Packaging and Production," January 1974, pp. 86-99.
11. Baudrand, D. J., *Electronics*, May 1983, Allied-Kelite Division, Witco Chemical Co.
12. Krzanowski, J., "Interfacial Reactions in Aluminum Wire Bonded Specimens Bonded to Ni and Au Plated Copper Substrates," Univ. of N. H., reported in *Materials Developments in Electronic Packaging, Conference Proceedings*.
13. Jackson, T., "Electroless Nickel Plating of Printed Wiring Boards for Solderability and Wire Bonding," Lexmark, in conjunction with J. Greene and Crotty of Allied-Kelite, American Electroplating and Surface Finishers Society, SUR/FIN '92.
14. Lovie, J., "Nickel and Gold Plating in Electronic Packaging," *Electronic Packaging and Production*, March 1991, pp. 56-60.
15. Endicott, J., and Nobel, "Effects of Gold Plating Additives on Semiconductor Wire Bonding," *Plating and Surface Finishing*, November 1981.
16. "Micro-Corrosion of Al-Cu Bonding Pads," *IEEE/CHMT Journal*, vol. CHMT-10, no. 2, June 1987.
17. Mayumi, et al., "The Effect of Cu Addition to Al-Si Interconnects on Stress Induced Open Circuit Failures," 1987 IEEE/IRPS CH 2388-7/87/0000-0015501.00.

18. Whitman, C., and Lip Wah Chung, "Thermomechanically Induced Voiding of Al-Cu thin films," *Journal of Vacuum Science and Technology*, A9 (4), July-August, 1991.
19. Puttlitz, Ryan, and Sullivan, "Semiconductor Interlevel Shorts caused by Hillock Formation in Al-Cu Metallization," IBM/Burlington, VT, IEEE Log No. 8931389, from *IEEE/CHMT Journal*, vol. 12, no. 4, December 1989.
20. Hosoda, Yagi, and Tsuchikawa, "Effects of Copper and Titanium Addition to Aluminum Interconnects on Electro- and Stress Migration Open Circuit Failures," Fujitsu, 1989, IEEE/IRPS CH 2650-0/89/0000-0202.
21. Ryan, J. T., et al., "The Effect of Alloying in Stress Induced Void Formation in Aluminum-based Metallizations," IBM/Burlington, VT., *Journal of Vacuum Science and Technology*, A8(3), May-June 1989.
22. Wesserman, C., Editor, "Cleaning 1991: The Picture Stabilizes," *Surface Mount Technology*, July 1991.
23. Rust, Doane, and Sawchyn, *IEEE/CHMT Transactions*, vol. 14, No. 3, Sept. 1991.
24. Harman, G., *Reliability and Yield Problems of Wire Bonding in Microelectronics*, National Institute of Standards and Technology, Published 1989 by the International Society for Hybrid Micro-electronics (ISHM, P.O. Box 2698, Reston, VA. 22090-2698. ISBN 0-930815-25-4.)
25. Krzanowski, J., et al., "A Transmission Electron Microscopy Study of Ultrasonic Wirebonding," University of N.H., IEEE/CHMT, vol. 13, no. 1, March 1990.
26. Selikson, and Longo, "A Study of Purple Plaque and Its Role in Integrated Circuits," *IEEE/Proceedings*, December 1964.
27. Haag, *Hybrid Circuits*, no. 26, September 1991.
28. Kashibara, and Hattori, "Formulation of Al-Au Intermetallics Compounds and Resistance Increase for Ultrasonic Al Wire Bonding," *Rev. Elect. Comm. Lab.*, vol. 17, pp. 1001-1013, 1969.
29. Maiocco, Sinyers, et al., "Intermetallics and Void Formation in Gold Wirebonds to Aluminum Films," Thayer School of Engineering, Dartmouth College, *Materials Characterization*, 24: 293-309, 1990.
30. Hu, et al., "Study of Temperature Parameter on the Thermosonic Gold Wire Bonding of High Speed CMOS," IEEE/CHMT, vol. 14, no. 4, December 1991, July-August, 1991.
31. Rice, C., "Reliable Bonding of High Lead Count SMDs," Lexmark, *Surface Mount Technology*, February 1992.
32. Davis, G., and P. Douglas, *Technical Report #2* "Metallurgy, Fabrication and use of Aluminum 1% Silicon Bonding Wire," American Fine Wire Corp.
33. Douglas, P., *Technical Report #7* "Bonding Wire: A Practical Guide for the Packaging Engineer," American Fine Wire Corp.
34. Douglas, P., *Technical Report #4* "Metallurgical Fundamentals of Gold Bonding Wire," American Fine Wire Corp.
35. Coxon, Kershner, and McEligot, "Transient Current Capacities of Bond Wires in Hybrid Micro-circuits," *IEEE/CHMT Transactions*, vol. CHMT-9, no. 3, September 1986.

36. Munroe, R. A., and J. O. Honeycutt, private communication to R. A. Christiansen, 1987.
37. Jacob, *Heat Transfer*, Vol. I, 1949 Wiley.
38. Training Staff, Kulicke and Soffa Industries, "Gold Wire Bonding," *Connection Technology*, July 1985.
39. Moore, A. H., "Understanding Ultrasonic Wire Bonding," Semiconductor Equipment Corporation, *Connection Technology*, July 1985.
40. Harman and Albers, "The Ultrasonic Welding Mechanism as Applied to Aluminum and Gold Wire Bonding in Microelectronics," *IEEE Transactions on Parts, Materials and Hybrids*, PHP-13, no. 4, December 1977.
41. Singer, P., "Techniques of Wire Bond Testing," *Semiconductor International*, July 1983.
42. "Measuring Pull Strength of Microelectronic Wire Bonds," American Society for Testing and Materials, Specification F459.
43. Mil-Std-883D, "Test Methods and Procedures for Electronics," Method 2011.7 (Destructive) and 2023.4 (Non-Destructive).
44. "The Destructive Bond Test," NBS Special Publication 400-18.
45. Charles, Romanesko, et al., "Hybrid Wire Bond Testing-Variables Influencing Bond Strength and Reliability," Johns Hopkins University, Applied Physics Laboratory.
46. "The Microelectronics Wire Bond Pull Test–How to Use It, How to Abuse It," *IEEE Transactions* CHMT, vol. CHMT-1, no. 3, September 1978.
47. Bilgutay, Li, and McBrearty, "Development of Non-Destructive Bond Monitoring Techniques for Ultrasonic Bonders," Drexel University, *Ultrasonics*, Vol. 24, November 1986.
48. Dunlap, and Adell, "Statistical Quality Control of Gold Ball/Wedge Bonding," Micra Corp., *Solid State Technology*, September 1987.
49. Charles, and Clatterbaugh, "Ball Bond Shearing—A Complement to the Wire Bond Pull Test," Johns Hopkins University, Applied Physics Laboratory.
50. Fitzsimmons, and Miller, "Brittle Cracks in Al-Si wire by the Ultrasonic Bonding Tool Vibration," *IEEE Transactions on CHMT*, vol. 14, no. 4, December 1991.
51. Vogel, T. R., "Rebonding: Bond on Bond Rework for Ultrasonic Wire Bonding," ASM 5th Electronic Materials and Processing Conferences, August 1992.
52. Malloy, and Kondounaris, "Contract No. F30602-78-C-0310," Hughes and Farrell of Rome Air Development Center (RADC).
53. Burkhart, Art, "Considerations for Choosing Chip-on Board Encapsulants," Dexter Electronic Materials, *Electronics*, September 1985.
54. Wong, C. P., "An Overview of Integrated Circuit Device Encapsulants," AT&T Bell Labs, *Journal of Electronic Packaging*, June 1989, vol. 11, pp. 97-107.
55. "The Seven Sins of Globbing," August 1990, Published by Dexter Electric Materials Company, Industry, California.
56. Benson, A., Senior Editor, "Dispensers Deliver," *Assembly*, April 1991.

8

Tape Automated Bonding Chip on Board and on MCM-D

Tom C. Chung, James Chang,
and Ali Emamjomeh

8.1 INTRODUCTION

Tape automated bonding (TAB) is an integrated circuit (IC) chip-level interconnect technology which is often compared to wire bonding technology.

TAB typically uses movie-film like tape material containing pre-patterned leads with rectangular cross-section (see Figure 8-1) instead of individually assembled round wires to distribute input/output (I/O) signals between IC chip and its next level interconnect which can be a plastic or ceramic package, or a substrate such as printed circuit board (PCB) or high density interconnect (HDI) substrate used in the multichip module (MCM) related applications.

TAB was originally conceived by General Electric in the late 1960s as a technology to replace traditional wire bonding technology. There were few activities and applications of TAB from the 1970s to the early 1980s. (See References 1, 2, and 3 for detailed descriptions of TAB history and introduction.) TAB related activities and applications did not increase to a significant level until the middle 1980s. It was due to several reasons, e.g. Microelectronics and Computer Technology Corporation (MCC), a research consortium located in Austin, Texas, consisting of more than 20 major electronics companies, successfully developed and demonstrated TAB as a viable semiconductor packaging technology in 1987;[4] several key TAB applications such as Honeywell's MCM with TAB,[5-7] NEC's SX super-computers,[8-9] Hewlett-Packard's applications of TAB,[10-13] and DEC's

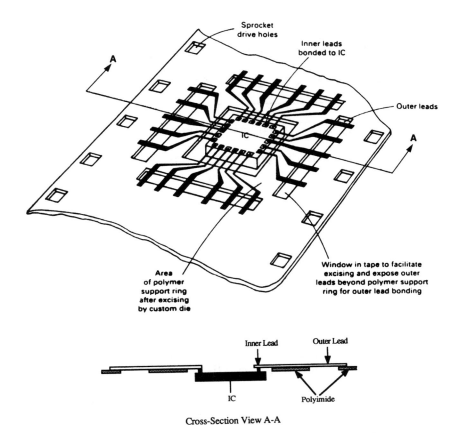

Figure 8-1 Typical TAB tape design.

VAX9000 computers[14-15] were successfully developed and published in the public domain between 1985 and the early 1990s.

TAB contains many key technology elements such as bumping, tape, encapsulation, inner lead bonding (ILB), chip-on-tape (COT) testing and burn-in, outer lead bonding (OLB), etc. The detailed description and discussion of each key TAB technology element can be found in Reference 1. It is not the authors' intention to discuss TAB technology in detail in this chapter. Rather, brief introductions of typical TAB manufacturing process flows are discussed with an emphasis on chip on board related applications. Key considerations of design, materials, assembly, and equipment for TAB on board (TOB) and on MCM-D related applications are discussed in detail. The issues, pros and cons, problems and solutions, and guidelines are provided to assist readers to examine a variety of applications. The examples of both TAB on board and on MCM-D applications are presented and discussed. An example is also used to explain the methodology of how to determine where TAB is a right technology for a COB application. In addition, advantages, disadvantages, and future trends of TAB technology are also discussed.

8.2 CLASSIFICATION OF TAB PACKAGING CONFIGURATIONS

TAB technology can be used to package semiconductor devices in the following three configurations:

- Application of TAB in a ceramic package. A typical manufacturing process flow of this type of TAB application is shown in Figure 8-2 which is rarely developed into volume production and seldom published. One example is the application of TAB in VHSIC (very-high-speed integrated circuits) package developed by MCC.[16]
- Application of TAB on board using pre-encapsulated chips. A typical manufacturing process flow of this type of TAB application is shown in Figure 8-3. One example is the TAPEPAK® developed by National Semiconductor.
- Application of TAB on board using COT devices. Many TAB applications fall into this category which can be further classified into TAB on board and TAB on MCM-D depending upon the selected substrate material and technology. While TAB on board refers to applications using conventional PCB materials and technologies, TAB on MCM-D covers specific MCM applications of COT devices assembled on substrate with thin film deposited copper (or other conductor material) and polyimide (or other dielectric material) interconnect structure. Figure 8-4 represents a typical manufacturing process flow for TAB on MCM-D. Since chip-on-board (COB) using TAB devices is the focus of this chapter, the issues, problems,

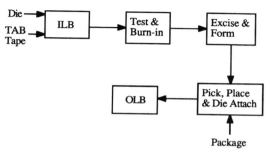

Figure 8-2 Typical manufacturing process flow for TAB in ceramic package.

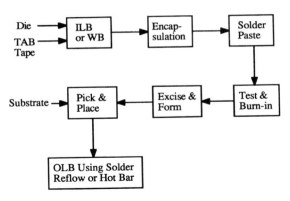

Figure 8-3 Typical COB manufacturing process flow using pre-encapsulated chips.

approaches, guidelines, and examples related to applications of TAB on board and on MCM-D are discussed from aspects of design, material, process and equipment, in the following sections.

8.3 DESIGN CONSIDERATIONS FOR TAB ON BOARD AND ON MCM-D

For the user of TAB on Board or TAB on MCM-D, the TAB "package" has the same meaning as any other package. It is defined to be the outside edge of the OLB polyimide (see Fig. 8-1) from where outer leads extend out. Therefore, the design considerations given here concentrate on TAB as a "package" and do not discuss the "inside" of the package.

In an attempt to reduce TAB packaging cost, various organizations, e.g. EIAJ (EE-13), JEDEC (JC-11), and ASTM (F1.16), have been active in working out standardization of TAB package sizes. Recently, one set of standards has emerged from these organizations[17] as: "Tape Automated Bonding (TAB) Package Family". The standards provide information for package size, carrier size, shipping medium (cups and tubes), tape format, etc., which are defined by a set of pre-determined parameters such as chip lead count, OLB pitch, and test pad pitch.

The specific package design is ultimately dependent upon the user's application, cost, performance and reliability targets. The following represents a generic TAB package design guideline.

Required information: Chip lead count (testable leads); OLB pitch (an assumption is needed to start); Test pad pitch (an assumption is needed to start).

Process: (See Design Flow Chart in Figure 8-5).

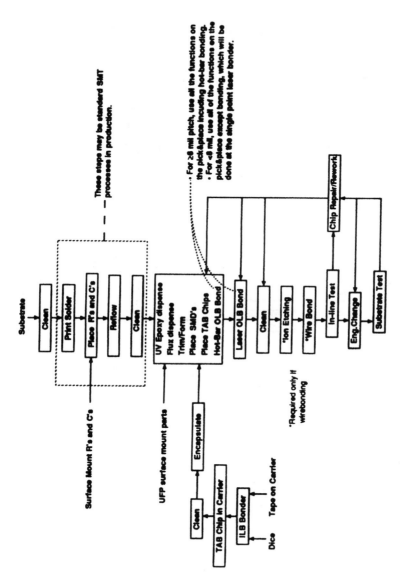

Figure 8-4 Typical COB assembly process flow using COT components.

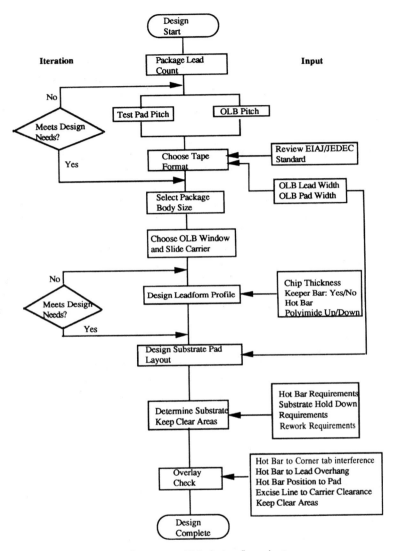

Figure 8-5 TAB design flow chart.

Step 1. Knowing the maximum number of leads that have to be tested and the test socket pad pitch, first determine whether the chip can be tested within the 35 mm, 48 mm or 70 mm tape formats using the JEDEC/EIAJ standard as a guide. Continue if the tape format meets the design and budget requirements.

Since the test/burn-in socket is a substantial investment for a TAB application, package and other layout designs are recommended to conform to any of the available (or soon to be available) sockets.

Example: 300 testable leads with 0.3 mm test pad pitch, OLB pitch of 0.25 mm (approximately 10 mils) requires a 48 mm tape format as designed in the JEDEC standard.

Step 2. In conjunction with the above, use the EIAJ/JEDEC standards as a guide (see Reference 17), look up the maximum number of leads based on the assumed OLB pitch and select the package body size. Continue if the body size meets the design and budget requirements. If not, a custom TAB design must be considered.

Example: The likely package body size for the above example is 26 mm × 26 mm which accommodates up to 326 testable leads.

8.3.1 OLB Window and Leadform Design[15,18]

Figure 8-1 shows a typical OLB window, it is the free-standing copper lead between the package body and the outside polyimide. Its purpose is to allow proper excise and lead forming (if needed) according to users' specification.

EIAJ/JEDEC TAB standards allow 2.25 mm opening in the OLB window for lead forming operation. This value has to be examined in order to determine whether it is adequate for a specific application. The following example explains what are the key geometric dimensions included in the TAB leadform profile and their mathematical relationship.

Example: Calculate the leadform profile based on the desired geometry. Required information:

- Chip or package thickness
- Die attach thickness
- Keeper bar: Yes/No
- Chip face up or down
- Polyimide up, down, or none

Figure 8-6 shows a typical TAB leadform profile. Various components of the leadform are related to each other according to the following equation:

$$L = A + 2(R + t/2)(\theta_1)(2p)/360 + B + C + D \qquad (8\text{-}1)$$

where
$\theta_1 = 90 - \theta_2$
$S = 2[R(1 - \cos(\theta_1)) + t] + (B)(\cos(\theta_2))$
$B = \{S - 2[(R)(1 - \cos(\theta_1) + t)]\}/\cos(\theta_2)$
$W = 2(R)(\sin(\theta_1)) + (B)(\cos(\theta_2))$
$T = A + W + C$

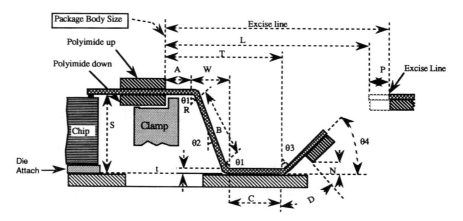

Figure 8-6 Typical TAB leadform profile.

Key dimensions and terminology associated with TAB leadform profile are explained as follows:

L or length of OLB window, is the key dimension which constrains the leadform profile and OLB footprint based on the given information of die thickness, die-attach thickness (if any), clearance between die and board surface, OLB pitch, leadform tooling dimensions, etc.

A or clamp area for form, is needed for proper clamping of the TAB lead prior to excising operation. Its value is important only in the case of polyimide "down" configuration. A typical clamp is shown in Figure 8-6. As it is shown, the clamp has to clear the polyimide, and be structurally robust. Typical values for this dimension range from 0.4-0.5 mm (15-20 mils). As is pointed out, polyimide down configuration is obviously at a disadvantage. In the case of polyimide up, most of the clamp area can be used for other purposes such as longer bond foot, larger stand-off, etc.

R or bend radii. The TAB tape has to be formed with a radius that not only provides the desired geometry, but minimizes stress concentration in the structure of the lead. The lead is basically stretched on its outer fibers and compressed on the inner fibers[18] without cracking the copper lead or any of the plating layers. Depending on the plating on the TAB leads, typical values range from 2.5-5 times the lead thickness.

t or lead thickness–TAB lead thickness in the OLB window.

B or length–the straight length along the incline.

θ_2 or bend angle is determined partially by excise and form tool design and fabrication limits and stress-relieving requirements. Typical values range from 5-20 degrees.

C or bond foot length is determined by the hot bar width and the OLB process and equipment requirements.

D or clamp area for excise. This is needed for proper clamping of the copper leads prior to excise operation. An argument similar to form clamp area can be made about this.

S or stand-off height is a sum of die thickness and die attach bond line thickness or clearance between die and board surface. It is usually a given dimension.

Keeper bar is a section identified for excise of the polyimide, with or without upform.

An iterative design cycle to determine appropriate TAB OLB footprint is explained with the following example:

Step 1. An assumption is made about the chip and die-attach thickness. In the case of no die attach, a clearance may be assumed between the chip and the substrate. A JEDEC standard OLB window length (2.25 mm) is assumed. See Table 8-1 for typical values for the form tool.

Step 2. Based on "typical" values for form dimensions, the package overall dimension is calculated. An important parameter to check is the bond foot length which depends upon the choice of hot bar technology and TAB rework requirements as explained in the following example:

Given: No keeper bar, polyimide down, standoff = S = 0.76 mm (30 mils)
            Standard OLB window = L = 2.25 mm (88.6 mils), θ_2 = 10 degrees
            R = 0.125 mm (5 mils), t = 0.035 mm (1.4 mils)
            Minimum bond foot = C = 0.76 mm (30 mils), needed for a special hot bar type
            Clamp area = A = D = 0.4 mm (16 mils)
Solution: From the given equations, bond foot length, C, is calculated to be 0.54 mm (21.44 mils). Obviously this value does not meet the design requirement. At this point, a number of options may be considered:

Table 8-1   Typical Dimensions of TAB Form

$A = D=$	Polyimide down: 0.4–0.5 mm (16–20 mils)
	Polyimide up: 0.125–0.254 mm (5–10 mils)
$t =$	0.035 mm (1.4 mils)
$R =$	2 (t) to 10 (t) = 70–350 μm (2.8–14 mils)
$\theta2 =$	0 to 20 degrees

(a) Reduce stand-off height: Use a thinner chip/package.
(b) Reduce bend radii: Use smaller angles.
(c) Reduce clamp areas: Use smaller A and D values.
(d) Re-evaluate the minimum bond foot requirement.
(e) Increase L: Use a non-standard OLB window length; a 2.5 mm (98.4 mil) window allows a 0.79 mm (31.3 mil) bond foot.

Step 3. Continue if the leadform geometry meets the design and budget requirements. If the result is not satisfactory, a different value for chip thickness, die attach or the OLB window length, etc. can be tried until the desired geometry is reached. The OLB window length modifications must be reserved until the end to avoid costs associated with custom tooling of TAB tape.

Step 4. Note the overall toe-to-toe is the package dimension if no keeper bar is used as shown in Figure 8-7. If keeper bar is part of the design, the overall package size has to include the keeper bar width (see Figure 8-7).

Typically, for a successful TAB design implementation, the final design geometry can only be accomplished through cooperation between the design team, process team, TAB tape vendor, supplier of chip-on-tape, and the outer lead bond equipment and tooling manufacturers.

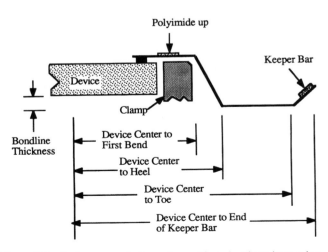

Figure 8-7   Package overall dimensions with and without keeper bar.

8.3.2 Keeper Bar Design

A keeper bar as shown in Figure 8-6 is a polyimide strip that is cut out of the TAB tape during the excise and leadform operation and becomes part of the TAB package. It is often included in the tape design when the OLB pitch shrinks less than 0.4 mm (15 mils). An "upform" shown in Figure 8-7 is used if polyimide is "down". This is done to clear the polyimide from the substrate and allow clearance for flux cleaning operation. The "upform" may not be necessary for polyimide "up" configuration.

The keeper bar width is typically determined based on its thickness. Polyimide thickness typically ranges from 50 to 125 µm (2-5 mils). Keeper bar width is recommended to be two to three times its thickness to ensure adequate rigidity in the structure.

The keeper bar length is typically determined by the position of the last lead. Figure 8-8 shows that the keeper bar extends 0.25-0.4 mm (10-15 mils) beyond the last lead to adequately allow for various tolerance stack.

A keeper bar relief as shown in Figure 8-8 can be designed into the TAB tape in order to facilitate manufacturing of the excise and form die set and provide one straight and clean cut on each side of the tape. The relief area is typically extended 125-250 µm (5-10 mils) beyond the desired keeper bar width in order to allow for tolerance stack.

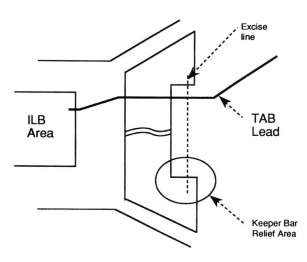

Figure 8-8   Keeper bar length and relief area.

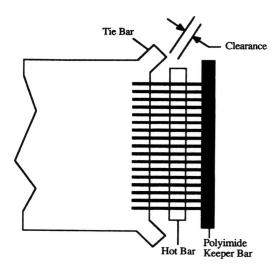

Figure 8-9   Corner tie bar to hot bar clearance.

8.3.3 Corner Tie Bar to Hot Bar Clearance

The corner tie bars shown in Figure 8-9 must be cut during the excise and leadform operation. The cut line is determined by the design requirements as well as OLB equipment and substrate layout rules.

An overlay to check possible interference is recommended. Clearance of 0.25-0.4 mm (10-15 mils) between the corner bar cut line and closest hot bar corner is recommended.

8.3.4 Hot Bar Selection[19-22]

As mentioned previously, different hot bar or heater bar technologies demand different geometrical limits. Here are a few examples of how the hot bar width and the bond foot varied depending on different hot bar technology:

Folded Metal: Hot bar is made from a "sheet" of metal which is formed and folded over to the desired geometry (see Fig. 8-10a). The hot bar width in the folded region in contact with the bond foot cannot be less than 0.41-0.5 mm (15-20 mils) due to the limits of forming operation. Therefore, this limit in addition to the pick and place equipment tolerances dictate the bond foot length when this type of hot bar is used.

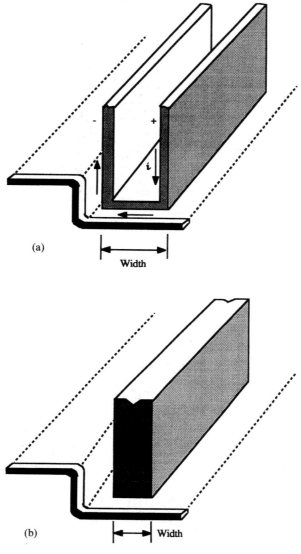

Figure 8-10   (a) Typical folded hot bar. (b) Typical solid hot bar.

Solid Metal: Hot bar is made from a solid "sheet" of metal with specified thickness cut to the desired geometry by Electric Discharge Machining (EDM) (see Fig. 8-10b). The advantage is that the hot bar can be made in virtually any thickness of the sheet metal material. The EDM process is independent of the material thickness. Therefore, shorter bond foot can be achieved using narrower hot bars, ranging from 0.125 to 0.25 mm (5-10 mils).

Solid Ceramic or Ceramic Coated Metal: Hot bar is made from a ceramic composite or an overlay coating on the metal. The manufacturing process for this type of hot bars cannot tightly control the ceramic geometry. Therefore, the bond foot must be made longer to accommodate the tolerances (similar to the folded metal type).

Hot bar length depends on the position of the last lead on each side of the TAB footprint. It is typically extended 0.6-0.7 mm (25-30 mils) beyond contact to the last lead in order to allow for positional and thermal tolerances. An interference check with the corner tie bars must be done before finalizing the hot bar design.

8.3.5 Hot Bar Location

Figure 8-11 shows a cross-section view of the outer lead, hot bar and the substrate pad. The hot bar is typically designed to be 0.4-0.7 mm (15-30 mils) away from the inside edge of the substrate pad. The hot bar to pad (inside) edge must be used in an iterative fashion with the clearance to corner tie bars and lower bend radius in mind to avoid any interference.

Hot bar blade must be placed on the bond foot with proper clearances to the inside lead bend radius, and to the lead toe (or the keeper bar if used). Added to the tolerance stack are the component placement and hot bar head manufacturing tolerances.

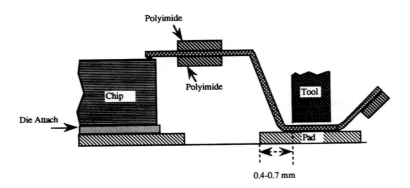

Figure 8-11   Hot bar position.

8.3.6 Carrier and Test/Burn-in Socket Selection

Although TAB component may be used in reel-to-reel or carrier formats, we limit our discussion in this section to TAB packages that are supplied in carriers.

A TAB component in a carrier can be considered as a singulated package in the family of fine pitch surface mounted packages, therefore it can be burned in and tested similarly.

EIAJ/JEDEC have suggested standardized carrier and test socket sizes, all in metric units, see Reference 17. The standards have fixed test pad row centerlines, with varying test pad pitch. Therefore an advance in test socket technology that reduces pad pitch will not affect the carrier selection and excise and leadform tooling. This minimizes the cost impact caused by the need for various sizes of test socket/carrier combinations.

Carriers are typically designed to handle "floating TAB tape". This feature allows the tape to move slightly while it is held in the carrier. TAB tape has to be located in various operations, e.g., ILB, test, excise and form, etc., through the sprocket holes. The coarsely "floating" design allows the carrier to be registered, then TAB tape can be fine registered by engaging locating pins through sprocket holes, allowing it to "float" and self-align.

The outer lead excise line must be checked to ensure proper clearance from the carrier inside opening. Depending on the tooling manufacturers, a clearance of 1 mm (40 mils) must be maintained whenever possible.

8.3.7 Substrate Pad Design

Substrate pad length and width (shown in Figures 8-6 and 8-21) are primarily functions of:

Pad length
  Bond foot length (hot bar length + tolerance)
  Hot bar clearance to inner pad edge
  Hot bar clearance to outer pad edge
Pad width
  Substrate pad pitch
  Outer lead width
  Minimum solder thickness

Lead and pad widths are dependent upon the user's design requirements, equipment capability and reliability needs. Table 8-2 shows a guideline for choosing the lead and pad width as a function of outer lead bond pitch:

Table 8-2 Guideline for Choosing the Lead and Pad Width vs. OLB Pitch

OLB pitch (mm)	0.5	0.4	0.3	0.2	0.1
Lead width (mm)	0.260	0.200	0.125	0.750	0.050
Pad width (mm)	0.325	0.250	0.200	0.125	0.055
Lead-to-pad ratio	0.80	0.80	0.62	0.60	0.90

The goal is to provide a reliable solder joint after all equipment, material, and process tolerances have been considered. For a wide process window, it is desirable that a TAB lead be as narrow as possible without sacrificing the rigidity and stability of the leads, while making the substrate pad as wide as possible without causing problems such as solder bridging, violation of substrate design rules, etc.

8.3.8 Substrate Keep Clear Areas

It is important to review outer lead bond assembly process requirements during the substrate layout design cycle.

Hot bar clear area–It is often necessary to use a device to flatten the substrate and hold it during the hot bar soldering process. This is especially true with FR-4 based PCBs that have components on both sides pre-soldered (through-hole and surface mounted technologies). Depending on the equipment vendor, the hold down device can be a rigid plate incorporated into the board support mechanism or can be a part of the hot bar head. Regardless of the design, OLB areas on substrate may need to be large enough to accommodate such a device to perform its function.

Rework clear area–This applies particularly to components with die-attach adhesive. It is often necessary to heat up the device and rotate it in order to break it from the adhesive. A designer must take the added swept area due to the rotation into consideration during the substrate layout. The TAB rework considerations are discussed in detail in the latter part of this chapter.

8.4 MATERIAL CONSIDERATIONS FOR TAB ON BOARD AND ON MCM-D

The important materials used in applications of TAB on board and TAB on MCM-D are substrate, TAB tape (in COT components), die attach material, flux, anisotropic conductive adhesive or epoxy, solder, and flux cleaning solvent. The key considerations of these materials are discussed in the following sections.

8.4.1 Substrate

PCB is the typical substrate technology used in TOB applications. The typical PCB is constructed with several layers of materials. Each layer is made of copper circuit pattern with epoxy resin and reinforcement material such as glass. The typical resin materials are flame-retardant epoxy (FR-4), high-performance epoxy, bismalimide triazine epoxy, cyanate ester, polyimide, and fluoropolymer.[23] Glass is the other type of substrate material used for TOB applications. This type of substrate material is typically used in the display industry.[24-25]

For applications of TAB on MCM-D, alumina, aluminum nitride (AlN), glass ceramic, beryllium oxide, silicon carbide, silicon, copper, and aluminum are the typical substrate base materials.[23,26-27] Silicon is relatively brittle with a low Young's modulus, beryllium is somewhat poisonous, alumina has a relatively low thermal conductivity and silicon carbide has a low dielectric strength. Aluminum nitride has a good thermal conductivity, a fairly close match to silicon's coefficient of thermal expansion, a fairly high Young's modulus with enough rigidity to minimize substrate warpage, a reasonably low dielectric constant; however, it also has a relatively high cost. See Table 8-3 for physical properties of various substrate materials. Most of the MCM-D substrates are built on the base material with thin film deposited metal conductors and dielectric material. The typical dielectric materials are polyimide, RO2800 (by Rogers), Teflon, benzocyclobutene, aliphatic resin, and silicon dioxide (SiO_2).[23,26-27] It is important to select the right substrate technology and material for a specific

Table 8-3 Physical Properties of Typical Substrate Materials

Substrate Material	Young's Modulus (Mpsi)	Coefficient of Thermal Expansion (ppm/°C)	Thermal Conductivity (W/m °K)	Dielectric Constant (Relative)	Dielectric Strength (kV/cm)
Silicon	29	2.5–3.5	120–200	12	100
SiC	83	2.5–3.8	150–270	40	0.7
Alumina	52	5.0–6.5	20–25	9.8	100
AlN	39	2.7–4.3	140–220	8.7	155
BeO	45	6.4–7.5	250–280	6.6	100

Table 8-4 MCM-D Substrate Technology

Manufacturer	Base Material	Dielectric Material	Conductor Material
DEC	alumina, aluminum, copper alloy	polyimide	copper
Hughes	alumina, silicon	polyimide	aluminum
MCC	alumina	polyimide	copper
NEC	alumina	polyimide	copper, gold
NTK	alumina, AlN	polyimide	copper, aluminum
NTT	alumina	polyimide	copper, aluminum
nChip	silicon	silicon dioxide	aluminum
Rogers	copper alloy	RO2800	copper

MCM application in order to meet specific application requirements. Table 8-4 shows the substrate material system used by different MCM-D vendors.[23,26-27] The detailed description and discussion of each vendor's substrate technology may be found in References 23, 26, and 27.

The most common conductor options on substrate are copper and aluminum.[28-30] A few companies use gold because it resists corrosion and has high electrical conductivity, but it has poor adhesion to polymer dielectrics, as well as being expensive. Aluminum has lower electrical conductivity than copper or gold, which is important because signal delays are directly proportional to the signal line resistance, but it does have good adhesion to polymers. Copper needs a barrier and adhesion layer of chrome, nickel, titanium or titanium-tungsten when used with polymer dielectrics such as polyimides. But copper has high electrical conductivity, good solderability, it can be chemically plated and it is inexpensive. The common dielectric options are various generations of polyimide.[28-31] The major drawbacks to the use of earlier polyimides were moisture absorption and a propensity to corrode (oxidize) copper. Third generation polyimides have reached dielectric constants of 2.9, CTEs of 3 ppm per °C and a moisture uptake of 0.5%.

The OLB bond pads on the substrate for PCB and MCM-D are typically copper pads plated with Au, Sn, or solder. For an Au-plated substrate, nickel is commonly used as the barrier material between copper and Au. In general, Au-plated bond pad provides a more reliable and less corrosive contact surface which is important for the demountable OLB. Solder plated OLB requires lower temperature and force during bonding process.

Die-attach pads on the substrate can be made with different surface metallurgies. In general, no die pad plating is needed if die-attach material is not required for the COT assembly. Typically, TAB on board application does not require die-attach material and process while TAB on MCM-D often requires plated die-attach pad especially for application of face-up TAB with requirement of through substrate cooling.[15,32] The most common die-attach pad metallurgies are Au, nickel, and solder. The compatibility of the die-attach material and bond pad surface condition must be taken into consideration to avoid epoxy bleedout.

8.4.2 TAB Tape

TAB tape can be classified as follows:[1,33,34]

1. Single metal layer TAB: 1 metal layer (no dielectric material); 2-layer (1 metal and 1 dielectric); 3-layer (1 metal, 1 adhesive and 1 dielectric)
2. Two metal layer TAB: 1 signal layer, 1 ground plane layer, 1 dielectric (with or without adhesive)
3. Three metal layer TAB: 1 signal layer, 1 ground plane layer and 1 power plane layer, 2 dielectric layers (with or without adhesive)

The most commonly used tape option is single metal layer TAB which offers potential improvement in lead inductance due to TAB lead's shorter length and rectangular cross-section compared to wire bonding. All TAB tape options except one metal layer (no dielectric material) offer capability of electrical pretestability and burn-in before assembly. Two metal layer TAB tape provides controlled impedance, reduced cross-talk between adjacent leads, and makes high-speed testing of the TAB IC functions possible. Three metal layer TAB tape offers all the advantages of two metal layer, as well as using a low inductance power plane layer to reduce simultaneous switching noise. Three metal layer TAB's advantages are especially suitable for IC technologies (like CMOS) with high impedance inputs and large simultaneous switching current spikes on the power supplies. It is important to note that due to the polyimide ring needed for multi-metal tape, the lead length has to be somewhat compromised. Therefore, to achieve the above capabilities one needs to evaluate the advantages of the shortest lead length vs. multi-metal layer tape. In addition, cost is the other key factor which needs to be considered. While three metal layer tape offers potential advantages over other tape options, it also has the highest cost.

Copper is the most common metal used for TAB conductor leads.[35] There are two types of copper: rolled annealed (RA) and electrodeposited (ED).

Each has its own advantages and disadvantages. In selecting which type of copper to use there are often conflicting requirements from various operations demanded. In addition, the choice of copper is often limited by the TAB tape vendor's preference. For example, it is shown that when TAB is used as the final package, such as in MCM-D, the thermal stresses developed during temperature cycling are higher in the copper leads than in the solder joint. This phenomenon is contrary to what was typically found in the surface mount package.[36-38] This signifies that RA copper is a better choice if high thermal stress in the copper lead becomes a concern. Ductility is also required in the thermocompression inner lead bonding process, yet the leads after exposure to high temperature burn-in may lack enough stiffness to stand the excise and form operation. In addition, too much ductility may limit yield in TAB tape manufacturing process.

Grain structure is another difference between the two copper types. RA copper has horizontal grain structure caused by various cold-working operations during manufacturing, ED copper has vertical grain structure produced during electroplating. Horizontal grain structure is potentially more resistant to microcracks during excise and form operation compared to vertical grain structure. But small grain structures in ED copper can eliminate the potential microcracks. On the other hand RA copper has a preferred rolling direction which may result in nonuniform leadform profile, compared to ED copper which is relatively homogeneous. Among other properties important in selection of copper are: low thickness variance, low surface roughness, low pinhole probability, low oxygen content, high tensile and yield strengths.

The two most common dielectric materials used in TAB tapes are Upilex-S® and Kapton®. Upilex-S® offers better flatness due to its high stiffness and closer CTE match to copper, but recent improvements to Kapton® show progress. Again, the choice may be dictated by the TAB tape manufacturer's preference.

The TAB tape ILB and OLB leads can be plated with Au, Sn, or solder. The selection of the ILB and OLB metallurgy depends on user's application requirements such as cost, performance and reliability. Tables 8-5 and 8-6 show the plating material and bonding options for ILB and OLB respectively.[39-41] Thermocompression (TC) gang bonding, thermode solder reflow gang and single point bonding, thermosonic (TS) single point bonding, and laser single point bonding are the common approaches for ILB bonding.

For some applications which do not require low OLB electrical resistance, anisotropic adhesive may be used to replace permanent metallic OLB. Electrically conductive metallic or metallic coated particles (usually gold, nickel, silver, or solder) are filled with epoxy adhesive to provide electrical interconnect in z-axis only between OLB leads and substrate pads.[42-44]

Table 8-5 ILB Plating Material and Bonding Options

Plating Material on Tape at ILB	Plating Material on IC Pads	Bonding Method
Au	Au	gang and single point bonding
Sn	Solder	thermode solder reflow gang and single point bonding
Au	Solder	thermode solder reflow gang and single point bonding
Sn	Au	thermode gang and single point bonding
Solder	Au	thermode solder reflow gang and single point bonding

Table 8-6 OLB Plating Material and Bonding Options

Plating Material on Tape at OLB	Plating Material on Substrate OLB Pads	Bonding Method
Au	Au	single point bonding
Au	Solder	thermode gang and single point bonding
Sn	Au	thermode gang and single point bonding
Sn	Solder	thermode solder reflow gang and single point bonding
Solder	Au	thermode solder reflow gang and single point bonding
Solder	Solder	thermode solder reflow gang and single point bonding

8.4.3 Die-Attach Material

Die-attach material is typically used in MCM-D applications where heat generated by COT device is dissipated through the substrate. The process uses thermally conductive adhesive as the medium to attach the COT device and dissipate the heat through substrate. The typical die-attach materials used to accomplish this function are eutectic solder, polymer, and silver glass.[45-48] The ideal die-attach material must exhibit a low residual stress during and after curing, be highly thermally conductive, and be either electrically conductive or non-conductive depending upon application requirements. In general, eutectic solder exhibits better heat transfer capability but exerts higher residual stress on device. Conductive particle filled polymer is more flexible in attaching and reworking device. Silver glass die-attach material provides high bonding strength and good reliability while higher residual stress can be a concern. The challenge of using polymeric adhesive is the issue of reliability. The selection of appropriate die attach adhesive requires a full understanding of application constraints and material behavior.[49,50] The comparison of polymeric and eutectic die-attachment materials is shown in Table 8-7.

8.4.4 Flux and Cleaning Solvent

Flux is commonly used as the medium of cleaning oxide, and transferring heat[51] during the OLB solder reflow. In general, OLB can be performed with or without flux (fluxless). For applications which require flux during OLB, selection of type of the flux will affect the manufacturing and cleaning process. Table 8-8 shows different types of the fluxes commonly used in the microelectronic industry.

Table 8-7 Comparison of Different Die Attachment Materials

Type of Die Attachment	Pros	Cons	Comments
Polymeric	flexible in material selection & assembly process development	lower thermal conductivity	easier to repair by using proper adhesive material
Eutectic	better thermal conductivity	higher residual stress due to CTE mismatch & high stiffness	well known in military application

Table 8-8 Options of Flux Materials

Flux Type	Description
R	rosin without activator
RMA	mildly activated rosin
RA	fully activated rosin
OA	organic acid flux
SA	synthetic activator

The flux may or may not need to be cleaned after OLB depending upon the flux material. In applications where cleaning is required, several different cleaning solvents are used to accomplish this function. In general, flux cleaning can use chlorofluorocarbons (CFCs) or CFC-free solvent. Since it is now considered that CFC cleaning solvents are environmentally hazardous,[52] the three remaining cleaning choices are:[53-56]

1. No-clean fluxes
2. Semi-aqueous cleaning with solvents (such as terpene or Axarel®) followed by water rinsing
3. Aqueous cleaning by high pressure and high volume sprays or by immersion.

No-clean fluxes would of course be the most elegant cleaning solution, but there is not sufficient understanding of the long-term reliability issues on such type of application at this time on fine-pitch MCMs. Any intention of using conformal coatings on the substrate could have serious long-term problems from lack of cleanliness. On the other hand, aqueous cleaning by high pressure, high volume and high energy impact sprays or by immersion can cause physical damage to delicate COT components.

8.5 ASSEMBLY CONSIDERATIONS FOR TAB ON BOARD AND ON MCM-D

8.5.1 TAB Mount Options

There are two alternatives in mounting the COT device to the substrate, i.e., face-up and face-down or flip-TAB.[57] Face-up TAB can have either a lead form to accommodate for the die thickness or straight leads (also called flat-mount TAB) where a cavity is cut in the substrate instead. Both designs offer the advantage of dissipating the heat from back side of the die through the substrate using thermal vias, pillars, or directly attaching the die to the base substrate. Flip-TAB, however, requires the heat to be conducted through the top of the module which adds to the challenges of an already complicated module assembly. On the other hand, flip-TAB also offers the advantages of having 100% routability under the die, as there is no loss of routing channels for thermal vias. Flip-TAB and flat-mount TAB offer the best electrical advantages brought about by the shortest lead length. Flip-TAB and flat-mount TAB also offer the smallest footprint due to their short lead lengths. A disadvantage of flat-mount TAB is that the cavity cut into the dielectric layers needed for direct connection of die to substrate significantly reduces signal trace routing and increases substrate cost.

8.5.2 Assembly Process Overview

The typical assembly process for COB using COT devices is shown in Figure 8-4. This assembly process can be further broken into several key areas such as die-attach dispensing, excise and form of COT, flux dispensing, component pick and place, outer lead bonding, cleaning, inspection, rework and repair. The use of die-attach material depends on the application requirements. In general, die-attach adhesive provides through-substrate heat dissipation capability. In most TOB applications, die attachment may not be needed. In the case where a fluxless process is used such as laser bonding in an inert environment or gang bonding with hot melted conductive adhesive, flux dispensing and cleaning processes can be eliminated.

The necessity of applying die-attach material on the substrate depends on the module design. The substrate using through-substrate thermal management technology requires a good thermally conductive die-attach material. The die-attach material can be applied on the substrate with printing, plating, stamping, or dispensing. Printing and stamping of die-attach pads give a higher throughput while dispensing provides a more flexible solution.

Prior to placement and bonding of COT, tape is excised and formed (E&F'd) to a proper configuration with precision E&F tooling. In general, the E&F'd COT device can be made in either gull-wing or straight lead configuration. The advantage of the straight lead tape gives a smaller footprint and lower tooling cost (no forming required). The disadvantage is higher residual stress on ILB and OLB under severe thermal environment due to CTE (coefficient of thermal expansion) mismatch. The good excise and form operation is important for the success of OLB assembly process. Keeper bars as shown in Figure 8-7 are commonly used for fine pitch (less than 15 mils) gull-wing face-up COTs. Leads extending from the support ring are formed down and outward to provide a downset of the lead frame and a mounting foot for OLB attachment. In order to reduce the OLB footprint on substrate and ease for cleaning, a "kicked-up" keeper bar is often used for fine-pitch MCM-D applications. The support polyimide ring sits on either top or bottom side of the leads.

An E&F'd COT device is 3-D in nature. The important criteria for evaluating quality of E&F'd COTs are: lead pitch, lead profile, lead coplanarity, lead skew, etc. The critical dimensions of E&F'd COTs as shown in Figure 8-7 can be used to verify capability of the E&F tooling and feeders as well as quality of E&F'd COTs. The dimensions of the E&F'd COT profile can be measured with tool microscope, optical comparator or other inspection system. Some inspection system can be built in with bonder to perform the inspection process before placement and bonding of E&F'd COTS.

Typical defects associated with E&F'd COTs are: (1) lead dislocation (skew), (2) burrs and contaminations, (3) coplanarity, (4) tooling marks, (5) lifted ILBs, and (6) uneven heel bend radius. Table 8-9 explains the causes for these defects.

Table 8-9   Typical E&F'd Defects and Their Causes

Defects	Causes
Lead skew	inaccurate tooling design and manufacturing
	excessive E/F stroke
	foreign substances or debris
	quality of the tape (copper, plating, support ring, and dimensions)
	improper encapsulation
Burrs and contamination	inaccurate tooling design and manufacturing
	excessive E/F stroke
	quality of the tape (copper, plating, and support ring materials)
	foreign substances and debris
Lead coplanarity problem	inaccurate tooling design and manufacturing
	foreign substances or debris
	quality of the tape (copper, plating, and support ring materials)
	improper encapsulation
Tool marks	inaccurate tooling design and manufacturing
	quality of the tape (copper, plating, and support ring materials, dimensions)
	improper encapsulation
Lifted ILBs	poor ILB quality
	inaccurate tooling design and manufacturing
	foreign substances or debris
	quality of the tape (copper, plating, and support ring materials, dimensions)
	improper or no encapsulation
	speed of E/F stroke
Uneven heel bend radius	inaccurate tooling design and manufacturing
	foreign substances or debris left
	quality of the tape (copper, plating, and support ring materials, dimensions)

The E&F'd COT device is typically picked up by a compliant vacuum cup, which is mounted on a quill co-axially assembled with the center of the thermode or hot bar assembly. For some applications, the vacuum cup with elastomeric O-ring material is used to provide compliant surface and stronger holding force on an unencapsulated device. In general, control of the Z-coordinates of tip of the vacuum cup and bonding surface of the thermode blades plays an important role in the final die-attach bondline thickness.

The next step is to place the E&F'd COT device onto substrate. A successful pick and place of the E&F'd COT device requires consistent and accurate excising and forming, thermode blades and vacuum tip set-up. Global and local fiducials are required to locate the reference point (corner or center) on substrate. Several types of fiducial shapes such as circle and bow-tie are frequently used.[58] In general, circle fiducials may be more convenient for an automatic vision system while bow-tie fiducials are usually used in manual placement systems. For an automatic OLB system, coordinates of all components need to be entered into the placement system's database. After locating the reference point, e.g. center of the substrate, coordinates of all the components on the substrate can be accordingly identified.

For an automatic OLB system, it is desirable to move the E&F'd COT device to pass a platform (inside the system) which is usually equipped with an up-looking camera to examine (right before placement) images of leads to determine the center of the device. The center of the COT is then placed onto the substrate and matches with the center of the OLB footprint. This approach of determining the center of the device before placement is considered to be a reliable way to pick, place and bond ultra-fine pitch devices. The local fiducials are also used to locate accurately the center of the bond site for placement. The procedure of locating local fiducials is similar to the one for locating global fiducials.

After the E&F'd device is placed, OLB is performed. Although OLB technology options are compared and discussed in detail in the following section, two OLB methods, i.e. thermode (or hot bar) gang bonding and single point laser bonding, are briefly reviewed and discussed. In general, thermode gang bonding is the most popular TAB OLB method and single point laser bonding offers promising features for ultra-fine pitch OLB applications. This is because single point laser bonding has demonstrated its flexibility and capability in MCM-D applications with 8 mils or less OLB pitch[32,59] as shown in Figures 8-12, 8-13, and 8-14 while thermode gang bonding is usually used in applications with more than 8 mils OLB pitch as shown in Figure 8-15.

For thermode gang bonding, there is an obvious need for uniform temperature control and the minimization of the heating current's voltage

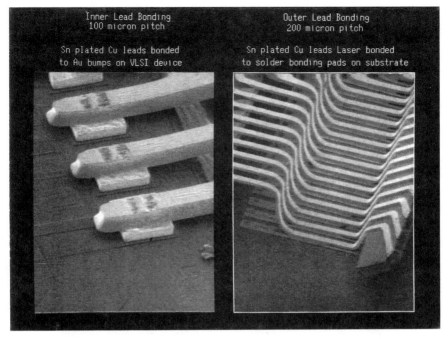

Figure 8-12   Example of single point laser ILB and OLB with an OLB pitch of 8 mils.

drop along the length of the thermode blade to insure the proper reflow of solder for each lead. This is necessary not only because of temperature drop along the length of the thermode blade, but also because of the different thermal loads on different leads due to their substrate connections. Most of the present thermode blade designs have a blade extension equivalent to 4 or 5 lead pitches beyond each corner lead of a component to minimize the cooling effects associated with the thermode corners. The problem with different thermal loads on different leads is a much more difficult problem to solve. Furthermore, the temperatures of different blades for bonding each side of the same component can have a significant difference from the intended uniform bonding temperature. This suggests that a practical thermode design should monitor and correct the temperature of each blade individually rather than feedback from one blade alone. In general, $+/-10°C$ or less temperature variation along each thermode blade is desirable to achieve good solder joints. Another major parameter that needs close control in thermode gang bonding is the downward force of each thermode blade on each side of the COT device. Not only must these forces be perfectly perpendicular to the leads on all sides of a potentially sagging substrate, but the forces need to be kept uniform on all sides of the

Figure 8-13   Example of single point laser OLB with a pitch of 4 mils.

component. This issue can be more challenging for TOB applications where board warpage is more severe than substrates used in MCM-D applications. What this means is that a practical thermode design should monitor and correct the direction and magnitude of the force vector applied to each blade of a thermode.

For single point laser bonding, the approach is to hold down individual TAB leads onto the bond pad one by one by a bonding tool as laser energy is delivered from the side to reflow the OLB joint.[59] The power density and duration of the laser are important factors to achieve good and reliable joints. Accurate placement of a good E&F'd COT device is a "must" for success of laser bonding. Without good placement, the insufficient solder reflow and burn marks can be easily made.

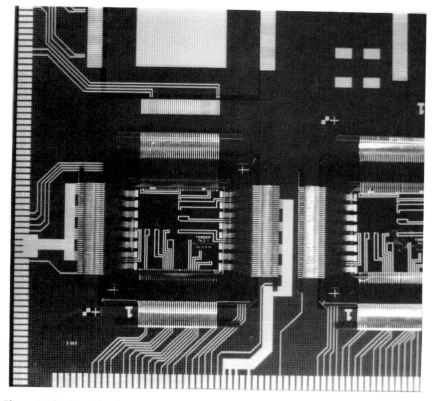

Figure 8-14   Example of single point laser OLB on MCM-D substrate with a pitch of 8 mils.

8.5.3 Outer Lead Bonding Options

Outer lead bonding can be either permanent or demountable.[1,13] The permanent OLB generally refers to conventional metallurgical bonds and the demountable OLB refers to outer lead bonds using demountable mechanical contact force to form interconnect between TAB leads and substrate pads. Generally speaking, the permanent OLB is more commonly used in TOB and on MCM-D while the demountable OLB technology provides the advantage of easy rework. The comparison of permanent and demountable OLBs is shown in Table 8-10.

The permanent OLBs are typically made by gang thermode (or hot bar bonding) or single point bonding. The gang thermode bonding has been used extensively at lead pitches of more than 8 mils. It becomes much more challenging in developing repeatable and reliable gang bonding processes at lead pitches of 8 mils or less. Using a single point bonder for outer lead

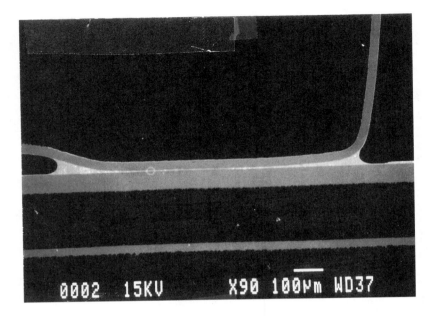

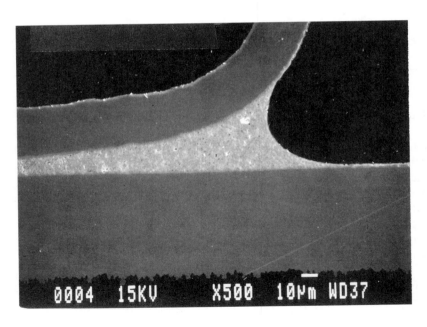

Figure 8-15   Example of gang OLB on PCB with a pitch of 10 mils.

Table 8-10 Comparison of Permanent and Demountable OLBs

Bonding and Material Options	Pros	Cons	Comments
Permanent OLB with solder-solder, solder-Sn, solder-Au	can be gang or single point, less expensive, easy to rework	in general, requires cleaning after bonding	well-known bonding material system in SMT applications
Permanent OLB with Au-Au	less (or no) cleaning required	requires higher local energy (thermosonic/ thermocompression), more expensive	can compensate for coplanarity problem with single point bonding
Permanent OLB with anisotropic conductive adhesive or epoxy	easy for rework, less expensive, gang bonding	higher contact resistance	reliability concern for higher operation temperature
Demountable OLB	easy for rework, demountable	requires clamp mechanism	minimum production experience & data

bonding offers advantages such as no need for oxide and residue cleaning of thermode blades, no need of expensive thermodes for each TAB footprint, more consistent and stronger OLB bond strength, etc. On the other hand, single point bonding also greatly slows down the outer lead bonding process for high pin count ICs. The total OLB thermode gang bonding cycle time typically takes about 20-30 seconds including excise and form, automatic alignment and solder reflow, regardless of lead count. Typical automated single point bonding cycle time is about 0.2-2 seconds[59] per lead for soldering only, excluding time for excise and form, etc. Single point bonding is also more forgiving in handling warped substrates and achieving uniform bond strength for fine pitch bonding applications. The comparison of gang and single point bondings is shown in Table 8-11. An example of single point laser OLB with a pitch of 4 mils is also shown in Figure 8-13. Preliminary pull test results of these laser bonds indicate that lead break failure mode can be achieved on all the bonds using optimized material system and bonding process.[32] For example, on 7 mils wide leads, the average strength was 145 grams with standard deviation of less than 10% where all leads broke during the pull test. The selection of one process over the other must also take OLB metallurgy into consideration. Thermode and single point laser are primarily used for tin, solder or gold/tin reflow

Table 8-11 Comparison of OLB Methods

Bonding Method	Pros	Cons	Comments
Gang	high throughput	less flexible, high equipment & tooling cost	suitable for simple footprint COT application
Single point	flexible, low equipment & tooling cost	low throughput	suitable for low volume & high mix applications, ideal method for ultra-fine pitch OLB

bonding, while single point thermocompression and thermosonic bonding are applied in the gold/gold metallurgical system.

Force Application Mechanism of Gang Thermode Bonding

Figure 8-16 shows one example of thermode used for heating the solder joints as well as applying the force on bonding leads. This thermode is designed in such a way that each set of thermode blades only fit in a particular TAB tape footprint. Four or two sides of thermode blades are commonly used during the bonding process. In this case, these thermode

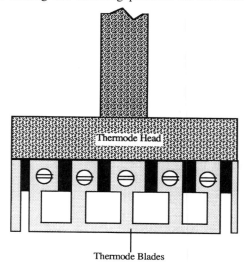

Figure 8-16 Example of OLB thermode.

blades are tightened to the metal block with mechanical screws. The force is applied from the machine to metal block to thermode blades and vacuum pick-up cup. Figure 8-17 shows the example of the operating sequence and force applied mechanism during an OLB process. The typical force application sequence for this thermode is:

1. Set-up a constant Z-axis spacing between the vacuum cup and thermode blades to allow the thermode to contact device surface and bond leads simultaneously.
2. Move thermode down until the back side of the die contacts the die-attach adhesive (if die-attach materials is used).
3. Lower thermode down with a preset bond force and temperature profiles.
4. Press die-attach adhesive until leads touch outer lead bonding pads.
5. Complete solder reflow, thermode cool down and vacuum shut off.
6. Lift the thermode.

Hot bar heads can incorporate a gimbaling feature. The gimbaling mechanism of the thermode provides a capability to overcome the minor coplanarity problem between the blades and the substrate surface. One of the commercially available gimbaling head designs allows the thermode head to gimbal simultaneously with four (or two) blades in the same direction. The problem of this gimbaling head design is that it may cause leads to shift during outer lead bonding. In addition, it is difficult for this type of gimbaling mechanism to accommodate a wide range of coplanarity existing between thermode blades and substrate pad surface. Other kinds of gimbaling thermodes can provide the gimbaling function on each blade individually. The comparison of different thermode gimbaling options is shown in Table 8-12.

Bonding Parameters

After successfully selecting a thermode, bonding parameters must be fully characterized. The key bonding parameters are: bonding temperatures/dwell time, bonding forces/dwell time, temperature and force ramp rates, stage temperature, flux activation temperature/dwell time, and amount and type of flux material.[58-60] Figure 8-18 shows the typical bonding temperature (on thermode blades) profiles and associated dwell times. The typical thermode blade heating/cooling sequence is described as follows:

1. Heat thermode blades to a standby temperature (T1).
2. Ramp up thermode blades to a flux activated temperature (T2).

Step 1 - Thermode Moves Down Before Contacting Die Attach Adhesive

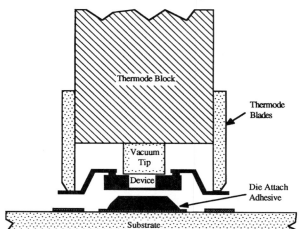

Step 2 - Thermode Moves Down after Contacting Die Attach Adhesive

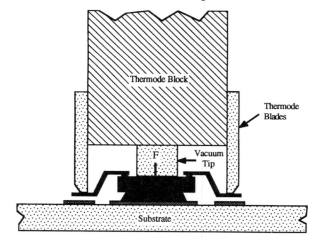

Figure 8-17   Example of thermode operation sequence and force applied mechanism.

Step 3 - Thermode Blades Apply Force on Bonding Pads and Vacuum Tip
Apply Force on Die Attach Adhesive

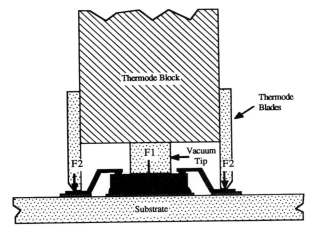

Step 4 - Thermode Blades Apply Force on Bonding Pads Only

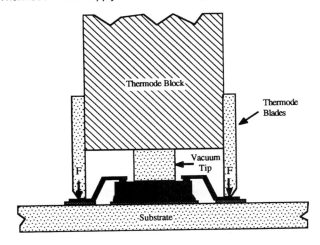

Figure 8-17   (Continued)

Table 8-12 Comparison of Thermode Gimbaling Options

Gimbaling Option	Pros	Cons	Others
No gimbaling	minimizes the lead shift due to gimbaling of the thermode head and blades	difficult to compensate for lack of coplanarity (especially for rigid substrate)	simple design
Thermode head gimbaling	can compensate minor coplanarity problem	lead shift caused by rotation of the head and blades	complicated design
Thermode blade gimbaling individually	can compensate more complicated coplanarity problem	lead shift caused by rotation of blades	complicated design

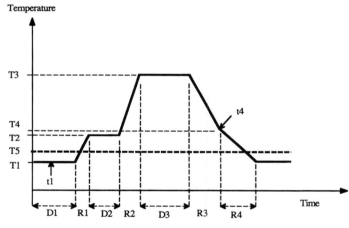

where

T1: standby temperature (°C)
D1: standby time (sec)
R1: ramp-up rate from standby to flux activated temperature (°C/sec)
T2: flux activated temperature (°C)
D2: flux activated dwell time (sec)
R2: ramp-up rate from flux activation to bonding temperature (°C/sec)
T3: bonding temperature (°C)
D3: bonding dwell time (sec)
R3: cool-down rate from bonding to thermode retrieve temperature (°C/sec)
T4: thermode retrieve temperature (°C)
R4: cool-down rate after thermode retrieved (°C/sec)
T5: stage temperature (°C)
t1: bonding activated time
t4: thermode retrieve time

Figure 8-18 Typical OLB temperature profile.

3. Hold at the flux activated temperature for a period of time (D2).
4. Ramp up thermode blades to a bonding temperature (T3).
5. Hold at the bonding temperature for a period of time, i.e. bonding dwell time (D3).
6. Cool down thermode blades to a specified temperature (T4) at which thermode head is retrieved.
7. Cool down thermode blades with natural air convection and maintain thermode blades at standby temperature (T1).

Force applied on the OLB leads must be high enough to overcome the coplanarity problem (caused by substrate, device, and thermode blades) without causing damage on substrate and device. The average stress on each bonding lead and compression force on the substrate can be calculated with Equation (8-2). As a rule of thumb, the maximum stress on leads or substrate must be less than the stress which lead or substrate material can withstand without any failure at the highest bonding temperature. This stress is also defined as material failure criterion. The failure criterion, in general, is material and structure dependent. For a brittle material (such as ceramic, glass, or silicon), fracture failure will be the better criterion to use. For a less brittle material (copper, FR4, or aluminum), yield criterion (maximum principal stress, Von Mises, or Tresca)[61] is the more appropriate failure criterion.

$$Sij = Fi/(nAj)(i,j = 1,3) \qquad (8\text{-}2)$$

where Sij = average stress in j-direction on i-plane
Fi = applied force in i-direction
n = number of leads
Aj = force applied area in j-direction

Figure 8-19 shows the typical force profile used simultaneously with the temperature profile to control the solder reflow bonding condition. The sequence of applied bonding force is:

1. Ramp up the force to a bonding force (F2) after thermode blades contact leads.
2. Hold at this bonding force during thermode heat-up, bonding, and cool down.
3. Retrieve thermode and release contact force after thermode blade temperature is below the solder re-solidify temperature (T4 shown in Figure 8-18).

Stage temperature is another important parameter affecting the solder reflow process. A higher stage temperature means a lower thermode

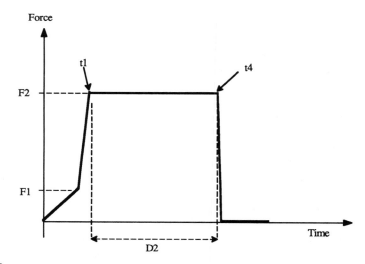

where
  F1: force which the bonding parameters activated (lbs)
  F2: bonding force (lbs)
  D2: bonding dwell time (sec)
  t1: bonding activated time
  t4: thermode retrieve time

Figure 8-19   Typical bonding force profile.

temperature. This is particularly important for substrates with higher thermal conductivity. The constraints of substrate heating (by stage) are driven by thermo-mechanical behavior and properties of materials (e.g., die-attach adhesive, substrate, solder, etc.) contained in the assembly.

Placement Accuracy

The placement accuracy of OLB bonder can be evaluated using different approaches such as: (1) chrome-on-glass placement accuracy and repeatability test, (2) place and bond E&F'd devices without die-attach adhesive, and (3) place and bond E&F'd devices with die-attach adhesive (for application with die-attach adhesive). The chrome-on-glass measurement can provide placement accuracy and repeatability of the machine. In general, die-attach adhesive material can introduce an extra factor which influences the OLB placement accuracy.

Figure 8-20 shows the graphic representation of the alignment accuracy of the component. According to this figure, the movement of the lead tip on the glass coupon to glass substrate can be calculated by the following equations:[62]

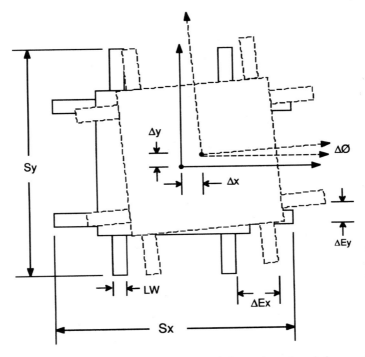

Figure 8-20   Lead tip misalignment due to translation and rotation of placement.

$$\Delta Ex = \text{Abs}(\Delta X) + (Sy/2)\sin(\Delta\theta) \qquad (8\text{-}3)$$

$$\Delta Ey = \text{Abs}(\Delta Y) + (Sx/2)\sin(\Delta\theta) \qquad (8\text{-}4)$$

where ΔEx = lead tip error in X-direction
      ΔEy = lead tip error in Y-direction
      ΔX = translation in X-direction
      ΔY = translation in Y-direction
      $\Delta\theta$ = rotation of component to substrate
      Sx = component lead span in X-direction
      Sy = component lead span in Y-direction

The high percentage (> 75%) of lead-on-bond pad area is important for a reliable solder joint and defect-free OLB. This is particularly true for a fine pitch OLB application. Figure 8-21 shows the substrate pad coverage by lead. The maximum percentage of lead coverage on pad area can be calculated by Equation (8-5).[62]

$$\text{MLOP} = 100 - \{[MLTE - (PW/2 - LW/2)]/LW\} \times 100 \qquad (8\text{-}5)$$

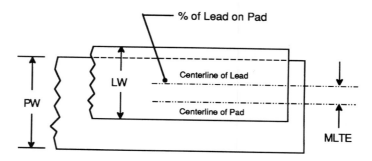

Figure 8-21   Pad coverage by lead.

where   MLOP   = maximum percentage lead on pad
         $MLTE$ = larger value of (ΔEx & ΔEy shown in Equations (8-3) and (8-4)
         LW = lead width
         PW = pad width

The COT device may be placed and bonded onto substrate with or without die-attach material. For applications without die-attach material, the overall placement accuracy is affected by a combination of device (tape manufacture and E&F), substrate, placement, and bonding. For applications with die-attach material, the overall placement accuracy is further affected by the properties of the die-attach adhesive. In addition to the overall placement accuracy, several other factors important for successful solder OLB gang bonding are: substrate and tape metallurgies, substrate materials and surface topology, bonding methods (environment, thermode blades, and bonder), bonding force mechanism, and bonding parameters.

Substrate materials and structure can play important roles in achieving reliable OLB joints. In general, substrates with lower thermal mass or heat capacitance are preferred. The high thermal performance substrate can transfer heat out faster than from thermode to substrate during OLB. This means a lower soldering temperature with the same bonding parameters.

8.5.4 Flux Cleaning

As the size of components utilized in the electronics industry continues to shrink, the problem of contamination and its effect on long-term system reliability becomes more important. Most modern assembly processes utilize flux to enhance the wetting properties of solder, oxide removal, and heat conduction improvement during the bonding process. After soldering flux residues composed primarily of primaric and abietic acids stay with the system. Over a period of time these residues will promote the formation of dendritic growths that can result in electrical and mechanical failures. Therefore, cleaning is necessary to ensure the long-term reliability of modern electronic systems assembled with fine-pitch TAB'd devices.

Cleaning consists of removal of all flux residues and other sources of ionic contamination from surface of substrate assembly. In order to remove these residues effectively, some type of solvent is used to enhance the ability of cleaning process which solubilizes and removes these contaminants. An effective cleaning process consists of three fundamental steps: (1) penetration of solvent into contaminated areas, (2) dissolution of contaminant, and (3) removal of dissolved contaminant from the contaminated ares.[63] There are three major non-CFC cleaning methods employed today. The first method is aqueous cleaning where removal of contaminants is accomplished with either straight water or water mixed with some type of detergent saponifier. This approach is limited by the ability of the water or water/saponifier mixture to solubilize and remove various sources of contamination. The second method is semi-aqueous cleaning where contaminants are dissolved by solvent and then removed by a rinse step followed by some type of drying procedure. The third method is isopropyl alcohol (IPA) cleaning which does not necessarily require any type of rinse due to the evaporation of alcohol at room temperature. The drawback of IPA cleaning lies in the low flash point (53°F). With both aqueous and semi-aqueous cleaning there is a basic problem of wastewater elimination. Most municipalities require wastewater to meet certain regulations before it is drained. Thus, the cleaning process and equipment used must be able either to meet these regulations or provide for some type of treatment of the outgoing water to meet the regulations. These water treatment methods include static separation, closed loop solvent reclamation, distillation and fine mesh filtration. An effective cleaning process can only be achieved with sufficient considerations and evaluations in areas such as solvent selection, cleaning process development, equipment selection, and waste treatment.

8.5.5 Inspection

The bond joints are typically inspected after OLB. The inspection approaches can be classified as non-destructive and destructive. The non-destructive inspections include visual inspection, X-Ray, laser, acoustic inspection, and non-destructive pull test. The destructive inspections include SEM and destructive pull test.[64]

Visual inspection can be used to inspect planar (2D) defects and OLB joint quality. The typical visual solder joint defects are solder bridge (short), solder ball, excess solder, misalignment, and insufficient solder, as shown in

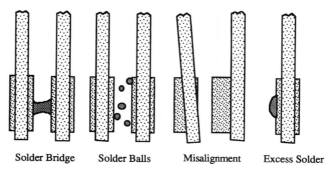

Solder Bridge   Solder Balls   Misalignment   Excess Solder

Figure 8-22 Typical solder OLB defects.

Figure 8-22. In general, the visual defects can be detected with either manual or automatic visual inspection techniques. The typical solder joint fillet (solder on the heel, toe, and between leads) can be inspected with this technique. Due to the relatively low set-up cost and ease of implementation, manual visual inspection is considered as the most popular method for solder joint inspection.

In order to inspect effectively the defects such as voids and hidden opens, nonvisual inspection techniques must be used. These inspection techniques are defined as the methods with measuring media other than conventional optical viewing. The typical techniques are: laser heat injection, X-Ray imaging, laser scanning imaging, scanning beam X-Ray laminography, and acoustic microscopy.[64]

The pull test is usually conducted either non-destructively or destructively. For gull-wing TAB design, the traditional pull test is adequate for ILB characterization but may mislead the results for OLB test (due to the failure mode of break in span). In order to test effectively (test to failure) the solder joint strength, modification of test fixture is needed. Figure 8-23 shows the set-up of this modified pull test fixture.

In Figure 8-23, the force applied on the solder joint is a combination of shear force and peel force. The shear force is related to the actual thermo-mechanical strength of the joint (due to coefficient of thermal expansion mismatch) while peel force is usually for characterization of strength of the interface. The force analysis is shown in Figure 8-24 and Equations (8-6) and (8-7).

$$Fs = F\sin(\theta) \qquad (8\text{-}6)$$

$$Fp = F\cos(\theta) \qquad (8\text{-}7)$$

where Fs = shear force component
       Fp = peel force component

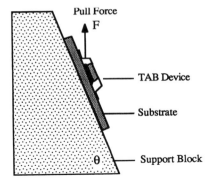

Figure 8-23   Gull-wing TAB OLB pull test set-up.

In order to analyze the effect of the pull force on the solder joints under test, the tape lead shown in Figure 8-24 is assumed as a beam. By beam theory,[65] Figure 8-25 shows the schematic of the force components and assumed boundary conditions. The force components on the OLB solder joint can be calculated with Equations (8-8) and (8-9). Figure 8-26 shows the normalized force vs. the distance of applied forces at $\theta = 70°$.

$$Fs(x) = (1 - x)Fs \qquad (8-8)$$

Figure 8-24   Pull force components

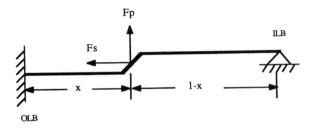

Figure 8-25   Force component and boundary conditions on the tape lead.

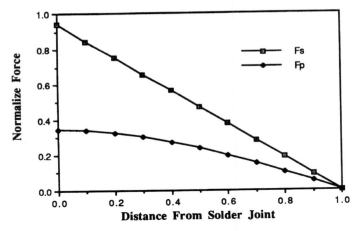

Figure 8-26 Normalized force vs. location of applied force at $\theta = 70°$.

$$Fp(x) = [(2 - 3x^2 + x^3)/2]Fp \qquad (8-9)$$

where $Fs(x)$ = shear force component on the solder joint
$Fp(x)$ = peel force component on the solder joint
x = normalized distance between location of the applied force and OLB solder joint

In order to obtain effectively a pull test result for OLB solder joints, two techniques are important for the test: (1) pull hook placed perpendicular to lead, (2) pull hook must be applied as close to OLB joint as possible to avoid lead span break.

8.5.6 Repair and Rework

Repair and rework are important areas in applications of TOB and TAB on MCM-D especially for applications using fine-pitch or high pin count COT devices. In general, "repair" refers to performing correction of local defects, e.g. lead skew, lead open or short, etc., detected on OLB'd COT devices while "rework" refers to replacement of a defective COT device which has unrepairable defects. In SMT applications, removing defective devices and remounting new devices for solder reflow has been practiced for many years.[66] The typical tools used in device removal are vacuum and mechanical pick-up tools. Some of the current rework equipment vendors and their capabilities are shown in Table 8-13. Companies such as Austin American Technology (ATT), Air-Vac, PACE, Semiconductor Equipment Corp.

Table 8-13   Current Equipment Vendors and Their Capabilities

| Vendor | Solder Reflow Heat Source | | | Device Removal Force | | |
	Hot Air	Focused IR	Thermode	Vacuum	Pick-up Tool	Thermo-plastic
AAT	X		X	X	X	X
Air-Vac	X			X	X	
Conceptronics		X		X	X	
Ismeca		X		X	X	
PACE	X			X	X	
SEC	X			X	X	
Seika	X			X	X	
SRT	X		X	X	X	
Ungar	X			X	X	
Wenesco		X		X	X	

(SEC), Seika, and Ungar utilize hot air to reflow the solder joints, while removing the device with either vacuum or manual pick-up tools. Conceptronics, Ismeca, and Wenesco utilize focused IR to reflow solder while manually removing the device. In general, the main difference between rework on TOB and SMT applications is that COT usually has finer OLB pitch and close device spacing which make rework on TOB more challenging. As for TAB on MCM-D, rework becomes even more difficult due to presence of the die-attach material. In order to achieve a successful TAB rework on MCM-D, the die-attach material must be removed in addition to desoldering of all the OLB joints. AAT located in Austin, Texas, has jointly worked with MCC and successfully developed a TAB rework system which uses a thermoplastic adhesive material reflowed and then solidified on the top surface of the defective die to facilitate removal of the defective COT device.[67,68]

Figure 8-27 shows the typical rework process of fine-pitch TAB'd devices on PCB or MCM-D substrate with die-attach adhesive. This rework process can be broken into three major areas: removal of the defective device, preparation of the substrate for device replacement, and the replacement with a new "known good" device. It is important to monitor the removal process closely to ensure that the substrate is not damaged at any stage of the rework process.

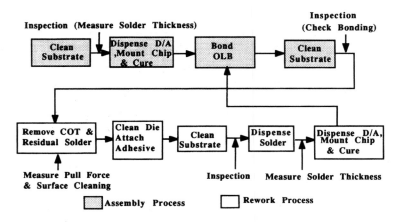

Figure 8-27 Typical rework process with die-attach material.

Removal of Defective COT Device

A globally defective device must be removed from the substrate. A successful removal process involves consideration of several areas such as substrate design, substrate material, adhesive material, and assembly process. The devices can be removed from substrate by means of pull, shear, or combined loads. The substrate heating temperature should be determined based on substrate structure, properties of adhesive and encapsulation materials. The combined action of pulling the die-attached device as well as reflowing/wicking the OLB solder joints can be completed in either one step or several steps.

For some applications, such as face-up TAB on board, heat dissipation through substrate is not needed, therefore, the use of die-attach adhesive is not required. The mechanism and process used for separation of the COT device from the substrate is much simpler for applications without die-attach material. In addition, the load required to pull the device from the substrate will be much less for the case of no die attach. It is because force is only required for the disengagement of the OLB solder joints. A vacuum force in most cases will be sufficient. A torsional force may not be needed for this type of die removal. The die lift motion can be accomplished by a vacuum tip. The typical temperature and force profiles for this type of die removal are shown in Figures 8-28 and 8-29.[67]

It is more challenging to remove the defective COT device with die-attach material from PCB or MCM-D substrate. Forces, temperatures, and times are key parameters to be monitored and controlled in this type of rework process. Combined torsional and tensile loads are typically used during the

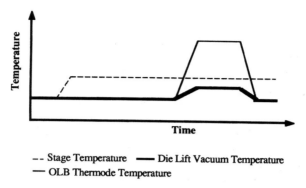

-- Stage Temperature ── Die Lift Vacuum Temperature
── OLB Thermode Temperature

Figure 8-28 Typical temperature profiles for COT removal without die-attach adhesive.

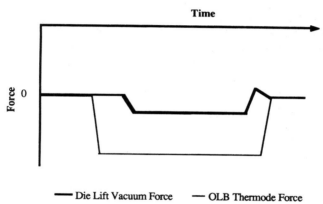

── Die Lift Vacuum Force   ── OLB Thermode Force

Figure 8-29 Typical force profiles for COT removal without die-attach adhesive.

removal process. It is important to make sure that the least amount of polymeric die-attach adhesive residuals is left on the substrate's die pad surface after completion of die removal. The typical temperature and force profiles for this type of die removal are shown in Figures 8-30 and 8-31.[67]

Cleaning

The need for residual die-attach adhesive material removal from the substrate depends on the die-attach adhesive material used and die bond pad surface condition. In general, the use of a thermoset die-attach adhesive for die attachment will require cleaning of remaining residue after the die is pulled from the substrate. If the die-attach adhesive material is thermoplastic, it can usually be reused without removing its residue. Typical

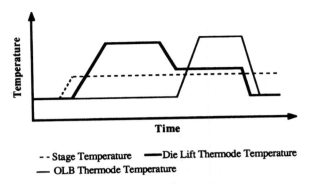

- - Stage Temperature   ——Die Lift Thermode Temperature
— OLB Thermode Temperature

Figure 8-30 Typical temperature profiles for COT removal with die-attach adhesive.

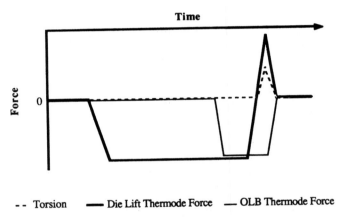

-- Torsion   —— Die Lift Thermode Force   — OLB Thermode Force

Figure 8-31 Typical force profiles for COT removal with die-attach adhesive.

methods for removing thermoset die-attach adhesive residue from the substrate are: mechanical (e.g., scrubbing), chemical (e.g., solvent), and combined mechanical and chemical approaches.

The amount of solvent applied to the adhesive residue must be controlled in order not to overflow to adjacent devices. The overflowing of the solvent may lead to failure of the die-attach adhesive joint on adjacent devices. A sufficient amount of solvent must be applied in order to cover the complete adhesive residual area. The reaction can also be speeded up substantially with the addition of controlled applied heat.

Flux used during removal of the defective device and OLB process along with other residues left on the substrate must be removed. The cleaning process is similar to the one used for OLB cleaning. It is a good practice to remove residual flux immediately after removal of the defective device and die-attach adhesive residues.

Inspection and Test

The cleaned substrate should be inspected to determine the condition of the substrate in order to verify its suitability for remounting a new device. In some cases, electrical testing may be needed to verify this suitability. Visual inspection with a high power microscope is the most common inspection approach. The amount of solder remaining on the OLB pad should be inspected and quantified to ensure the proper amount of solder replenishment. It is particularly important for a very fine pitch OLB. The amount of the solder left after solder wicking can be measured by a surface profilometer.

Replenishment of Solder and Die-Attach Adhesive

After removal of the defective COT device, fresh solder has to be replenished on the solder pads in order to make up the difference for the solder removed during the solder wick-off process especially for fine-pitch device. Typical methods used to replenish the solder are solder-plated tape, solder dispensing on the lead joints,[69] solder plating on the OLB leads, and solder carrier (such as Solder Quik® from Raychem). A comparison of these solder replenishment methods is shown in Table 8-14. One of the key

Table 8-14   Comparison of Solder Replenishment Methods

Method	Pros	Cons	Comments
Solder plated tape	higher throughput, standard OLB process	higher cost for tape and tooling	suitable for high volume production
Solder dispensing on the pads	flexible	higher equipment cost, low throughput	suitable for development and low quantity COT rework
Solder plated on the pads	—	not well defined	more development will be required
Solder carrier	easy to apply	cannot handle ultra fine-pitch application	Raychem patent

advantages of using the solder dispensing approach is that it provides a flexible way of replenishing new solder and the solder dispenser can be also used for die-attach adhesive dispensing. For fine-pitch devices, a well-controlled volume dispenser is required to replenish the correct amount of solder.

Replenishment of new die-attach adhesive material may be required before the new device is remounted. The die-attach adhesive material can be applied in the same fashion as the one used in the original assembly process.

Remounting

The new "known good" device is mounted on the substrate to complete the rework process. The process for mounting the new device is the same as is used in the original assembly process. The reasons are: (1) the functions required for component replacement can be performed by the original OLB system, and (2) the expenses involved in purchasing a second complex system for replacement would be quite high. By using a flux containing solder paste during solder replenishment, fluxing operation is not required for remount process.

Curing of Die-Attach Adhesive Materials

The thermoset die-attach adhesive mateial must be cured before post-bond flux cleaning. The cure temperature and its associated ramp-up rate must be designed in such a way that both thermal stress and shock are minimized. In addition, mechanical and thermal damage to adjacent devices must be avoided. For 63/37 eutectic solder-plated substrates, the maximum curing temperature is typically recommended to be less than 150°C.

Repair

The typical local defects which can be repaired are single lead open, single lead solder bridge, solder balls, etc. In general, a manual process is used for repair of single lead defects. A fine-tip soldering iron and solder wick material are used for repair of solder bridging, open leads, and lead skew/misalignment. For a very fine pitch device, a semi-automatic single point bonder is useful for rebonding open leads (such as single point bonder).

Three most common types of lead solder bridge defects are usually found after OLB bonding. One of the solder bridges is usually located at the heel of the leads, another between the bonded leads at the section pressed by the thermode, while the third is located at the toe area. The causes of these solder bridges are quite complicated. In general, excess solder, uneven thermode forces and bonding temperatures, poor alignment, and poorly excised/formed devices are the key reasons for these defects.

Due to improper bonding temperatures/forces or poor quality of original solder, solder balls may be formed after bonding. The approach for repairing this defect involves both manual and equipment cleaning efforts. The manual cleaning tool (such as a soft brush) is designed in such a way that a "soft" force is applied to the solder balls, without damaging any other device on the substrate.

8.5.7 Assembly Equipment Selection

The success of the assembly process is definitely affected by selection of the right equipment mix to meet assembly requirements. The key equipment required for assembly of COT devices on PCB and MCM-D include fine pitch pick and place equipment (or OLB bonder), rework equipment, flux and/or die-attach dispenser, cleaning equipment, etc. The following list represents typical procedures to investigate, locate, evaluate, and test/accept a complicated assembly equipment.

- Equipment vendor survey
- Define the key components and requirements of the equipment.
- Write equipment specification and acceptance test requirements.
- Establish the equipment selection procedure and criteria.
- Send out request for quote (RFQ) to vendors.
- Review the equipment RFQ specification and acceptance tests with potential vendors.
- Final review of potential equipment vendors.
- Select equipment vendor and place purchase order.
- Conduct equipment acceptance test and training.
- Final acceptance of the equipment.
- Update progress and capability of current equipment vendors.

The equipment vendor survey starts with gathering equipment specifications and technical data sheets from all the potential vendors, followed by discussions with user(s)/customer(s) of the vendors, engineers and experts about their experience and knowledge on the evaluated equipment. Trips to visit the equipment vendors and observe demonstrations of the equipment are important to understand the capability of vendors and equipment in detail. Table 8-15 shows an example of evaluation matrix for fine-pitch OLB bonders.

Table 8-15 Example of Fine Pitch OLB Equipment Vendors Evaluation

Alternatives	A	B	C	D
Placement accuracy	+/ − 0.5 mil	+/ −1.2 mil	+/ −2.0 mil	+/ − 2.0 mil
Bonding method	Two-stage hot bar	Laser	One-stage hot bar	One-stage hot bar
Lead inspection	yes	no	no	no
Maximum board size	24 in. × 24 in.	20 in. × 15 in.	18 in. × 18 in.	10 in. × 10 in.
Minimum pitch	0.008 in.	0.008 in.	0.008 in.	0.008 in.
Max. number of components	6	8	6	1
Flux dispensing	yes	yes	yes	custom
Adhesive dispensing	yes	no	yes	custom
Cycle time (sec)	18	25	20	>60
Approximate cost	$400K	$500K	$300K	$200K

It is important to define the key components and requirements of the equipment, which are usually used as key criteria for evaluation of the surveyed equipment vendors as shown in Table 8-15. The equipment specifications and acceptance test requirements must be written in detail especially for equipment with advanced or new capabilities, e.g., ultra-fine pitch OLB or cleaning, die-attach dispensing, etc. The specifications must be precise enough to cover all the application requirements and at the same time flexible enough to accept any new idea. The acceptance criteria must be capable of verifying the repeatability, accuracy, manufacturability, and reliability of the equipment.

For complicated equipment, the equipment selection procedures and criteria must be established and used as the guideline for selecting the most appropriate equipment. The typical criteria used for fine pitch OLB equipment vendor selection are: excise/form capability, placement accuracy and repeatability before and after OLB, OLB reflow soldering, engineering support, delivery schedule, system cost, etc.

The equipment RFQs are then sent to all the potential vendors for quotations. After reviewing with these vendors, the final RFQ is written and the vendor selection is completed. The overall vendor capability is then rated according to the criteria defined previously. The rating may be weighted using the similar approach described in Section 8.6 depending upon the application requirements. The vendor with the highest score is then selected as the supplier.

8.6 HOW TO DETERMINE IF TAB IS THE RIGHT TECHNOLOGY FOR A COB APPLICATION

It always causes argument when TAB is compared with other chip level interconnect technologies such as wire bonding[69] (WB) and flip-chip or controlled collapse chip connection[70] (C4) for a COB application. The typical advantages of TAB are: TAB can offer better electrical performance than wire bond due to its shorter interconnect leads with rectangular cross-section, TAB'd devices can be pre-tested and burned-in prior to assembly, can be easily automated, and can have finer pitch bonding pads on chip, etc., while its disadvantages are mainly: TAB requires extra wafer processing (bumping), requires extra tooling (ILB and OLB), and it requires large capital investment, etc. The detailed discussions on advantages and disadvantages of TAB can be found in References 1, 2, and 3. It is the authors' opinion that each interconnect technology has its unique advantages and disadvantages. It is very difficult to obtain the optimal packaging technology selection for a COB application without a detailed trade-off analysis among key packaging objectives such as cost, performance and reliability. In this section, instead of repeating statements of TAB's advantages and disadvantages, an example is used to explain the methodology of qualitatively determining if TAB is the right technology for a COB application.

Example: Let us assume the application under consideration has a chip with 500 I/Os and maximum 20 watts of power dissipation. The packaged device is mounted on a conventional PCB. Three candidate interconnect technologies, i.e. wire bonding, TAB, and flip-chip or C4, are evaluated. An evaluation matrix generated using the described methodology is shown in Table 8-16. The first column represents the packaging attributes which are considered "must" by the user. The attributes used in this example are electrical and thermal performance, pre-test and burn-in, cost, repairability, etc., as shown in Table 8-16. Other attributes such as system compatibility, design simulation, etc., can be easily added into this table depending upon application requirements. A scoring system using "2" for yes, "1" for maybe, and "0" for no is applied in this example. The unweighted scores are assigned to each cell in the columns of WB, TAB and C4 by user based on features and capabilities of each technology. The next column, "weight", is optional. In some cases, user may find some attributes such as pre-test and burn-in, and repairability are much more important than others. Then, a higher percent of weight may be assigned to those attributes. The sum of all the cells in this "weight" column must equal 100%. The scores shown in the cells of the last three columns, "weighted WB, TAB and C4", are obtained by multiplying the cells in the "weight", and "unweighted WB, TAB and C4" columns respectively. The total scores, both unweighted and weighted, for each technology are then summed up for each column and show in the bottom row of Table 8-16. The higher the total score the more suitable the technology is for the application under consideration.

The example described in Table 8-16 provides general methodology of selecting the right chip-level interconnect technology for a typical COB application. In order to make a right and successful decision, it is important to distinguish the "must" attributes from the "wants". In addition, the scores and "weights" must be objectively provided and approved by persons with enough packaging and interconnect knowledge and experience.

8.7 EXAMPLES OF TAB APPLICATIONS

Applications of TAB technology can be found in many segments of the industrial, consumer, and military markets[71] such as computers, automotive, telecommunication, consumer electronics, avionics, etc. In general, TAB can be used in either performance-driven computer-related applications[5,6] or cost-driven consumer electronics-related applications.[72] The COT

Table 8-16 Example of Evaluation Matrix for Chip on Board Assembly (Assume: Board Material: FR4; Chip I/O count: 500)

Packaging Attributes	Criteria	WB	TAB	C4	Weight %	Weighted Scores		
						WB	TAB	C4
Electrical performance	75 MHz clock frequency	2	2	2	10	0.2	0.2	0.2
Thermal performance	$\theta_{jc} \leq 1$ C/Watt	1	2	0	15	0.15	0.3	0
Package footprint	≤ 3 in sq.	2	1	2	5	0.1	0.01	0.1
Pre-test & burn-in capability	Required	0	2	2	20	0	0.4	0.4
Assembly cost	≤ 1 Cent/pin	2	1	2	5	0.1	0.01	0.1
Rework	≤ 2 Per site without failure	0	2	2	15	0	0.3	0.3
Infrastructure	Required	2	1	0	5	0.1	0.05	0
Reliability	High	1	2	2	10	0.1	0.2	0.2
Production experience	Required	2	0	0	10	0.2	0	0
Production throughput	Cycle time ≤ 40 seconds	0	2	2	5	0	0.1	0.1
Total		12	15	14	100	0.95	1.65	1.4

• Yes = 2, Maybe = 1, No = 0

devices used in the TAB related applications can be either face-up or flip TAB design. The pros and cons of face-up and flip TAB designs and associated examples are shown in Table 8-17 followed by examples of typical TOB and TAB on MCM-D applications.

8.7.1 TAB on Board

TAB on board (TOB) applications can have metallurgical, anisotropic conductive adhesive or epoxy, or demountable OLB. While metallurgical, conductive adhesive, and epoxy OLBs are considered as the permanent bonding, demountable TAB (DTAB)[13] offers many features which cannot be overlooked. The examples of TOB applications are:

Metallurgical OLB

HP have demonstrated a special TAB application for manufacturing their calculators.[75] The characteristics of this application are:

- Design
  SMT compatible TAB manufacturing line
  Using 35 mm TAB tape
- Material
  Tin plated TAB tape
  Gold plated bumps on IC pads
  Solder paste printed on PCB OLB pads
- Assembly
  Thermode gang bonded ILBs
  Reel-to-reel COT
  High volume TAB'd devices with 20 mils (170 leads) and 22 mils (104 leads) OLB pitches assembled on PCB
  Using plastic clamp to hold down the COT device during mass solder reflow
  Using infrared to reflow OLB solder
  Using trichlorotrifluoroethane and 1,1,1,-trichloroethane as defluxing solvent

The production yield data associated with the application using TAB technology was found to be better than the one using SMT (i.e. 32-mil-pitch plastic quad flat packs with solder-dipped gullwing leads).

Table 8-17 Comparison of Face-up and Flip TAB

	Face-up TAB	Flip TAB
Pros	less complicated mechanical assembly structure	potentially higher heat removal capability, higher silicon to substrate area ratio, better electrical performance
Cons	require die-attach material to provide interface to conduct heat away from back side of die, less thermally efficient due to more thermal interfaces involved, difficult to rework	complicated mechanical assembly structure, tighter mechanical tolerance, require compliant material to provide die surface contact
Examples	DEC's VAX9000 MCM[14-15] (cavity-mounted face-up), –Tandem's MCM[32,73,74] (surface-mounted face-up), –HP's SMT Compatible[75] (TAB)	Siemen's MCM,[76] HP's demountable TAB[13] (DTAB), NEC's SX-2 and SX-3[8-9] (Flip TAB Carrier)

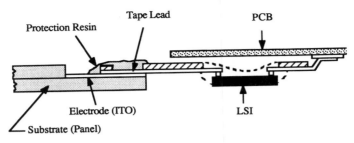

Figure 8-32 Cross-section of LCD assembly structure.

Anisotropic Conductive Adhesive or Epoxy

Anisotropic conductive adhesive and epoxy have been used to joint the TAB OLBs with electrodes which are made of indium tin oxide (ITO) for LCD display-related applications. This anisotropic conductive adhesive or epoxy can be cured either with heat or light. Figure 8-32 shows the typical cross-section of LCD assembly structure[72] where epoxy is used as the primary material to hold interconnect between TAB outer leads and electrodes on substrate. The characteristics of this application are:

- Design
  ITO with OLB pitch of 200 μm or 280 μm
  With asymmetric TAB tape
- Material
  Gold plated tape leads
  With PCB for input leads OLB and liquid crystal display panel for output leads OLB
  Using transferred bump TAB to make ILBs
- Assembly
  Input lead solder bond on PCB
  ITO and output OLB bonded with insulated resin
  Light-cured epoxy to provide pressure contact force to form OLBs

Demountable OLB (Development)

The demountable TAB (DTAB) developed by HP is described in detail in Reference 13. The concept of DTAB involves application of reworkable and demountable OLB technology instead of conventional (permanent) OLB. The key features of DTAB are:

- Design
  A flat-mounted (no leadform) flip-TAB configuration

Bumpless ILB and pressure contact OLB

Alignment of OLBs achieved by precisely drilled holes located on PCB

Area array OLB directly interconnected with pads on PCB with contact pressure

Using composite spring system to provide controlled contact force

Using heat spreader to conduct heat from device to heat sink

- Material

Both 48 mm (432 I/O) and 35 mm (284 I/O) TAB tape designs successfully demonstrated on PCB

- Assembly

Using die-attach material to attach device to heat spreader

Using demountable fixture for board assembly

- Others

Easy to rework

A minimum of tooling required for OLB

Requiring fixture for provision of OLB contact force

Requiring accurately fabricated alignment holes on PCB

Application of DTAB-432 demonstrated with the following features: I/O: 432; footprint: $2'' \times 2'' \times 0.165''$ (w/o heat sink); chip size: 15 mm max.; chip power: 40 watt max.

8.7.2 TAB on MCM-D

The TAB on MCM-D applications have been recently demonstrated by several companies such as DEC,[14-15] Tandem,[32,74,75] and MCC.[77] Figure 8-33 shows a cross-section of a typical face-up TAB on MCM-D assembly including key components such as excised and formed face-up TAB device, HDI with through-substrate thermal pillars, module heat sink, off-substrate connector assembly, and mother board. In TAB on MCM-D related applications, the COT devices are either directly mounted on HDI[32,74,75] as shown in Figure 8-34 or cavity-mounted on substrate as shown in the famous DEC VAX9000 MCM-D module. The characteristics of DEC's TAB on MCM-D application are:

- Design

Substrate size of $4'' \times 4''$

Through substrate air cooling; using face-up tape automated bonding (TAB) technology to assembly ICs on a high density interconnect substrate, with the heatflow path being from the back of the die through die-attach material to the underlying substrate, then through the substrate to a heat sink.

Figure 8-33   Cross-section of typical face-up TAB on MCM-D.

Pin fin heat sink with impingement air cooling to dissipate up to total 300 watts of module heat.
- Material
  Copper polyimide high density interconnect (HDI) supported by aluminum or copper-based metal plate
  Tin-plated TAB tape OLB'd to solder plated pads on HDI substrate
- Assembly
  Face-up COT devices are mounted in the cut out area (or cavity) on the substrate and directly die-attached onto the substrate support plate as shown in Figure 8-35.
  Gang solder OLB bonding with a minimum of 8-mil pitch.

8.8 FUTURE TRENDS OF TAB TECHNOLOGY

The need for an effective packaging and interconnect technology has become overwhelming. The challenges are in bridging the gap between incredibly fine IC geometry (less than a micron) and relatively coarse substrate design rules (5 μm to several mils), and meeting electronics

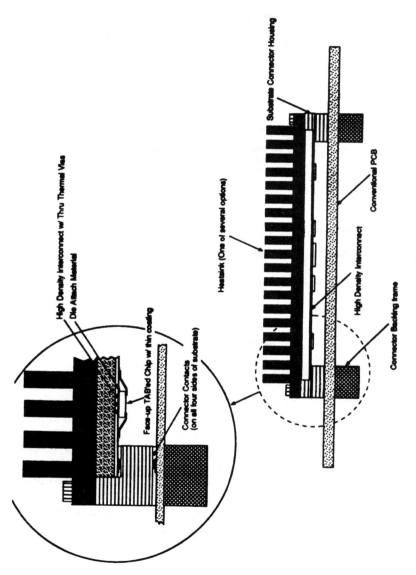

Figure 8-34   Example of face-up TAB on MCM-D.

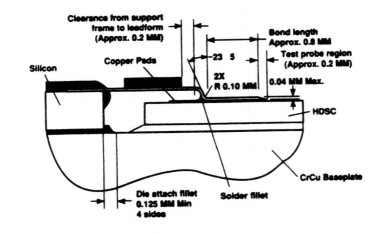

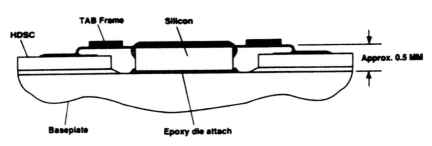

Figure 8-35 Cross-section of DEC's VAX9000 MCM-D with TAB.

packaging trends, i.e. lower in product cost/performance ratio, and lighter, thinner, shorter, and smaller in product physical dimensions. TAB has been promoted and demonstrated as a promising technology to meet these challenges. Partly due to TAB's inherited advantages such as smaller footprint, reduced weight, lower profile,[78] and superior electrical and thermal performance. More and more applications of TAB, e.g. LCDs, ASICs, functional IC cards, Notebook PCs, and MCMs driven by either higher performance and/or lower cost, are being introduced into the markets. It is expected that the share of the TAB related applications will increase at a faster pace than other chip interconnect technologies due to TAB's growing infrastructure and technical advantages while wire bonding remains to be the key technology for the medium or low pin count applications. There will be more demands on the following TAB related technologies:

- Higher density–examples are 80 μm or less ILB, 100 μm or less OLB, array TAB, etc.
- Higher performance TAB–examples are multiple metal layer tape (controlled impedance), flip-TAB (smaller footprint), etc.
- More flexible applications–examples are surface mount compatible TAB,[75] demountable TAB (DTAB),[13] tape with selective plating, etc.
- Higher performance materials and equipment. Materials and equipment used in TAB technology will be continuously improved in order to meet requirements of applications with higher density and better performance. The examples are low stress die-attach and dielectric materials, highly accurate ILB, OLB, and bond inspection equipment, etc.

REFERENCES

1. Lau, J. H., *Handbook of Tape Automated Bonding*, Van Nostrand Reinhold, New York, 1992, pp. 1-43.
2. Tummala, R. R., and E. J. Rymaszewski, *Microelectronic Packaging Handbook*, Van Nostrand Reinhold, New York, 1989, pp. 409-435.
3. *Electronic Materials Handbook*, Vol. 1, *Packaging*, ASM International, Materials Park, Ohio, 1989, pp. 274-296.
4. Kroger, H., et al., "Tape Automated Bonding for High Lead-count, High Performance Devices," *IEEE International Reliability Physics Symposium, Tutorial Notes*, pp. 2.1-2.39.
5. Belcourt, F. J., "Electrical Issues Associated with High Density Packaging," *Proceedings 3rd ASM Conference on Electronic Packaging: Materials and Processes and Corrosion in Microelectronics*, 1987, pp. 71-77.
6. Griffin, N., "High Lead Count TAB in a Multi-Chip Environment," *NEPCON West*, February 1988, pp. 569-575.
7. Jensen, R. J., "Copper/Polyimide Thin Film Multilayer Interconnections for High Performance Packaging," *Proceedings 3rd ASM Conference on Electronic Packaging: Materials and Processes and Corrosion in Microelectronics*, 1987, pp. 25-31.
8. Watari, T., and H. Murano, "Packaging Technology for the NEC SX Supercomputer," *IEEE Electronic Components Conference*, 1985, pp. 192-198.
9. Akihiro, D., W. Toshihiko, and N. Hideki, "Packaging Technology for the NEC SX-3/SX-X Supercomputer," *40th IEEE Electronic Components and Technology Conference*, May 1990, pp. 525-533.
10. Wong, S., "TAB Inner Lead Bonding for High-Lead-Count, High-Volume Applications," *NEPCON West*, March 1989, pp. 198-205.
11. Altendorf, J. M., "SMT-Compatible, High-Yield TAB Outer Lead Bonding Process," *NEPCON West*, March 1989, pp. 233-249.

12. Deeney, J. L., D. B. Halbert, and L. Nobi, "TAB as a High-Leadcount PGA Replacement," *International Electronics Packaging Conference*, September 1990, pp. 660-669.
13. Rajendra, D., et al., "Demountable TAB–A New Path for TAB Technology," *ITAB*, 1992, pp. 9-24.
14. Deshpande, U., S. Shamouilian, and G. Howell, "High Density Interconnect Technology for VAX 9000 System," *International Electronics Packaging Conference*, September 1990, pp. 46-55.
15. Joseph, J. P., and M. A. Kniffin, "Designing TAB Interconnect for the VAX 9000 Computer," *International Electronics Packaging Conference*, September 1990, pp. 683-695.
16. Hargis, B., "The Use of TAB Tape in High Performance Packaging," *NEPCON West*, March 1989, pp. 457-472.
17. JEDEC Standard–Tape Automated Bonding (TAB) Package Family, *JEDEC* document 95.
18. Kleiner, T., "TAB Excise and Form Tooling Principles," *Proceedings of the Second International TAB Symposium*, pp. 62-76, February 1990.
19. Roberts, L., "Developments In Hot Bar Reflow," *Circuits Assembly*, pp. 56-64, August 1991.
20. Zimmer, G., "Using Advanced Pulsed Hot-Bar Solder Technology for Reliable Positioning and Mounting of High-Lead-Count Flat Packs and TAB Devices," *Proceedings of the Second International TAB Symposium*, pp. 230-249, February 1990.
21. Sovinsky, J., "Outer-Lead Bonding TAB Packages–Your Soldering Options," *Technical Proceedings, EXPO SMT*, Las Vegas, 1989.
22. Evans, H. E., J. P. O'Hara, and P. Viswanadham, "Hot Bar Attachment: TAB Applications," *Proceedings of International Electronics Packaging Conference*, pp. 1108-1127, 1989.
23. Lau, J. H., *op cit.*, pp. 489-514.
24. Resor, G., "Lithography for Flat Panel Video Display," *Solid State Technology*, February 1988, pp. 103-107.
25. Tummala, R. R., and E. J. Rymaszewski, *op cit.*, pp. 1021-1086.
26. Hartenett, M. W., et al., *Worldwide Status and Trends in Multichip Module Package*, TechSearch International Inc., 1989.
27. Gary, R. M., and G. Geschwind, *Multichip Modules for the VLSI and ULSI ERA*, Rose Associates, 1990.
28. Rossi, R. D., "Multichip Modules: An Overview of Design, Materials and Processes," *Surface Mount Technology*, August 1990, pp. 17-20, Vol. 4, no. 8.
29. Balde, J. W., "Multichip Modules: The Constructions in Use Today," *Proceedings of the Technical Program of NEPCON WEST*, February 1990, pp. 965-974.
30. Mather, J. C., and J. K. Hagge, "Material Requirements for Packaging Multichip Modules," *Proceedings of the Technical Program of NEPCON WEST*, February 1990, pp. 1135-1142.

31. Craig, J. D., and W. J. Lautenberger, "New Polyimides for High Density Interconnect Packaging," *Proceedings of the Technical Program of NEPCON WEST*, February 1989, pp. 925-926.
32. Roe, K., et al., "Challenges in Multichip Module Substrate Assembly with TAB'd Components,"*NEPCON West*, 1991, pp. 1130-1144.
33. Duncan, P. M., "Noise Characteristics of High Density TAB Configurations," *Proceedings of the Technical Program of NEPCON WEST*, February 1990, pp. 638-649.
34. Lockard, S. C., et al., "Multimetal Layer TAB for High Performance Digital Applications," *Proceedings of the Technical Program of NEPCON WEST*, February 1990, pp. 1113-1122.
35. Doss, K., "Materials Issues for the TAB Technology of the '90s," *Proceedings of the Technical Program of NEPCON WEST*, February 1990, pp. 944-956.
36. Liljestrand, L. G., "Bond Strength of Inner and Outer Leads on TAB Devices," *Hybrid Circuits, Journal of the International Society for Microelectronics–Europe*, pp. 42-48, 1986.
37. Lau, J. H., et al., "Thermal Stress Analysis of Tape Automated Bonding Packages and Interconnections," *Proceedings of the 39th Electronic Components Conference*, pp. 456-463, 1989.
38. Emamjomeh, A., and M. J. Sandor, "Evaluation of Material and Geometric Parameters in a TAB-on-Board Package Using Finite Element Analysis," *Second International TAB Symposium Proceedings*, pp. 172-198, 1990.
39. Lau, J. H., *op cit.*, pp. 200-242.
40. Anderson, S. W., "Solder Attach Tape Technology (SATT) Inner Lead Bond Process Development," *Proceedings of 4th ITAB*, 1992.
41. Lau, J. H., *op cit.*, pp. 329-337.
42. Crea, J. J., and P. B. Hogerton, "Development of a Z-axis Adhesive Film for Flex Circuit Interconnects and TAB Outer Lead Bonding," *Proceedings of Technical Program, Nepcon West*, Volume 1, 1991.
43. Wong, M., "Anisotropically Conductive Adhesive Interconnects–A Case Study," *Proceedings of the Technical Program, Surface Mount International*, 1991, pp. 308-310.
44. Hatada, K., *Introduction to TAB Technology* (in Japanese), Kogyo Chosakai Publishing Co., 1990, pp. 221-247.
45. Kearney, K. M., "Trends in Die Bonding Materials," *Semiconductor International*, June 1988, pp. 84-88.
46. Davey, N. M., and F. W. Wiese, Jr., "Adhesive Mechanisms in Silver/Glass Die Attachment of Gold Backed Die," *Proceedings of the 1986 International Symposium on Microelectronics*.
47. Bourdelaise, R. A., "Solderless Alternatives to Surface Mount Component Attachment," *Proceedings of 4th International SAMPE Electronics Conference*, June 1990, pp. 1-10.
48. Ahukla, R. K., and N. P. Mencinger, "A Critical Review of VLSI Die-Attachment in High Reliability Applications," *Solid State Technology*, July 1985, pp. 67-74.

49. Chang, J., et al., "Repairable Die-Attach for Multi-Chip Modules," *Proceedings of the 7th Electronic Materials and Processing Conference*, August 1992, pp. 19-22.
50. Lee, C. J., and J. Chang, "Reworkable Die Attachment Adhesives for Multichip Modules," *Proceedings of International Symposium on Microelectronics*, 1991, pp. 110-114.
51. Hwang, J. S., *Solder Paste in Electronic Packaging*, Van Nostrand Reinhold, New York, 1989, pp. 33-70.
52. "Montreal Protocol for Substances That Deplete the Ozone Layer," September 1987, UNEP, Montreal, Canada.
53. Felty, J. R., "OZONE/CFC Cleaning Issues: Alternatives Testing Results," *3rd International SAMPE Electronics Conference*, June 1989, pp. 1011-1022.
54. Chow, M., "Aqueous Cleaning under Large Low-Clearance Parts," *Surface Mount Technology*, December 1990, pp. 27-33.
55. Hayes, M., "Chlorinated and CFC Solvent Replacement in the Electronic Industry," *Nepcon East*, 1988.
56. Rubin, W. R., and M. Warwick, "A No-Clean-Flux Review," *Surface Mount Technology*, October 1990, pp. 43-46.
57. Lea, C., *A Scientific Guide to Surface Mount Technology*, Electrochemical Publications, 1988, pp. 119-120.
58. Lau, J. H., *op cit.*, pp. 338-357.
59. Hammond, R., "Laser Ultrasonic Tape Automated Bonding," *Surface Mount Technology*, September 1990, pp. 25-31.
60. Messner, G., et al., "Thin Film Multichip Modules," *International Society for Hybrid Microelectronics*, 1992, pp. 282-330.
61. Timoshenko, S. P., *Theory of Elasticity*, McGraw-Hill.
62. Morris, J. E., *Electronics Packaging Forum–Parameterization of Fine Pitch Processing*, pp. 395-436.
63. Hayes, M. E., "Physiochemical Aspects of Electronics Assembly Cleaning and Their Implications for Halogen-Free Solvent Selection," *3rd International SAMPE Electronics Conference*, pp. 998-1010.
64. Lau, J. H., *op cit.*, pp. 358-386.
65. Gere, J. M., and S. P. Timoshenko, *Mechanics of Materials*, 2nd edition, PWA Publishers, 1984.
66. Prasad, R. P., *Surface Mount Technology–Principles and Practice*, Van Nostrand Reinhold, 1989.
67. Chang, J., and C. Spooner, "Rework of Multi-Chip Modules—Die Removal," *NEPCON WEST*, 1992, pp. 512-522.
68. Bertram, M., "Repairable Method for Solder Reflow TAB OLB," *Proceedings of 2nd ITAB*, 1990, pp. 147-164.
69. Tummala, R. R., and E. J. Rymaszewski, *op. cit.*, pp. 391-408.
70. Tummala, R. R., and E. J. Rymaszewski, *op. cit.*, pp. 366-391.
71. Khadpe, S., "Worldwide TAB Status–1992," *Proceedings of 4th ITAB*, 1992, pp. 352-354.
72. Lau, J. H., *op. cit.*, pp. 539-570.

73. Wesling, P., V. Dobroff, and T. Chung, "Thermal management for an MCM assembly with TAB'd Components," *NEPCON WEST*, 1992, pp. 502-511.
74. Wesling, P., et al., "CMOS Multichip Module Test Vehicle with TAB'd Components," *41st Electronic Components Technology Conference (ECTC)*, May 1991, pp. 712-718.
75. Lau, J. H., *op. cit.*, pp. 593-618.
76. Hacke, H. J., and H. H. Steckhan, "Micropack Packaging Technology," *Siemens Research and Development Reports*, Bd. 17 (1988) Nr. 5.
77. Pan, J. T., C. Hilbert, and B. Weigler, "Multichip Modules for High Performance Computer Applications," *Thin Solid Film*, 193/194, 1990, pp. 886-894.
78. Hatada, K., *op. cit.*, pp. 30-32.

9

Solder Bumped Flip Chip Attach on SLC Board and Multichip Module

Yutaka Tsukada

9.1 REQUIREMENTS FOR PACKAGING

The requirement for low-end computer packaging has various aspects. Cost always comes first. Since there are many alternatives for the solution of low-end packaging, the one with the lowest cost gets the first priority in most cases. At the same time, small and light requirements have more attention in these days because of the down-sizing of computers and other electronic products. In particular, the spread of portable computers and the advancement of consumer product electronics are driving this trend. Car electronics is another high potential area in the near future.

The performance of low-end computers is getting better and closer to high-end computers. In the past, only four or five years separated the performance of high-end computers and low-end computers. Today, the emergence of a high-end computer using a microprocessor as its engine indicates that there is almost no interval between them as far as the hardware is concerned. Also the density of the package is high due to the integration of semiconductors and miniaturization of the package. Reliability is another area which requires firm and wide consideration. The products are used in very different conditions. There is great heat or cold outside, and frequent power on/off or vibration and shock due to handling are not unusual conditions for low-end products.

Surface Laminar Circuit (SLC) and Flip Chip Attach (FCA) are the technologies which were developed to satisfy these various requirements. SLC is a new printed circuit board technology with construction which is similar to today's semiconductor wiring structure.[1] As a result of photo processed via hole, SLC has twice the wiring density of the conventional PCB (Printed Circuit Board). FCA is a chip packaging method which makes it possible to attach a bare chip on the epoxy base carrier with flip chip bonding.[2]

9.2 SLC TECHNOLOGY

9.2.1 Structure of SLC

The SLC structure has two parts. One is a substrate and the other is surface laminar layers for the signal wiring. An ordinary glass epoxy panel is used for the substrate and the surface laminar layers are sequentially built up with the dielectric layer made of photo sensitive epoxy and the conductor plane of copper plating. Figure 9-1 shows a typical cross-section of SLC with the explanation of solutions which SLC provides to the limitation of conventional FR4 glass epoxy laminate.

The overall construction is a build-up structure. The dielectric layer has photo-via holes with $125\,\mu m$ (5 mil) diameter for the signal wiring connections. Not only is it small in diameter compared with a plated thru hole, but photo-via hole is also able to connect each signal wiring without interfering with the other wiring planes. Copper plating can provide a thinner conductor plane which is the most important parameter to achieve the fine pitch wiring. The ordinary glass epoxy laminate has copper plating for the inside of the thru hole which automatically adds thickness over the laminated copper foil for the conductor plane.

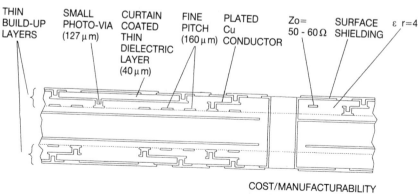

Figure 9-1   Solution by SLC.

The characteristic impedance is designed to be 50-60 Ω to maintain easier control for the noise of a high performance chip. At the same time, the thickness of dielectric layers may be varied easily by adjusting the apply tool dimension. This allows the creation of various electrical conditions which are adaptive to the applications. Since SLC has been developed consistent with the objective of low-end product use, all materials and toolings are the ones available in the printed circuit board industry. Because of this strong intention, low-cost manufacturing has become possible.

Figure 9-2 shows the variations of the cross-sections. SLC1.0 is the very first version of SLC which has two SLC build-up layers with two different

SLC 1.0

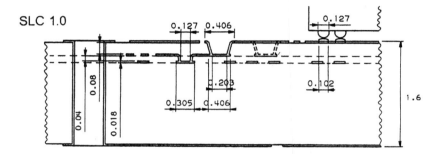

SLC 1.1

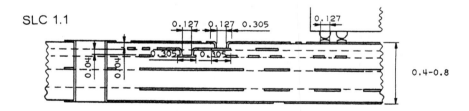

SLC 1.5

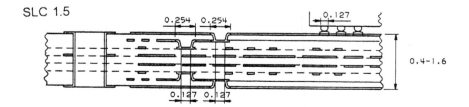

Figure 9-2 SLC cross-sections.

dielectric thicknesses for the characteristic impedance matching for the two signal planes which are located in micro strip conditions. SLC1.1 has a tri-plate structure and is used for the PCMCIA format IC card. The thickness of SLC card is 0.4 to 0.8 mm (15.8 to 31.6 mil) to achieve thinner package which must be placed in the IC card. SLC1.5 is a double-sided version with the two signal layers at each side. Both sides have tri-plate constructions. In every SLC type, though the thickness and configuration may vary, the ground rule of SLC is basically the same. Only the part of the base FR4 changes without the major change of the electrical characteristics in the SLC layer.

The strength of SLC is driven by the via density. Conventional PCB uses LPC (Lines Per Channel) as an indication of the density. However, the wide utilization of SMC (Surface Mount Component) brought change in the carrier density utilization. All the I/O of the components are once located at the surface of the carrier and then distributed inside it. Due to the density of I/O which has to be handled, the high density plated thru holes are required around the component. Therefore the density of the carrier is highly dependent on the via density. SLC uses the parameter called VPSG (Via's Per Square Grid) as an indication of its density. This explains the numbers of via per square grid which is the area of 2.5 mm (100 mil) square.

Figure 9-3 shows the comparison of the wiring density with this VPSG for the maximum wirability. The conventional FR4 glass epoxy laminate has 4 VPSG with the drill size which is the minimum for the optimum production use. The first version of SLC had 8 VPSG by the effect of photo-via hole utilization. It is extended to 12 VPSG utilizing mini-land via holes. With the landless vias, SLC provides 20-28 VPSG. Figure 9-4 shows the maximum wiring density for the various carriers. SLC has a wide range of wiring density and bridges from the conventional FR4 glass epoxy laminate to silicon carrier which has 25 μm (1 mil) line pitch fabricated in the semiconductor process.

9.2.2 Manufacturing Process of SLC

SLC has been planned and developed to achieve low cost for low-end packaging. Industry available materials and toolings are utilized to satisfy this objective. The processes are carefully balanced by adjusting the process windows for the reliable utilization of these. Figure 9-5 shows the basic process steps of SLC. SLC is processed by 495 mm × 600 mm (19.5 in. × 24 in.) large panel size.

The starting material is ordinary FR4 glass epoxy laminate as a base. The figure shows a non-internal plane FR4, but the multi-layer FR4 can be used without any problem. The surface conductor plane of FR4 is circuitized

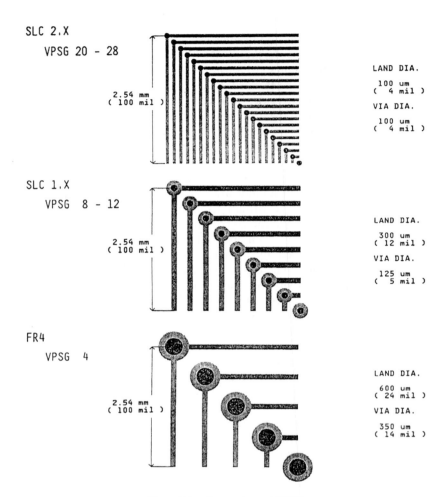

Figure 9-3 Comparison of VPSG.

using the ordinary subtractive etching process. This circuit is to be the bottom part of SLC layers. The first dielectric layer is applied to this surface by using curtain coat process where the liquid epoxy resin is flowing in the narrow slit to form the curtain and the work runs through under this nozzle to leave a thin layer of epoxy resin on the surface. The thickness variation is $\pm 10\ \mu\text{m}$ ($\pm 4\,\text{mil}$). The photo-sensitive epoxy resin is exposed and developed to form the via hole. After curing the resin, its surface is etched in the permanganate bath to provide the roughness for the copper plating adhesion.

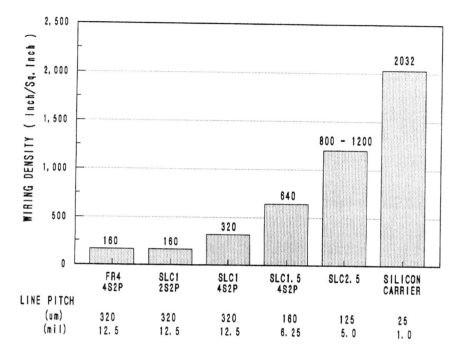

Figure 9-4 Wiring density.

Electroless plating makes the seed layer and the electro plating adds the necessary thickness. The nominal thickness is 18 µm (0.7 mil) which is equivalent to 1/2 ounce copper foil. A thinner conductor layer which can provide finer lines is easily achievable by simply plating with less time. After the plating, the second circuit is made by the same etching process. The same steps are followed until the required SLC layers are built up. After finishing to apply the top layer dielectric, the through holes are drilled for the power pick-up and the functional connections between the front and back layers. Through holes and top conductor layer are plated at the same time. Apply the solder protective coating to finish the fabrication. The material properties which are used for SLC fabrication are shown in Table 9-1. The photosensitive epoxy for the dielectric layer has very low modulus compared with glass epoxy laminate which is used for the base. The dielectric constant is 4.0 because it is almost pure epoxy. CTE is higher compared with the FR4 base.

The circuit lines in each SLC layer are inspected by the optical tester. The software of the tester is modified to adopt the unique role of SLC because the FCA area has lines such as open end which do not appear in the

1. CIRCUITIZE

2. COAT DIELECTRIC/MAKE PHOTO-VIA-HOLE

3. Cu PLATING/CIRCUITIZE

4. DIELECTRIC/PHOTO-VIA-HOLE/DRILL THRU-HOLE

5. Cu PLATING/CIRCUITIZE

6. SOLDER PROCOAT

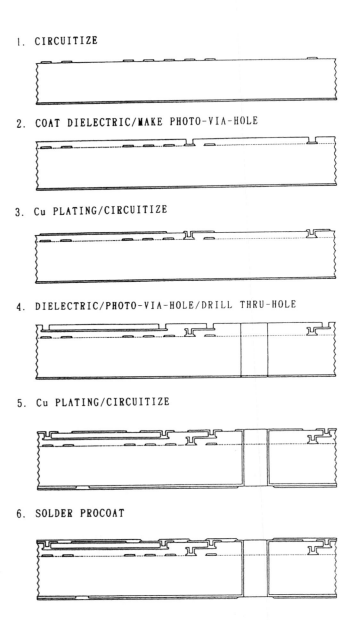

Figure 9-5   Process steps for SLC manufacturing.

Table 9-1 Material Properties

Material	CTE (×10E-6)	Young's Modulus Kgf/mm²	(MPa)	T_g °C (K)	Er
Glass Epoxy	X 15.0	2100	(20580)	138 (411)	4.7
	Y 13.2	2250	(22050)		
	Z 55.6				
Photo Sensitive Epoxy	69.0	482	(4724)	132 (405)	4

ordinary PCB pattern. At the end of process steps, SLC is tested by the electrical open/short tester. If the board is large, FCA area is tested separately due to the difference of required pitch for the test probe. FCA area is probed by 250 µm (10 mil) pitch sleeved spring probe.

9.2.3 Quasi Module

Due to the high surface wirability, SLC can fan out the high density I/O lines from the bare chip and distribute into its inside for the wiring. One of the major difficulties of the bare chip attach package is to control the physical and electrical reliability of the chip. In the existing system and environment, the gap between the chip and carrier is significant. In SLC/ FCA technology, the concept called "quasi module" is introduced to bridge this gap smoothly. Figure 9-6 is a typical quasi module which is provided for 95K gate VLSI logic chip with 372 pads. Quasi module is considered as an imbedded module in SLC.

Since the components other than the semiconductor chips have the larger I/O lead pitch, the narrow pitch I/O lines from the FCA pads are expanded to the natural wiring pitch of the board. The quasi module in Figure 9-6 has the FCA pad with minimum 250 µm (10 mil) pitch in triple row staggered pattern. The lines which escape from the edge of the chip have a pitch of 176 µm (7 mil) for each layer, while the other area of the board has a pitch of 317.5 µm (12.5 mil). The fan-out lines connect this difference of the pitch. The pads for the test and engineering change wiring are at the peripheral edge of the quasi module. When SLC/FCA design is evaluated, the quasi module is put to the test vehicle which has the same cross-section as the application board. The test vehicle passes the same process as the application board for FCA assembly and is put into the stress test for the

Figure 9-6   Quasi module for VLSI logic chip.

physical reliability evaluation. Also the electrical performance is tested by this test vehicle assembly.

The design of quasi module is registered in the component data base file in the design system. The board designer uses the quasi module as if it is one of the components. The structure of quasi module depends on the I/O count and the electrical performance requirement of the chip. Figure 9-7 shows SLC fan-out capability of I/O lines to determine the structure of the quasi module. Figure 9-8 shows a cross-sectional view of the high performance quasi module which has plated through holes right under the chip for power pick-up which goes directly inside the chip area. The inductance of the power line for this quasi module is 0.1-0.15 nH.

9.3 FLIP CHIP ATTACH (FCA) TECHNOLOGY

9.3.1 Flip Chip Bonding

Flip chip bonding of the semiconductor chip using small solder ball as joint material has the following advantages over other types of package.

- High density I/O connections by area array joint
- Superior electrical performance due to small bump for the joint

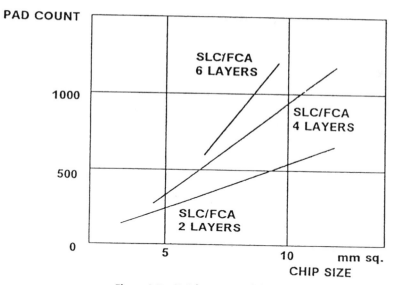

Figure 9-7   SLC fan-out capability.

Figure 9-8   High performance quasi module.

- Multichip can be attached by the single reflow of bump solder
- Life of the joint is estimated simply as the fatigue of solder.

Because of these advantages, it has been used as a main packaging technology in large computers for almost three decades. However, since life of the joint is dominated by CTE mismatch between the chip and substrate materials,[3] high-cost material such as ceramic has been used and not considered as an alternative for the low-cost technology.

To utilize this superior technology in low-cost packaging, two issues have been resolved in FCA technology. One is to lower the reflow temperature so that material such as epoxy can be used. This was achieved to provide the eutectic solder bump on FCA pad in the quasi module on SLC. This bump is reflowed with the temperature around 200°C and forms the joint with the high temperature solder bump on the chip side. By this method, it becomes possible to use the low-cost material such as epoxy for the carrier material. Figure 9-9 shows the eutectic solder bump on the SLC. The chip side bump is high temperature solder which is provided at the end of the semiconductor

Figure 9-9   Eutectic solder bump on SLC.

line. It works to allow the stress on the bump during the assembly process until the completion of the encapsulation process.

The next issue is the large CTE mismatch between the chip and carrier. The CTE of chip is normally 3.5 ppm/°C. The one for alumina substrate is 7-9 ppm/°C which has been used as a typical material for a flip chip bonding. However, SLC which uses the ordinary glass epoxy laminate as a base has CTE about 15 ppm/°C which is twice that of alumina CTE. To overcome this large CTE mismatch, the encapsulation technique is applied. After the joint is formed, the space between chip and SLC is filled with epoxy resin to disperse the stress from the bump. Figure 9-10 shows the cross-section of complete SLC/FCA package.

In Figure 9-10, the carrier is SLC with 0.7 mm (28 mil). SLC layer is thin, locating at the top surface. The joints with three rows staggered pattern are located at the peripheral of the chip. The encapsulant fills the space and forms the fillet at both edges of chip. Figure 9-11 shows a magnified view of the joints. The chip bump which is the dark part due to higher lead content is joined with the carrier eutectic bump which is the light part due to higher tin content. The chip bump does not melt during reflow because of the lower reflow temperature which is lower than its melting point. Instead, the carrier

Figure 9-10 SLC/FCA cross-section.

eutectic solder bump melts and wets the surface of the chip bump. There is a slight diffusion of each solder at the interface area. The complete joint shows a deep bowl shape where the eutectic solder is working like a cap to hold the chip bump. The figure also shows the excellent filling condition of the encapsulant even in the high density bump area.

As a result, this package has significant improvements in the reliability compared with the conventional flip chip bonding. More than 20,000 cycles in 100°C delta temperature thermal cycle test and over 2000 hours in Temperature/Humidity/Bias test with 85°C/85%/5V are achieved.

9.3.2 Stress Analysis of FCA Joint

The stress condition of this package is analyzed by Finite Element Method. Figure 9-12 shows the model for this analysis. The chip is a memory chip of 10.7 mm × 5.7 mm (421 mil × 224 mil), and the thickness is 0.6 mm (23.6 mil). The joint is 180 μm (7.1 mil) in diameter with 100 μm (4 mil) height. The maximum DNP (Distance from Neutral Point) of the bump for this chip is 5.51 mm (217 mil). The bumps are located along the shorter edge

Figure 9-11   FCA joint.

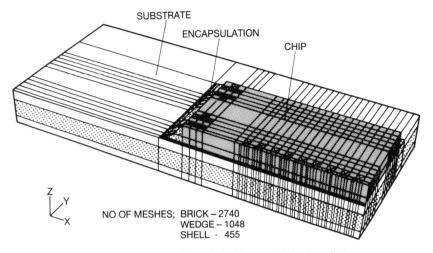

Figure 9-12   Model of Finite Element Method analysis.

and at the center in the longer direction. SLC has 1.6 mm (63 mil) thickness. The model which is shown in Figure 9-12 is a quarter size to reduce the mesh numbers for shorter calculation time.

The material properties which are described in Table 9-2 are put into the calculation. The result is shown in Figure 9-13, which shows the Von Mises strain relative to the chip bump location. The bump location on the chip is also shown in the upper left corner of the figure where the bumps are numbered from the center of the chip.

When there is no encapsulation, the strain increases significantly in relation to the distance from the center of the chip. It reached 4.5% on the bump at the far end. Under this level of strain, the mean cycle to failure by open joint in the stress test of 0-100°C thermal cycle is a little over 100 cycles which compares with 3000 cycles in the case of using ordinary alumina substrate. This significant difference is caused by the large CTE of SLC which is 14 ppm/°C and twice that of alumina. However, when the space between the chip and SLC is filled with epoxy and the encapsulation is completed, the strain drops drastically and becomes less than 1%. In this condition, the failure of the joint does not occur even if the cycle exceeds 20,000 cycles in the same 100°C delta temperature thermal cycle test.

The reason for this significant difference of the strain and resultant joint life extension is explained by the condition shown in Figure 9-14. This shows the calculated deformation of the package for both "without encapsulation" and "with encapsulation" which is exaggerated 25 times in Z-direction. If the package is "without encapsulation", the large difference in CTE simply causes the major deformation of the bump in shearing direction. In the case

Table 9-2 Material Properties for Modeling

Material	Modulus (MPa)	CTE (ppm/°C)	Poisson Ratio
Silicon (chip)	169540	3.5	0.07
Glass Epoxy Substrate	21560	X,Y-14	0.20
Photo-Sensitive Epoxy	4724	69	0.39
Solder (63/37)	6270	24	0.40
(5/95)	2940	29	
Encapsulant	4900-6370	26-60	0.36
	(0-100°C)	(0-100°C)	

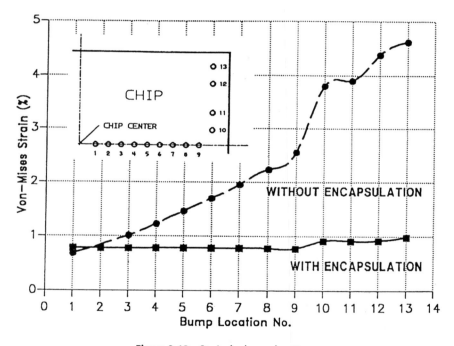

Figure 9-13 Strain for bump locations.

of "with encapsulation", there is no major deformation in shearing direction because all parts of chip and carrier are tightly adhered and the joints are protected by the encapsulant. As a result, the strain caused by the CTE mismatch is converted to the deformation in Z-direction and leaves no major strain in the joint. This is the same effect which we see in the bi-metal thermometer. This deformation depends on the CTE and modulus of each

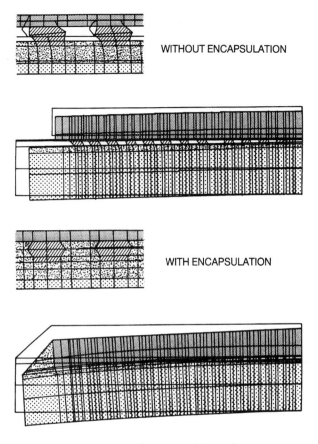

WITHOUT ENCAPSULATION

WITH ENCAPSULATION

Figure 9-14   Deformation of the package.

material. The measured actual deformation is about 5-7 mm (200-275 mil) for the chip of 10 mm (394 mil) square.

The other significant phenomenon of the encapsulation effect is shown in Figure 9-15, in which the strain under the chip is shown in different chip sizes when the encapsulation is performed. The area of each chip is 1/4 with the center at the lower right corner. The number indicates the level of the strain at each location. The strain level is very low and flat in the entire area under the chip and the same in the different chip sizes. There is a slight increase of the strain at the edge of the chip which is caused by the encapsulant fillet, but still lower than 1%. This indicates that the low strain is independent of the chip size when the encapsulation is applied. Theoretically, we can allow limitless chip size to attach on SLC if the

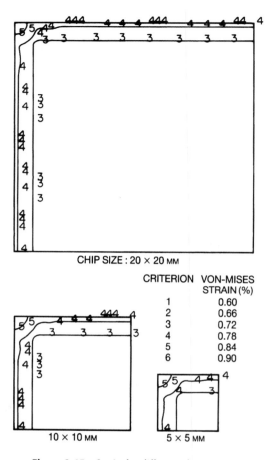

Figure 9-15   Strain for different chip size.

deformation curve is allowed and it means that we are free from "DNP" which has been the limiting factor of chip size in the conventional flip chip bonding.

In the further research of the stress conditions in detail, it is indicated that the reduction of the strain is shared with this deformation by 60% and the rest of the strain is absorbed by the low modulus SLC layer. This stress absorption of the SLC layer contributes more when the carrier deformation is restricted. For example, when we attach the chip on both sides of the carrier symmetrically, the strain increases more than with free carrier condition, but the life of the joint is still well maintained by the effect of SLC build-up layer absorption.

9.3.3 Manufacturing Process of FCA

Figure 9-16 shows the manufacturing process flow of FCA. The chip that comes into FCA process is tested at wafer level. It is temporarily attached to the burn-in carrier which is made of SLC and has the same quasi module as the application card. After the burn-in and test on this carrier, the chip is removed from the carrier for FCA assembly process. The process and tooling are basically the same ones used in the FCA process and the carrier recycles more than 20 times to achieve the low-cost burn-in.

The FCA process has three major sectors. One is eutectic carrier bump build, the second is chip join and the third is encapsulation. Eutectic carrier bump build is done by solder injection tool. Figure 9-17 shows the concept and operation sequence of this tool. In Figure 9-17: (a) the head of this tool

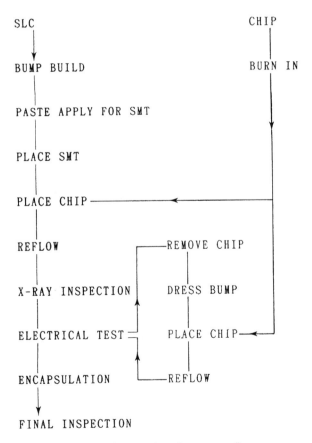

Figure 9-16   FCA manufacturing process flow.

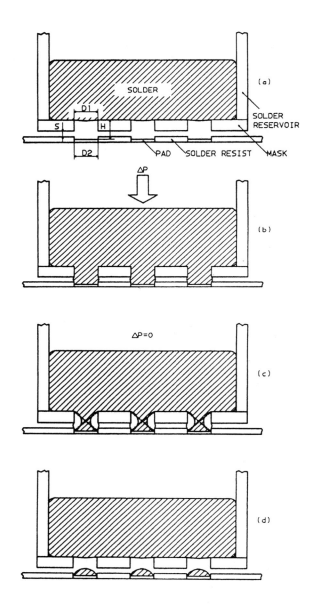

Figure 9-17 Solder injection process.

has a solder reservoir and a metal mask is attached to the bottom. The mask has small holes of which the pattern is the same as the pad pattern in the quasi module on SLC. (b) The head comes down to touch the quasi module after the alignment is adjusted. When the reservoir is pressurized, the solder in the reservoir is squeezed out from the holes in the mask and the solder touches the pad. After the solder wets the pads, the increased pressure in the reservoir balances with the sum of the surface tensions by the depressed solder columns between the mask and carrier. (c) When the pressure in the reservoir is released, the solder in the columns is pushed back into the holes by the surface tension of the molten solder itself. (d) The solder columns are split when the condition reaches the limit to keep its form. The lower part remains on the pad and forms the carrier bump. The upper part is drawn back into the reservoir.

This method can leave the larger solder amount more consistently than any other method. The volume deposited on the carrier pad is determined by the shape at the point of split by the surface tension. Since the precise calculation requires the detailed parameters which are difficult to obtain, the approximation by equation (9-1) can be used.

$$V = \frac{\pi}{12} \times H \times \frac{D_2^3}{D_1 + D_3} \qquad (9-1)$$

where V = bump volume
H = height from carrier surface to upper surface of mask
D_2 = carrier pad diameter

After applying the paste solder for other components, SMC and chip are placed for the required location and reflowed in the infra-red oven. Since the reflow temperature is 200°C, nitrogen purge in the oven is not required. The integrity of the joint is inspected by X-ray inspection. It is possible to use X-rays because SLC is epoxy based and does not block X-rays as in the case of ceramic. When the chip is confirmed as good by the electrical functional test, the space between the chip and SLC is encapsulated by epoxy resin. The encapsulation is simply done by the capillary action, however, it is sensitive to bump pattern to fill the entire space without a void.

Because the chip which comes into FCA process has already passed the burn-in and test, there is no major fall-out at the test in FCA process. However, when the defective chip is found, it is simply removed by heating up to reflow the eutectic solder. The new chip is placed and follows the same sequence of the first time join. It is very rare to find defective chip after the encapsulation, but if it is found, the replacement is done by the mechanical rework process. The chip is ground off by the milling machine, dressing the bump by solder injection tool and a new chip is placed for following the same step of normal chip joining. Figure 9-18 shows this process.

In Figure 9-18, (a) shows the joint after the chip and about a half height of encapsulant are ground off, (b) shows the eutectic chip deposited over the ground joint by solder injection process, and (c) shows the complete reworked joint which looks like a double-decker joint. The high temperature solder and eutectic solder are still not completely diffused. This rework method is effective for the other encapsulated flip chip joint.

9.4 ADVANTAGES OF SLC/FCA PACKAGE

By the combination of flip chip bonding and high density SLC carrier, this package has various advantages shown in Figure 9-19.

In Figure 9-19, (a) shows a comparison of the size. The chips which have similar signal I/O count are selected. QFP (Quad Flat Pack) has a lower number of power I/Os because of the limited availability of the lead, while the ceramic PGA (Pin Grid Array) and SLC/FCA have a larger number because of the area array joint connection by flip chip bonding. This difference influences the electrical performance due to switching noise, but it is simply compared with the same signal count here. The size is 7:5:1. In (b) is shown the noise level with adding the MLC (multi layer ceramic) package to the comparison.[4] SLC/FCA has the equivalent performance with the MLC package which has been used for the high performance large computer. The comparison of the cost is shown in (c). This figure is calculated including all the cost elements of the package such as chip burn-in and test, carrier bump build, chip assembly and test on the carrier, and occupied carrier space. The chip cost itself is excluded. The cost of each package is indicated relative to the index which regards SLC/FCA as one. A conventional QFP is 1.5 and QFP which has an interposer to improve the noise by providing a ground plane is about 2. A conventional ceramic PGA is 3 and a multi-layer ceramic is out of scale. The thermal performance is shown in (d). In SLC/FCA package, 60% of heat goes to the air and 40% into the carrier through the I/O fan-out lines. Since the back of chip is directly exposed to the air, the thermal resistance is very low to achieve a good cooling condition compared with QFP which has higher thermal resistance due to plastic mold. However, since epoxy is not a good heat conductive mateial, SLC/FCA is not effectively used for high power chip which dissipates heat more than 10 watts.

As a summary, SLC/FCA is considered as a package with high performance, low cost, small and light and the chip power dissipation below 10 watts. These conditions are the best fit to low-end computers, portable products and consumer products.

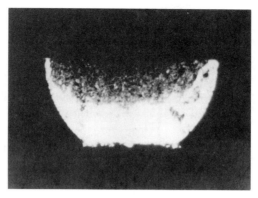

(a)

(b)

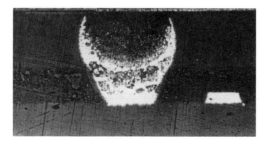

(c)

Figure 9-18 Mechanical chip rework process.

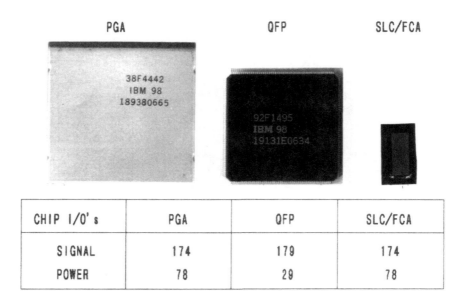

(a)

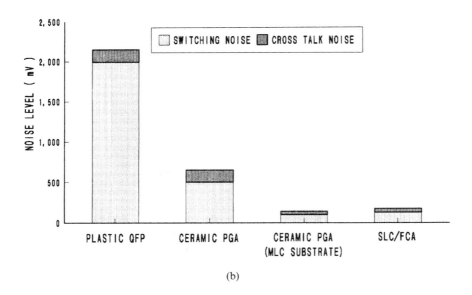

(b)

Figure 9-19   SLC/FCA advantages.

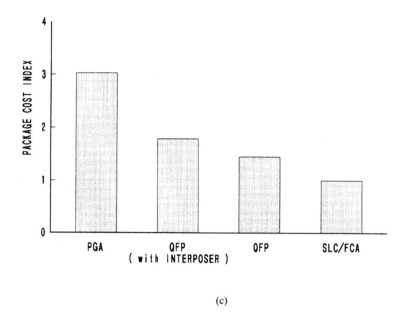

(c)

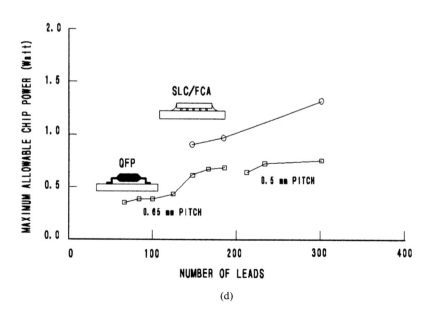

(d)

Figure 9-19 (continued)

9.5 SLC/FCA APPLICATIONS

9.5.1 Coverage of Applications

Figure 9-20 shows a chronological chart for the area of SLC/FCA applications. SLC and FCA were developed separately at the initial phase. SLC without FCA and memory chip FCA on normal FR4 were the first products qualified at the end of 1989, and brought into the field since 1990. In 1991, SLC with FCA has been qualified. The expansion of the application was started in 1992. Utilizing the small and light nature of SLC/FCA, the various IC cards which have the format adopted to PCMCIA are being manufactured. The detail of the example is introduced in the next section. The excellent electrical performance is used to develop the optical link adapter card which is operated with 500 MHz clock cycle. A large package heat sink provides a good cooling performance which becomes possible due to the very low thickness of SLC/FCA package. The large numbers of FCA are attached to the processor card of the workstation which is operated by 75 MHz clock and provides the excellent cost performance. The performance with low cost will be well shown for the application to MCM-L which is introduced in Section 9.5.3. SLC/FCA technology can be adapted to most of the electronic package, achieving a good performance and low cost.

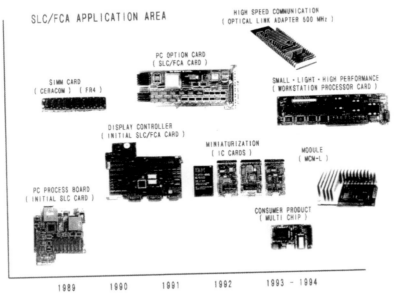

Figure 9-20 SLC/FCA applications.

9.5.2 Token Ring LAN Adapter Card

Figure 9-21 shows the Token Ring Adapter card in two types: (a) shows the outside appearance of PCMCIA IC card, and (b) shows the package of this card. The card shown in (c) was designed as PC option card and was available before PCMCIA card appeared. The PCMCIA card has the size of 85.6 mm × 54.0 mm (3.4 in. × 2.1 in.) with 5 mm (0.2 in.) thickness. At the center of the card, there is one FCA of Token Ring LSI chip which has various functions. The chip is 40K gates equivalent custom chip with analog part and I/O count is 248. Just above Token Ring chip, there is the PCMCIA interface chip which is added by this release. SLC used to this card is six layers with 0.67 mm (26.4 mil) thickness. The quasi module has single layer fan-out with 176 μm (7 mil) pitch with 70 μm (2.75 mil) width lines. Compared with the IC card by SLC/FCA package, the card shown in (c) has 140 mm × 88 mm (5.5 in. × 3.5 in.) with 18 mm (0.7 in.) thickness. The PCB is four layers FR4 and Token Ring chip was packaged in 36 mm (1.4 in.) PGA. The card is inserted into the option slot in a desk top PC by the card edge connector.

The SLC/FAC PCMCIA IC card is inserted into the PCMCIA type II slot and the user of a portable PC can connect his equipment to Token Ring LAN at any place where the IEEE802.5 Token Ring LAN is available.

9.5.3 Extension to MCM-L

The MCM-L described here is a prototype package of IBM RISC System/6000. The RISC system/6000 is partitioned into several functional units which are an instruction-cache unit (ICU), a fixed-point unit (FXU), a floating-point unit (FPU), a data-cache unit (DCU), a storage-control unit and an input/output interface unit (IOU). Each unit is implemented in CMOS VLSI logic as described in Table 9-3. A complete processor contains 9 chips. This prototype is a cost-reduced version which has two DCU chips rather than four.[5] The table also describes the chip sizes which are the largest chip size in IBM CMOS logic and high chip I/O which provides the large amount of simultaneous switching capability.

SLC substrate cross-section for this MCM-L is shown in Figure 9-22. FR4 base is four layers and five SLC layers are built up on it. Signal layers have a line ground rule of 114 μm (4.5 mil) pitch with 38 μm (1.5 mil) line width and 22 VPSG density. Four signal layers are required in this application to complete the wiring within 75 mm (3 in.) square function area. 15 mm (0.6 in.) is added to allow the utilization of low cost pins for the module I/O connection. To obtain a low noise performance, the signal planes are located in the triplate structure by adding the reference planes. If

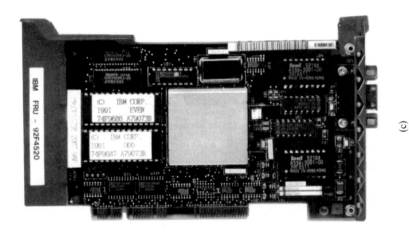

(c)

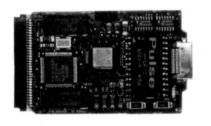

(b)

(a)

Figure 9-21 Comparison of Token Ring Adapter card.

Table 9-3   RISC System Chip Set

Chip	Transistor		Die Size (mm)	Signal I/O	Total Pads
	Logic	Memory			
ICU	200k	550k	12.7 × 12.7	252	818
FXU	250k	250k	12.7 × 12.7	256	818
FPU	360k	60k	12.7 × 12.7	224	818
DCU ×4	175k	950k	11.3 × 11.3	184	799
SCU	230k	—	11.3 × 11.3	255	799
IOU	300k	200k	12.7 × 12.7	293	703
Total	2040k	4860k	1284 mm^2	2016	7152

Note: 9 chip set implementation.

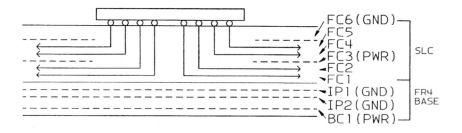

SLC/FCA substrate structure

Figure 9-22   Cross-section of MCM-L.

the signal layer count is increased to seven layers, the size of substrate will be 54 mm (2.1 in.) square. However, 90 mm (3.5 in.) square is defined to have an optimum balance of the cost, performance and size allowable for the requirements. Figure 9-23 shows each layer of wiring design on 90 mm substrate which are five layers of SLC build-up part and four layers located in the FR4 base.

To reduce the switching noise, the power to ground effective inductance is designed to have less than 100 pH which leads to less than 800 mV noise at full operating frequency. A low cost surface mount chip capacitor is applied and reflowed together with the VLSI chip. The interconnection to mother board for the signal and power I/O is a significant matter for any MCM from the point of view of performance, cost and manufacturability. One of major advantages of MCM-L is an ability to provide plated thru holes to accept pins. The low cost pins in peripheral plated thru holes are applied.

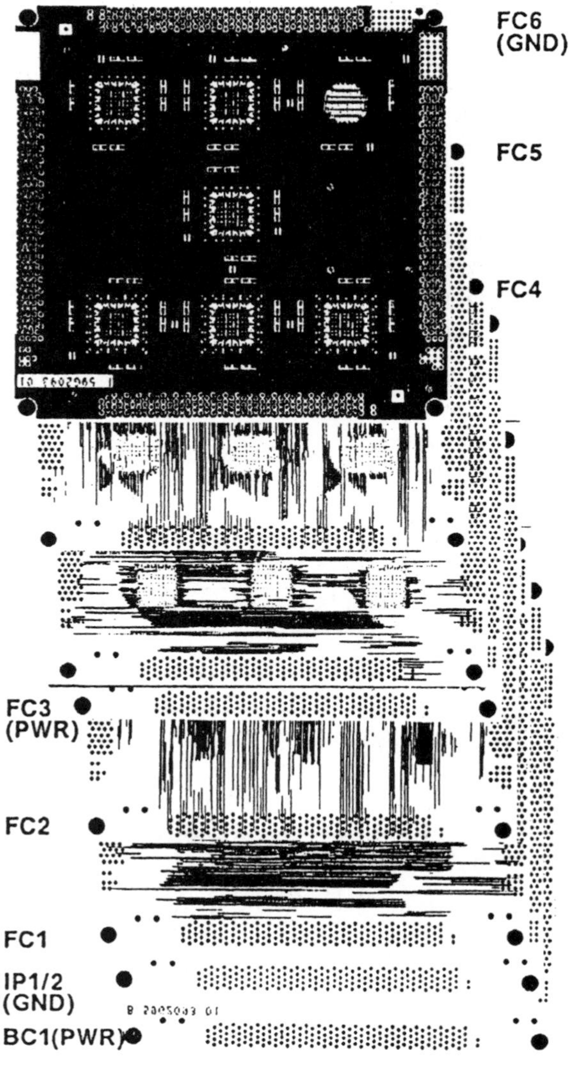

Figure 9-23 Wiring planes of MCM-L.

The heat sink of this MCM-L is very simple because of FCA technology, the chip requires no hermetic sealing which increases the thermal resistance. The individual heatsink is directly attached to the top surface of each chip with thermal conductive adhesive. The heatsink is designed to maintain the chip junction temperature below 85°C in 40°C environment for 3 W dissipation per chip in 1 m/sec air flow.

Table 9-4   Comparison of MCM Package

Properties	SLC	SOS	SCM/PCB
Substrate			
Size (mm × mm)	90 × 90	54 × 54	156 × 175
Chip to chip (cm)	2.3	1.5	5.5
Pkg efficiency (%)	15.9	44.0	4.7
Critical path (cm)	8.9	6.4	25.4
Line width (μm)	38	13	127
pitch (μm)	114	25	317.5
Via land dia. (μm)	100	8	610
Dielectric constant	4.0	3.8	4.4
Chip			
Number of chips	7 (max. 9)	7 (max. 9)	7 (max. 9)
Total joints/chips	5554	5554	5554
Module			
Number of signal	512	512	256/SCM
Vdd/GND	86/86	86/86	24/20/SCM
Heat sink size (mm)	20 × 20 × 25	60 × 64 × 20	40 × 38 × 19
numbers	7	1	7
Signal line			
Resistance (ohm/cm)	0.600	10.25	0.024
Capacitance (pf/cm)	1.36	2.97	0.75
Zo (ohm)	53	21	89
Delay (psec/cm)	66.6	62.4	66.4
Noise coefficient (mV/V)	114	12	364

Table 9-4 describes the physical dimensions and electrical parameters comparison of other package implementations which are SCM (Single Chip Module) package and MCM-D by SOS (Silicon on Silicon) package. The data for the SOS package were referred to and calculated from the technical papers.[6,7] SLC occupies 90 mm × 90 mm (3.5 in. × 3.5 in.) which is almost 30% of the area for the SCM case. The package efficiency defined as the area occupied by chips relative to the total package area is 15.9% in the SLC case which is three times that of the SCM case. SOS has the highest efficiency due to the highest wirability by utilizing semiconductor processes. The chip to chip distance is defined as the distance between the chip centers. Other significant differences are in the line resistance and the cross-talk noise coefficient. SOS signal line has 10.25 Ω/cm resistance which is 17 times greater than 600 mΩ/cm of SLC. This large resistance influences the signal transmission capability in high frequency range. On the other hand, SOS has the smallest noise coefficient of 12 mV/V which is 10 times lower than 114 mV/V of SLC. SLC noise level is still significantly better than that of single chip value.

Figure 9-24 shows the size of each package and a critical path for the package performance is shown. With this critical path length, the comparison of signal propagation performance is shown in Figure 9-25. The circuit configuration for ASTAP (Advanced Statistical Analysis Program) simulation is shown at the top and the result is shown by three different frequencies. The circuit has assumed an ideal driver which has a matched series output impedance to input a signal. The amplitude is 5 V, and 5 pF capacitance is placed at the end of the line as a receiver input impedance. When the frequency is 20 MHz which has 50 nsec cycle time, there is no problem in all package types. When the frequency increases to 100 MHz, the delay time in SCM becomes almost 2 nsec which is 20% of the cycle time. This may cause problems such as a timing skew and waiting cycle added, and degrade the performance. In 200 MHz, this possibility becomes greater. SCM performance limit comes from the large area size. The high level width which is defined as the pulse width above 3.5 V becomes significantly shorter in SOS application in 200 MHz. There is only 1.86 nsec which is 70% of an ideal case. This trend becomes worse with the increase of

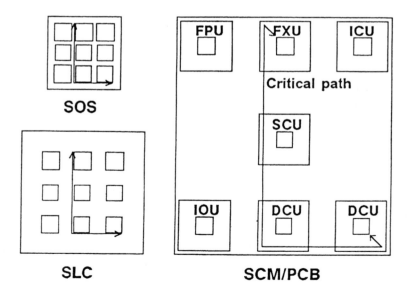

Figure 9-24   Package size and critical path.

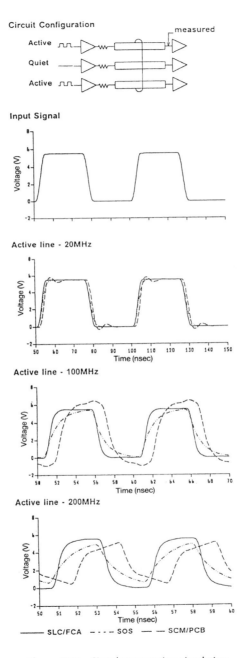

Figure 9-25   Signal propagation simulation.

the frequency. It may be impossible for a signal to propagate on the line to switch the receiver status in a higher frequency. This limitation comes from the high signal line resistance. It is shown that 200 MHz signal can be transmitted on SLC without adding any circuit element in terms of the chip to chip delay and the signal integrity in this simplest circuit. Although the cross-talk noise itself on SLC is not so great, care must be taken in the design phase to avoid a noise problem which may occur by the accumulated noise with other noise sources such as simultaneous switching and reflection.

9.6 SUMMARY

As described in the earlier part, the requirements for the low-end computer and low-cost product are in the various aspects. There has not been a package to provide a single solution to satisfy all such items. The emergence of SLC/FCA technology gives one of the answers for this issue. SLC is a high density low cost carrier for the surface mount technology components and FCA which is the ultimate edge of it provides us with a tremendous opportunity to use various materials for a chip on board carrier. SLC/FCA also resolves the many other items which become technical issues for a chip on board technology. For example, low cost burn-in, lower reflow temperature, encapsulated chip rework and quasi module concept. The expansion of SLC/FCA technology provides more applications in the high performance and low-cost area. The wide spread of this technology in industry is expected to drive us down this road further for future packaging technology.

REFERENCES

1. Tsukada, Y., and S. Tsuchida, "Surface Laminar Circuit, A Low Cost High Density Printed Circuit Board," *Proceedings of Surface Mount International Conference*, Vol. 1, pp. 537-542, August 1992.
2. Tsukada, Y., Y. Mashimoto, T. Nishio, and N. Mii, "Reliability and Stress Analysis of Encapsulated Flip Chip Joint on Epoxy Base Printed Circuit Board," *Proceedings of ASME/JSME Joint Conference on Electronic Packaging*, Vol. 2, pp. 827-835, September 1992.
3. Norris, K. C., and A. H. Landsberg, "Reliability of Controlled Collapse Interconnections," *IBM Journal Research and Development*, Vol. 13, No. 3, pp. 266-271, May 1969.
4. Venkatachalam, P. N., and W. F. Shulter, "Figures of Merit for SMT and Pin Grid Array ASIC Packages," *Proceedings of IEPS Electronics Packaging Conference*, Vol. 1, pp. 444-453, September 1989.

5. Bakoglu, H. B., G. F. Grohoski, and R. K. Montoye, "The IBM RISC System/ 6000 Processor: Hardware Overview," *IBM Journal Research and Development*, Vol. 134, No. 1, pp. 12-22, 1992.

6. Bregman, M. F., "Multi Chip Module for Advanced Workstation Applications," *Proceedings of 7th International Microelectronics Conference*, pp. 30-36, June 1992.

7. Kimura, A., T. Tsujimura, K. Saitoh, and Y. Kohno, "Fabrication and Characterization of Silicon Carrier Substrate for Silicon on Silicon Multichip Modules," *Proceedings of 1st International Conference on Multichip Modules*, pp. 23-27, April 1992.

8. Tsukada, Y., Y. Maeda, and K. Yamanaka, "A Novel Solution for MCM-L Utilizing Surface Laminar Circuit and Flip Chip Attach Technology," *Proceedings of 2nd International Conference on Multichip Modules*, pp. 252-259, April 1993.

10

Micron Bump Bonding Chip on Board

Kenzo Hatada

10.1 INTRODUCTION

The more compact design of modern electronic equipment devices is keenly desired today, and the inevitable tendency of larger chip size, narrower pitched multipins, and smaller pad area is eminent. Thus the conventional semiconductor packaging, methods such as the QFP (Quad Flat Package) or SOP (Small Out Line Package) of wire bonding methods, TAB (Tape Automated Bonding), or FC (Flip Chip) method are not applicable any more to satisfy such requirements.

In contrast to the above, the presently developed micron bump bonding (MBB)[1] method is a new packaging method which is capable of solving all the requirements shown above, and the practical feasibility was confirmed through reliability tests where LSI (Large Scale Integration) chips have an electrode pitch of 10 μm and the number of electrodes of 2320 were assembled.

A module was developed by using this new micron bump bonding method together with a dedicated bonder for this. This chapter reports briefly on the assembling process for this module.

10.2 OUTLINE AND FUTURES OF NEW TECHNOLOGY

Figure 10-1 is a schematic cross-section of an LSI chip after the new bonding process. The LSI chip is connected face down on the substrate. The

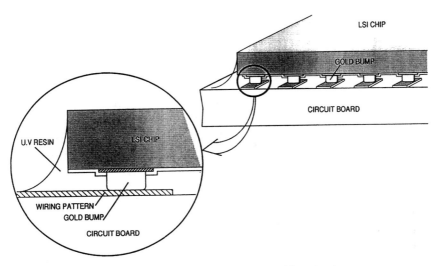

Figure 10-1   Cross-section of new assembly technology.

main characteristic of this new method is the use of a light setting insulation resin for bonding the chip to the substrate. By taking advantage of the inner stress contractile force, the resin develops during the setting time, and the LSI and substrate electrodes are connected under pressure. Unlike the existing solder bump method (Ex: C4 method), the electrodes are not bonded completely. In other words, the fixing of the LSI chip and bonding of the electrodes are separate processes. The attachment of the LSI chip is due to the adhesive power of the resin while the bonding between the electrodes is accomplished by the contractile force developed during setting. These two different processes are accomplished by one kind of resin. For this reason, the electrode pitch can be reduced to a low limit, while at the same time a higher thermal shock reliability can be achieved.

The following are other characteristics:

1. Connection at a micro-order pitch is possible. Since the process uses insulation resin, there is no restrictive element toward the electrode pitch. As far as reliability is concerned, the process can guarantee down to a 10 μm pitch with 2320 pins. The bonding is assured down to 2-3 μm pitch.
2. Due to the mechanical pressure bonding between the electrodes of the LSI chip and the substrate, the structure is flexible and more resistant to thermal and mechanical stress, resulting in less breakage in the bonding area.

3. The LSI chip is easily replaced. With the use of a special solution, the LSI chips are easily removed and replaced. After 30-40 replacement cycles, the substrate electrodes remained unharmed.
4. Bonding costs and overall material costs are low and the process is relatively simple, compared with TAB, for a wire bonding method.
5. Since there is no need for a heat application cycle, the structure is never exposed to high temperatures and this results in less thermally induced stress. For the substrate electrodes, ITO (Indium Tin Oxide) or Au-treated material can be used. Furthermore, glass, ceramic, and resin substrate material can be used.

10.2.1 Method of Process

Figure 10-2 explains the processes of this bonding method. The process is started with the dropping of light setting resin on the substrate by means of a dispenser for an amount corresponding to approximately one-half of the total volume of LSI to be bonded.

The LSI electrodes are the substrate electrodes that are aligned next, and the pressing force is then applied from the LSI chip side against the substrate, and the UV (Ultra Violet) light is irradiated on them. The

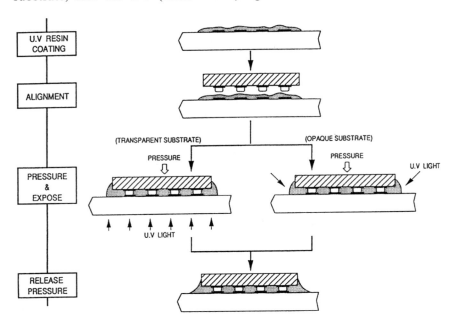

Figure 10-2 Process of micron bump bonding method.

pressing force and the amount of UV irradiation in this case are determined for the minimum bonding resistance and the bondability of 100%, and the realistics are determined from the reliability test results.

The pressing force is necessary to align the electrode planes with each other, between the LSI and the substrate, and the force was set to cause the gold bump deformation of 0.5-1.0 μm. However, the amount of force required depends largely on the flat area, hardness, and the shape of the bump, and the smaller size bump and lower hardness bump naturally require lower forces. For example, the bump with 5 μm² by 5 μm² flat area requires 1.5-3 g bump force.

The UV irradiation time depends on the light source intensity, but if a light source of 1000 mW/cm² was used, the setting would be completed within 3 s. The irradiation was made from the rear side in case of a transparent substrate, and is made from the LSI side for a nontransparent (ceramic or plastic) substrate wherein only the rim area was hardened first and the nonirradiated area was naturally set after the pressing force was removed.

The light setting insulation resin is selected with the following three elements in consideration: thermal stress, adhesive force, and compression stress.

The applied resin was the acrylic type, and a mixture of about eight types of resin was used. This resin is very important material for this method. Key technology of this method is preparedness of the composition for this resin.

10.2.2 Principle of this Technology

Figure 10-3 shows both the distance between the LSI chip and the substrate, and the thickness of the resin is determined by the height of the bump and the amount of pressure applied during the hardening process. The hardening of the resin results in the contraction or compression stress W affecting the LSI chip. The compression stress W has to be roughly equal to the amount of resin contraction Δh. α is the adhesive force between the LSI chip and the resin, β is the adhesive force between the substrate and the resin. The chip and substrate both are affected by the counter force of the compression stress W. In order for the chip to be connected securely, both α and β must individually be greater than W. P is the thermal stress.

Supposing that the surrounding temperature increases and the resin expands. This expansion will be in a direction which tries to separate between the substrate and the chip, causing thermal stress P. Therefore, if the relationship of α, $\beta > W > P$ could be maintained by the proper choice of the resin composition throughout the practical temperature range, high bonding reliability should be assured.

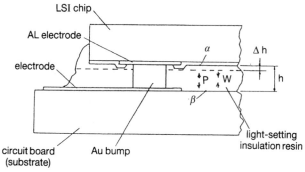

Figure 10-3   Principle of micron bump bonding method.

Figure 10-4 shows the relationship between compression stress W and thermal stress P over a range of temperature applied to the resin. The adhesive force β, between the resin and substrate, and adhesive force α, between the resin and the chip have to be greater than compression stress of

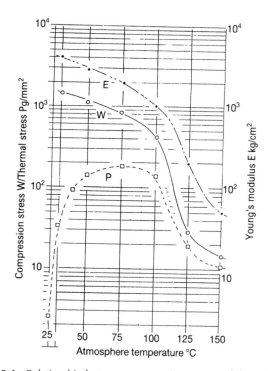

Figure 10-4   Relationship between compression stress and thermal stress.

the resin W. Furthermore, the composition of the resin is chosen to make compression stress W larger than thermal stress P.

At higher temperatures, both the Young's modulus of the resin E and thermal stress P decrease. Thermal stress P is obtained by calculations based on the thermal expansion coefficient and Young's modulus of the resin E. Compression stress W is obtained by a new calculation method based on the amount of warping of the substrate and other elements.

10.2.3 Electrical Characteristics

Electrical characteristic measurements and the reliability tests were carried out for the specimen with 2320 electrodes at the pitch of $10\,\mu m$. The height of the gold bump was $3\,\mu m$, and the glass substrate on which Cr-Au (Cr: 900A. Au:2000A) wiring pattern were formed is shown in Figure 10-5. Figure 10-6 shows the enlarged bonding area of $10\,\mu m$ pitch pattern formed on a glass substrate.

As shown by Figure 10-7, the contact resistance of the bonded area was $8\,m\Omega$ constant. The V-I characteristics between the chip and the substrate electrodes showed a linear relationship until $10^{-5}V$, as shown by Figure 10-8, and this proved that no electrical barrier layer was formed in it.

Figure 10-5   Specimen with 2320 electrodes at the pitch of $10\,\mu m$.

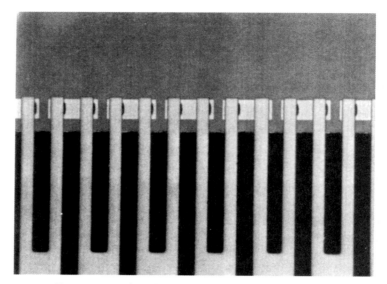

Figure 10-6 Enlarged bonding area of 10 μm pitch pattern.

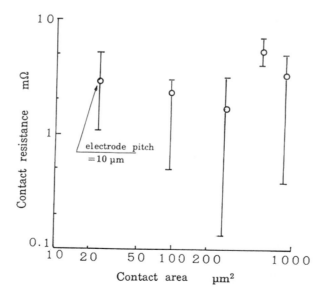

Figure 10-7 Contact resistance of the bonded area.

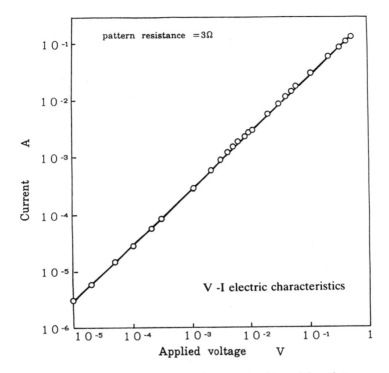

pattern resistance = 3Ω

V -I electric characteristics

Current   A

Applied voltage   V

Figure 10-8 V-I characteristics between the chip and the substrate.

The maximum permissible current for $2000\,\mu m^2$ contact area was then measured, and the perfect connection was achieved even at 1A current, in spite of the temperature rising to a certain extent due to the joule loss in the wiring resistance, which exceeded the contact resistance between the electrodes.

The insulation resistance between adjacent electrodes is measured from 5 to $100\,\mu m$ (Fig. 10-9). The insulation resistance depends on that of the light setting insulation within resin and the range of the measured electrode pitch.

10.2.4 Reliability

Thermal shock tests were considered the severest tests among the reliability tests; the complete contacts were kept until a very high temperature. For the low temperature region, on the other hand, cracks took place in the resins with greater hardness, thus attention has to be paid to resin hardness control. The physical shape of the resin after the setting affected the thermal shock test results.

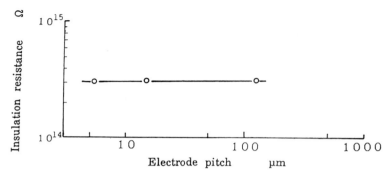

Figure 10-9   Electrode pitch dependency of insulation resistance.

Table 10-1 shows the tabulated results of three reliability tests including the 125°C high temperature storage test, the 85°C/85% RH tropical storage test, and the −55°C/125°C thermal shock test. Each point represents the average of 20 specimens.

Another thermal cyclic shock test (−55°C/125°C), and an acceleration test (10G) was also conducted, and satisfactory results were attained with acrylic resins.

Other reliability experiments, by using the practical LSI chip specimen which has the bump dimension of 50 μm by 50 μm at 100 μm pitch, mounted on the glass and the ceramic substrates on which ITO and gold writing patterns were formed respectively, were conducted and the results also proved the practicality of this bonding method.

Table 10-1   Result of Reliability

Test Type	Test Condition	Variation of Resistance Value	Existence of Defective Connection
High temperature storage	150°C 2000H	±5%	0%
High temperature High humidity	85°C/85% 2000H	±5%	0%
Heat shock	−55°C/+125°C 1000C	±5%	0%
Thermal cycle	−55°C/+125°C 1000C	±5%	0%

10.3 TYPICAL APPLICATIONS OF MBB METHOD[1]

This chapter is to describe some of the typical applications of the MBB method including the applications to the LED (Light Emitting Diode), thermal head modules, and MCM (Multi-Chip Module).

The LED module is a readout sensor fabricated on a glass substrate, while the thermal head is a printing device fabricated on a ceramic substrate. Both are incorporated in facsimiles. MCM is a memory module incorporated in personal computers.

10.3.1 LED Array Module

Construction of the LED Array Module[2-4]

Figure 10-10 shows a constitution of the LED array module wherein both LED and driver LSI chips are disposed. A heat-resistant glass plate of 246 mm by 14 mm in size was employed as the circuit substrate, on which a circuit wiring pattern was formed, by conducting a photolithographic process on the gold layer deposited on the entire substrate surface at a thickness of 2.0 μm. The LED consists of a GaAs chip of 4.06 mm by 0.5 mm, having an electrode pitch of 63.5 μm, the number of electrodes is

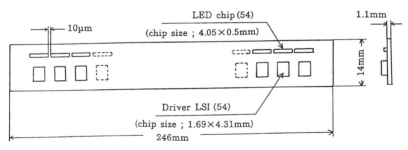

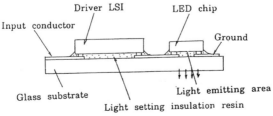

Cross-sectional of LED array module

Figure 10-10   Constitution of LED array module.

126, and the LED dot number is 64. The P-N (Positive type-Negative type) junction thereof is a coplanar construction in order to facilitate the face down bonding of LED chips on the glass substrate. The chip size of driver LSI is 1.69 mm by 4.31 mm, and 110 electrodes are provided on the chip at a pitch of 100 μm, and the size of formed gold bump is 30 μm in diameter with a height of 7 μm (Table 10-2).

Table 10-2   Dimensions of Chip

	LED Chip	Driver LSI
Size of Chip	4.06 × 0.5 mm²	1.69 × 4.31 mm²
Pad's Pitch	63.5 μm	100 μm
Pin Count	126	110
Dot Count	64	—

The LED light emission is externally available from the reverse side of the substrate glass plate, and by using a heat dissipating sheet, a heat radiating fin having a dimension the same as the glass substrate, Si (Silicon) is attached on the reverse size of the LED and driver LSI chips.

LED Array Module Fabrication Process

Figure 10-11 shows summarized LED array assembling processes among which the LED bonding process requires particular attention. This is explained in the following steps:

1. One of LED chips is vacuum-picked by a collet while light setting insulating resist is coated on the electrode side surface of LED.
2. The alignment work between the electrode of LED chip 1 and the electrode of the circuit substrate is then carried out, and these are pressed together by using a bonding tool. Thus by means of this application of force, the bumps on LED array electrode are pressed and deformed against the electrodes of the circuit for making an improved contact between these electrodes. The plastic deformation of bump is within an order of 10-20%, and the resin exited between these electrodes is expelled leaving nothing between at this time. UV light is then irradiated on the half of chip 1 which is remote from chip 2 to be bonded in the next step. Thus only the resin on the left chip is light set leaving the resin on the right unset as shown in Figure 10-11.

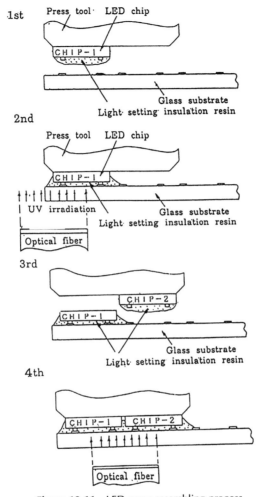

Figure 10-11   LED array assembling process.

3. The alignment between resin-coated chip 2 and the corresponding wiring pattern on the substrate is then performed, keeping the distance from the neighboring chip 2 at 10 μm.
4. A force is applied on chips 1 and 2 simultaneously by means of a tool while the UV light irradiation is made on both the right side of chip 1 and the left side of chip 2 for setting the resin coated on them. Thus the complete electrical connection between chip 1 and substrate is attained. The remaining chips are then bonded in the same way until the last chip is bonded.

A particular light setting resin having a main constituent of denatured acrylate resin with several additives was developed for this purpose. It has a viscosity of 2000 CPS, a thermal expansion coefficient of $8 \times 10^{-5}/°C$, and a Young's modulus of $3000\,kg/cm^2$ at 25°C.

If a chip failure is found after the completion of chip assembly, the chip can be removed by either swelling the resin by applying a solvent by a gauze patch, or heat application on the chip from the reverse side of the substrate until the chip is heated above its glass transition temperature Tg for loosening. The resin left behind can be removed completely by a special solvent, and a new chip is mounted on it for completing the chip replacement. The reliabilities were confirmed for such repair works, not only on the replaced chip, but on the neighboring chips on the left and right.

Evaluation of Electrical Characteristics

The electrical characteristics of modules assembled under the previously described conditions are evaluated and described below.

The V-I characteristics of LED before and after the bonding are shown in Figure 10-12 which shows virtually no difference, and thus no effects on the contact resistance from the pressing.

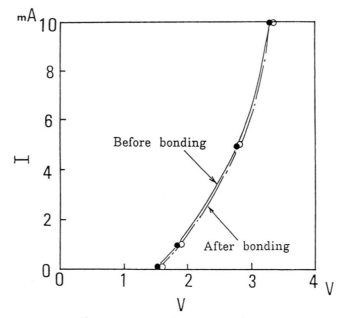

Figure 10-12 V-I characteristics of LED.

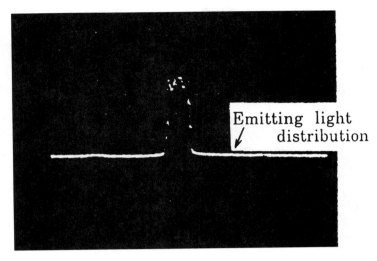

Figure 10-13   Light emitting characteristics of LED array modules.

Figure 10-13 is measured emitting light distribution showing very little light spreading, and this also shows very little light scattering during the light transmission through the light setting resin and glass substrate.

The temperature rises when all 54 LEDs mounted together with 54 driver ICs on an A4 size LED array head are energized, and a current of 3 mA is fed through each of the 3456 bonding dots at the same time and is measured, and found to be less than 40°C, which is considered satisfactory.

Experimental fabrication of A4 size LED array modules shown in Figure 10-14 was conducted under the above described process conditions, and bonding yield of 99.8% was obtained after approximately 2000 LSI chips were bonded. Analysis then found that the environmental dusts were entirely responsible for these bonding failures. It is then essential to perform a fine bonding work in the environment of improved cleanliness.

The results of bonding reliability tests and the aging effects of light emission characteristics were found satisfactory as shown in Table 10-3.

10.3.2 Thermal Head Module[5,6]

Construction of Thermal Head Module

A cross-section of the thermal head fabricated by the MBB method is shown in Figure 10-15. The heating elements are disposed on a glazed layer formed on a ceramic substrate in order to save the generated heat from

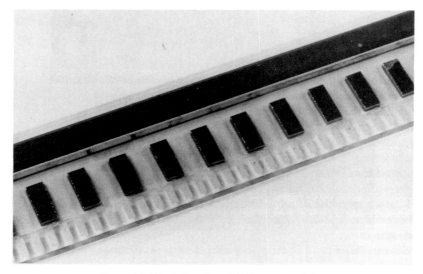

Figure 10-14   Entire view of LED array module.

dissipation, and gold wiring for interconnecting the heating elements is disposed on it.

The driver ICs with gold bumps are bonded on the wiring by means of the MBB method. In order to obtain stable contacts of driver ICs against the wiring, wiring consisting of a gold layer electroplated on the patterned Cr-Cu layer which is deposited on the substrate in vacuum, is used.

An acrylate resin having a viscosity of 850 cp is employed for the bonding. The resin composition is determined by considering the resistances against the humidity, heat, and the heat-cycles. In order to obtain the thermal heads of the thinnest construction, no protection coating is applied after the IC bonding is performed, so that the reliability has to be secured only by the light-setting insulating resin employed for bonding.

The resin is a single liquid type, and the setting of it is performed by applications of both ultraviolet rays and heat.

Table 10-3   Result of Reliability Tests for Array Modules

Test Items	Condition	Failures
High Temperature Storage	125°C 100H	0/240
THB	85°C/85% 1000H	0/240
Thermal Shock	−55°C/+125°C 1000c/s	0/240

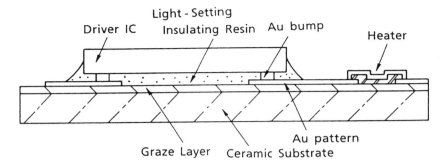

Figure 10-15   Cross-section of thermal head module.

Bonding Process

The bonding process developed for this case is shown in Figure 10-16. Though the number of ICs to be bonded is determined by the head size and bit construction of IC, 32 ICs have to be bonded to construct a B4-size, 400 DPI (Dot Per Inch) head.

In order to attain a higher production efficiency, the pressure application and UV irradiation of the chips are performed at an independent stage after the completion of IC bonding on the substrate. By employing this method, the bonding and the applications of pressure and UV irradiation can be performed simultaneously, and a production speed twice the one obtained by the conventional method is attained.

Characteristics

The characteristics of the thermal head fabricated by using the above-shown construction are given in the following:

Contact Resistance   Figure 10-17 shows the contact resistances attained between the gold wiring and the gold bumps. These are about $6\,\text{m}\Omega$ on average, which is sufficiently low.

High-Temperature Operation Test   As the heating element consumes the highest driving current, the highest current goes through between the IC bump and the wiring electrode. Furthermore, since the heat generated by the heating elements is spread throughout the substrate, the temperature of IC chips may go up to as high as $80°C$.

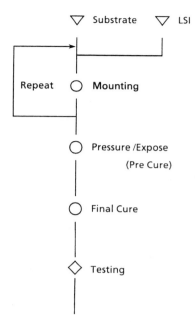

Figure 10-16   Process flow of thermal head module.

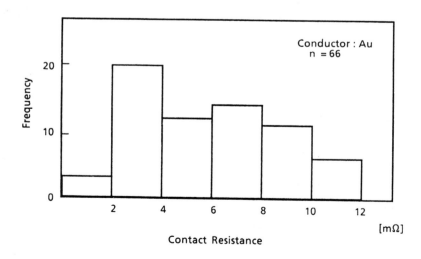

Figure 10-17   Contact resistance between the electrodes of substrate and the electrodes of driver' IC.

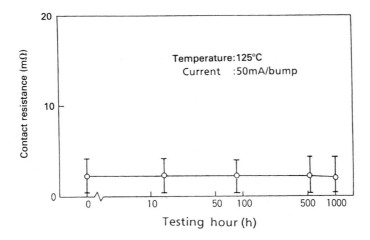

Figure 10-18   Result of high temperature and current stress testing.

In order to simulate such a thermal head proper condition, the assembled head is subjected to a high temperature and high driving current test. Figure 10-18 shows a result of the test conducted at a temperature of 125°C and a driving current of 50 mA/bump, showing the change in the contact resistance after 1000 hr, and is considered satisfactory.

Gang Bonding of ICs Having a Terminal Pitch of 50 μm

In order to attain a higher dot density, lower costs, and higher reliability at the mass production level, a head assembly employing IC chips having a terminal pitch of 50 μm has been tested.

Although the possibility of bonding the chip having a terminal pitch of 10 μm had been experimentally confirmed previously, problems of IC chip inspection and the alignment accuracy have to be solved before the mass production of heads assembled at a 50 μm pitch becomes possible.

The IC chip developed for this purpose has a dimension of $1.4 \times 4.2\,mm^2$, 208 electrodes disposed at a pitch of 50 μm, and 128 output bits. In order to match with the 50 μm pitch, a gold bump diameter of 25 μm is employed. Figure 10-19 shows an SEM photograph of the bump.

Since the direct probing of the bumps having a size less than 25 μm is considered impossible, a barrier metal layer of zigzag pattern is disposed on the active area of IC in order to make the probing of 50 μm pitch IC possible.

Figure 10-19 SEM photograph of the bump.

Since the chip is reduced to 58% of the conventional size, a substantial cost reduction of IC chips is accomplished also.

An MBB bonder dedicated to mass production, having an alignment accuracy less than $+10\,\mu m$ which is necessary to realize $50\,\mu m$ pitch bonding, has been developed.

The developed MBB bonder shown in Figure 10-20 is equipped with a Θ compensation mechanism having a resolution of $1/400°$ and a precision alignment method employing a camera having two viewing fields.

The optimum conditions for the suction collet, mounting speed, and the applied pressure have also been experimentally determined. As a result of these, an alignment accuracy of $\pm 5.4\,\mu m$ at 3σ is attained clearing the target value.

Mass-Production of 400 DPI Thermal Heads

An external view of the fabricated 400 DPI thermal head is shown in Figure 10-21. By using the MBB method, compared with the conventional TAB assembly method, substantial reductions of head size and assembly costs and improvements of yields and reliability at the mass-production level have been successfully accomplished.

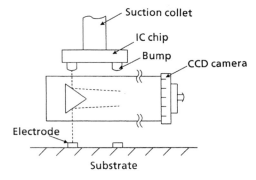

Figure 10-20   Alignment system of developed bonder.

Evaluation of Reliability

A result of the evaluation tests of 400 DPI thermal heads fabricated by the above, is shown in Table 10-4. This shows satisfactory results in every aspect of the specifications.

10.3.3 MCM Module[7]

Construction of MCM Module

A computer-use MCM module on which LSI chips are assembled by the MBB method is described in the following. Figure 10-22 shows a cross-section of the MCM module assembled by MBB.

Table 10-4   Result of Reliability Tests for Thermal Head Modules

Test items	Test conditions	N	Result
Thermal cycle	$-45°C \rightleftharpoons RT \rightleftharpoons 85°C$	28	300 cycle OK
High temperature and High humidity storage	85°C 85% RT	16	1000H OK
Low temperature storage	$-45°C$	4	1000H OK
High temperature	85°C	4	1000H OK
THB	85°C 85% RT 24V/5V	3	1000H OK
High frequency vibration	$5Hz \rightleftharpoons 100Hz \rightleftharpoons 5Hz$ 1G	3	OK
Compound test	$T/C \rightarrow TH$	2	1000H OK
Compound test	$TH \rightarrow T/C$	2	300 cycle OK

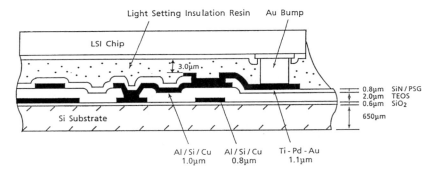

Figure 10-21   400 DPI thermal head by micron bump bonding method.

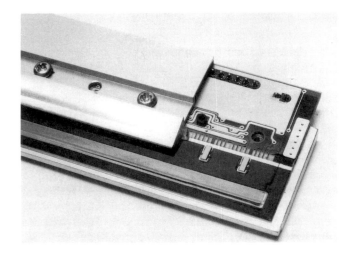

Figure 10-22   Cross-section of developed MCM for computer.

A silicon substrate is used as a wiring substrate on which a triple layered wiring made of SiO_2 and aluminum layers is formed. In order to reduce the contact resistance, the top electrode is of a triple layered construction of Ti-Pd-Au layers, while gold bumps are formed on the LSI chips.

An SiO_2 layer is deposited on the silicon substrate, and an aluminum layer as the first layer is formed on it. The thickness of the SiO_2 layer deposited between the first and the second aluminum layers is about 2 µm, and Si and Cu elements are introduced in the aluminum wiring material in order to prevent the electromigration between the wirings.

The third wiring layer has a triple layered construction made of Ti (2000A)-Pd (5000A)-Au (1000A) layers, so that the Ti layer is contacted against the aluminum electrode, and the chip bumps are contacted through the gold layers.

The wiring pitch is 60 μm with an inter-space of 30 μm, and the pierce hole diameter is 30 μm. The lowest insulation layer is formed by thermal oxidation, and the interlayer insulation layer is an SiO_2 layer formed by a CVD method. In here, the first and second wiring layers act as the signal and ground lines, and the third layer acts as the electrode pads for connection.

The bump formation on LSI chips could be performed by either of the two methods. One of the methods is to form a metal multi-layer on the LSI chip and the bumps are formed by electroplating on them afterward. The other is a so-called bump transfer method[8] by which the bumps are directly formed on the aluminum electrode of LSI.

The MCM module construction shown in Figure 10-22 is an example fabricated by the bump transfer method. Since no particular additional process is required for the LSI chips, and bumps are formed only on the accepted chips, the bump costs could be lower than those obtained by the former. Figure 10-23 shows the bumps transferred on the aluminum electrodes of LSI. The bump material is gold and has a height of 10 μm and a diameter of 30 μm. The chip is bonded on the aluminum electrode by means of Au-Al alloying in this case.

The LSI chips are mounted on the circuit substrate by means of the previously described MBB method. The designed inter-chip distance is

Figure 10-23 Transferred bumps on the electrodes of LSI chip.

0.5 mm, and the ratio of chip area to the circuit substrate is 0.8 which means a considerably high-density assembly.

Process

Figure 10-24 shows an employed MCM process. The circuit substrate is diced into a determined dimension after the circuit is formed. The LSI chips are tested at a state of wafer, and bumps are formed by the bump transfer method after the wafer is diced into individual chips.

A light-setting insulation resin is coated on the substrate area on which the LSI chip is to be bonded, and the chip and the circuit electrodes are aligned and pressed onto each other, and the resin is set while this condition is maintained by introducing the UV irradiation from the side of the chip.

The placing and the alignment of chip and the UV irradiation are automatically performed by the dedicated bonder. The interchip distance is 0.5 mm, and after the finishing of all the chip bondings, the electrical

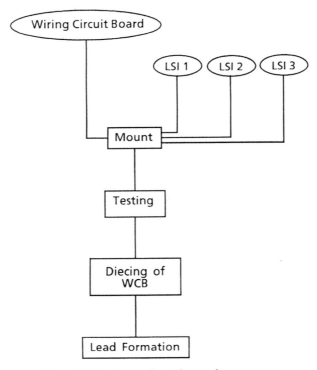

Figure 10-24 Process flow of MCM for computer.

characteristics of the module are tested. If a chip failure is found, the resin is softened and the chip is heated on the back side of the chip by heater block. The chip is removed from the circuit substrate by this action.

The particular area from which the failed chip is removed is cleaned by a specified solvent, and a new tested chip is bonded on it, and the module is tested again. Then the lead-wires for external connection are mounted on the peripheral region of the module.

The final function test is conducted after the lead connections are finished. The chips remain in this condition and no further protection coating is applied. A satisfactory reliability can be secured as it is.

Evaluation of Reliability

Figure 10-25 shows an assembled MCM module consisting of three first memory chips and two second memory chips and one logic LSI assembled on a 20 mm square silicon circuit substrate. Reliability tests shown below are conducted for this module which is operated at a frequency of 60 MHz.

Figure 10-25   Developed MCM of computer.

150°C high-temperature storage test for 2000 hr.
85°C, 85% RH tropical storage test for 2000 hr.
85°C, 85% RH bias test for 2000 hr.
PCT for 1500 hr.
−55°C/150°C thermal shock test for 1000 cycles.

As the result of these tests, no failures of modules are observed.

10.4 SUMMARY

The process of a new chip assembly method utilizing the shrinkage stress of light-setting resin by which LSI chips are press-bonded against the electrodes provided on a substrate, i.e., the MBB method, is developed. Its effectiveness to assemble electronic devices and their reliabilities are experimentally confirmed. This development and the reliability are made possible by the development of a new light-setting resin.

Thus, it is expected that the MBB method would be applied to all types of MCMs of electronic equipment, and that new applications of this would be produced in future, so that this is expected to be a core technology to assemble bare chips.

REFERENCES

1. Hatada, K., et al., "A New LSI Bonding Technology–Micron Bump Bonding Assembly Technology," *IEEE/CHMT Symposium Proceedings*, pp. 23-27, 1988.
2. Hatada, K., et al., "LED Array Modules by New Technology Micron Bump Bonding Method," *IEEE/CHMT Transactions*, Vol. 13, No. 3, pp. 521-527, September 1990.
3. Hatada, K. et al., "Applications of New Assembly Method–Micron Bump Bonding Method," *IEMT '89*, Japan, pp. 47-50, 1989.
4. Hatada, K., et al., "Micron Bump Bonding Assembly Technology," *ISHM '89 Proceedings*, pp. 245-248, 1989.
5. Hatada, K., et al., "A New LSI Bonding Technology–Micron Bump Bonding Technology," *39th ECC Proceedings*, pp. 45-49, 1989.
6. Fujimoto, H., et al., "High Density Thermal Head by Micro Bump Bonding Technology," *4th Microelectronics Symposium (MES '91)*, pp. 25-28, Tokyo, 1991.
7. Hatada, K., et al., "Assembly of Consumer Electronics Equipment by MCM," Japan Technology Transfer Association, *The 33rd High Density Assembly Technology Committee*, pp. 1-5, May 1992.

8. Hatada, K., et al., "Bump Transfer Technology to Al Electrodes of LSI Chip," *The Institute of Electronics, Information, and Communication Engineers*, VLD 88-71/ICD 88-89, pp. 55-58, October 1988.

11

Chip on Board Encapsulation

C. P. Wong, John M. Segelken, and Courtland N. Robinson

11.1 INTRODUCTION

The demand for small, lightweight, low-cost, high performance electronic devices and subsystems is becoming an ever-increasing dominant driving market force. Multichip module (MCM) technologies, and chip on board technology, and the combination of these two technologies are providing design and manufacturing solutions to the marketplace.

The electronics packaging industry has seen emerging MCM technologies in three categories:

1. MCM-C Typical multilayer co-fired ceramic – sometimes referred to as hybrid ceramics.
2. MCM-D Thin films deposited structures on a substrate structure–silicon, ceramic, metal, even epoxy-glass.
3. MCM-L Laminated Printed Circuit Board (PCB) or flex circuit constructions.

Over the past 20 years various forms of COB have continued to evolve into today's MCM-L technologies

We defined this technology as MCM-Ls manufactured with COB–"a device or subsystem with two or more unpackaged encapsulated integrated circuit (IC) dies (chips) assembled by a common interconnect that utilizes high density wire bondable PC or flex circuit material as the basic substrate comprising the functional circuit unit or module."

These modules have been manufactured for some years, in high volumes used in numerous commercial and industrial products such as watches, calculators, electronic games, cameras, memory cards, cellular phones, all manner of handheld personal computers, and small electronic devices.

Some competitive advantages gained through the use of MCM-L COB are:

- Low cost material base
- Proven "small" form factor part affecting assembly processes and materials
- Easily combined with through-hole and surface mount technologies
- Low profile, memory card format for notebook/palmtop computers
- Various size and form factor (shape) requirements can be satisfied
- Shorter signal path distances resulting in good electrical performance.

The design engineer is always faced with developing a packaging strategy that satisfies and optimizes manufacturability and market needs. The following factors (at the very least) need to be addressed:

- Cost
- Reliability
- Circuit density, speed, I/Os
- Power distribution
- Power dissipation, thermal management
- Testability, yield
- Desired shape, size, form factor
- Manufacturability
- Product life
- Time to market.

Market demands and high performance chip technologies continue to create challenges and continued innovation in electronic packaging.

MCM-L and COB as the industry has defined it will certainly continue to evolve with MCM-C and MCM-D solutions as well. Also, chip level interconnection will be not only wire bonded (Chapter 3) but Tape Automated Bonding (TAB) (Chapters 4 and 8) and Flip Chip [sometimes referred to as C4 (IBM-controlled collapse chip connection)] (Chapters 5 and 9) will also find their market/performance uses.

This chapter will focus on the encapsulation issues related to these challenging design, manufacturing, and reliability COB solutions. The approach of the chapter utilizes a roadmap or structural framework by first establishing purposes and goals and defining material requirements. Potential encapsulants are discussed and screening tests and procedures are presented. Appropriate manufacturing processes are also examined and reviewed. A list of extensive and useful references is provided.

11.2 ENCAPSULATION OF COB

Encapsulation is the key in COB operation. Prior to our discussion in encapsulation of COB, we have to address the purposes and goals for this operation, the material requirements, potential encapsulants and screening tests and analyses of the COB encapsulants. These are discussed as follows:

11.2.1 Purposes and Goals for Encapsulation

The purpose of encapsulation is to protect the electronic components from adverse environment (from $-65°C$ to $150°C$ for Mil. Spec. 883, thermal shock, temperature cycle during actual life applications, etc.), and increase the long-term reliability of the electronics with reasonably low-cost materials. High performance and low-cost materials are the major driving force in the 1990s.

Electro-oxidation (corrosion) and metal migration are attributed to the presence of moisture. Generally speaking, when enough moisture diffuses through the encapsulant to form a continuous water path, with the presence of mobile ion(s) and under electrical bias condition, electro-corrosion begins. Pure crystals, metals, glass, and other inorganics are the best moisture barriers; however, most organic compounds are quite permeable to moisture. Figure 11-1 shows the permeability of various materials. Pure crystals and metals are the best materials as moisture barriers. Glass (silicon dioxide) is an excellent moisture barrier, but is slightly inferior to pure crystals and metals. Organic polymers, such as fluorocarbons, epoxies, and silicones are a few orders of magnitude more permeable to moisture than glass. (Silicone materials which have the highest moisture permeability of most polymers, are one of the best device encapsulants. The reason for this is the subject of our discussion later in this section). Obviously, gases are the most permeable to moisture of all materials as shown in Fig. 11-1. In general, for each particular material, the moisture diffusion rate is proportional to the water vapor partial pressure and inversely proportional to the material thickness. This is accurate when moisture diffusion rates are in steady-state permeation. However, moisture transient penetration rates (perhaps more important because they determine the time it takes for moisture to break through) are inversely proportional to the square of material thickness.

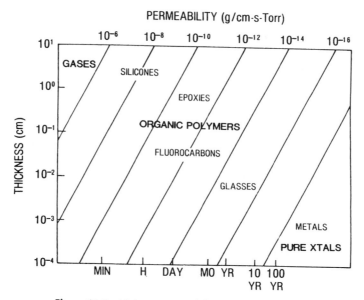

Figure 11-1   Moisture permeability of various materials.

Mobile ions, such as sodium and potassium, tend to migrate to the p-n junction of the IC device where they acquire an electron, and deposit as the corresponding metal on the p-n junction. This consequently destroys the device. Furthermore, mobile ions will also support leakage currents between biased device features, which degrade device performance and ultimately destroy the devices by electrochemical processes such as metal conductor dissolution. For example, chloride and fluoride ions, even in trace amounts (at ppm level), could cause the dissolution of aluminum metallization of complementary metal oxide semiconductor (CMOS) devices. Unfortunately, CMOS is likely to be the trend of the VLSI technology and sodium chloride is a common contaminant. The protection of these devices from the effects of these mobile ions is an absolute requirement. The use of an ultra-high purity encapsulant to encapsulate the passivated IC is the answer to some of these mobile ion contaminant problems.

Ultra Violet-Visible (UV-VIS) light radiation can cause damage to light sensitive opto-electronic devices. UV-VIS protection can be achieved by choosing an opaque encapsulant. However, impurities in an encapsulant, such as low levels of uranium in the ceramic or plastic packages can cause appreciable alpha particle radiation. So can cosmic radiation in the atmosphere. The alpha radiation can generate a temporary "soft error" in operating dynamic random access memory (DRAM) devices. This type of alpha particle radiation has become a major concern, especially in high-density memory devices. Good encapsulants must have alpha radiation levels less than 0.001 alpha particles/cm²/hour, and be opaque in order to protect a device from UV-VIS radiation. Since the alpha particle is a weak radiation, a few micrometers thickness of encapsulant will usually prevent this radiation damage of the DRAM devices.

Hostile environments, such as extreme cycling temperature (values from $-65°C$ to $+150°C$ in Mil. Spec. 883), high relative humidity (85% to 100%), shock and vibration, and high temperature operating bias are part of the real life device operation. It is critical for the device to survive these operation-life cycles. In addition, encapsulants must also have suitable mechanical, electrical and physical properties (such as minimal stress and matching thermal expansion coefficient, etc.) which are compatible with encapsulated IC devices and the underlayering substrates.

11.2.2 Material Requirements

In addition to the above requirements of low moisture permeability, excellent mobile ions barrier, good UV-VIS and alpha particle protection, excellent mechanical, electrical and physical properties, the encapsulant must have a low dielectric constant (less than 3.5) to reduce the device propagation delay, and excellent thermal conductivity to dissipate heat generated by IC devices. Furthermore, the encapsulant must be ultrapure material, with extremely low ionic contaminants (less than a few ppm level). Since the encapsulation is the final process step and some of the devices are expensive, particularly in the high density multichip modules (MCM), it must be easy to apply and repair in production and service. With the proper choice of encapsulant and process, the encapsulation can enhance the reliability of the fragile IC device, improve its mechanical and physical properties, and its manufacturing yields.

11.2.3 Potential Encapsulants

There are many polymeric materials that could be used as electronic components. However, only a few with suitable electrical, chemical, and physical properties, are widely used today.

Epoxies

Epoxies, as a class of polymers, have long been attractive candidates for silicon device encapsulation in COB applications. These materials received serious consideration once the level of ionic impurities was reduced to less than 25 ppm. The epoxy systems most widely in use for COB applications are liquid systems based on epoxy resins cured with either novolacs or anhydrides (both monofunctional and difunctional). Reactions typifying these systems are shown on Fig. 11-2. The resulting cured polymers are materials characterized by a highly crosslinked network structure. As such, they exhibit high Young's modulus ($E \approx 10^5$ psi) and high fluxial strength $\sigma_b \approx 2 - 10 \times 10^3$ psi) along with outstanding solvent and other chemical resistance. They also exhibit very good thermal stability with glass transition temperatures, $T_g \geq 150°C$. In addition to their very good physical and chemical properties, epoxies also exhibit excellent adhesion to a wide range of surfaces. When epoxies are used as chip encapsulants in COB applications, their natural properties give rise to several issues.

Most important is the need for the encapsulants to have a low thermal coefficient of expansion (TCE). This is necessary in order to minimize thermal stresses due to the TCE mismatch between the encapsulant and the

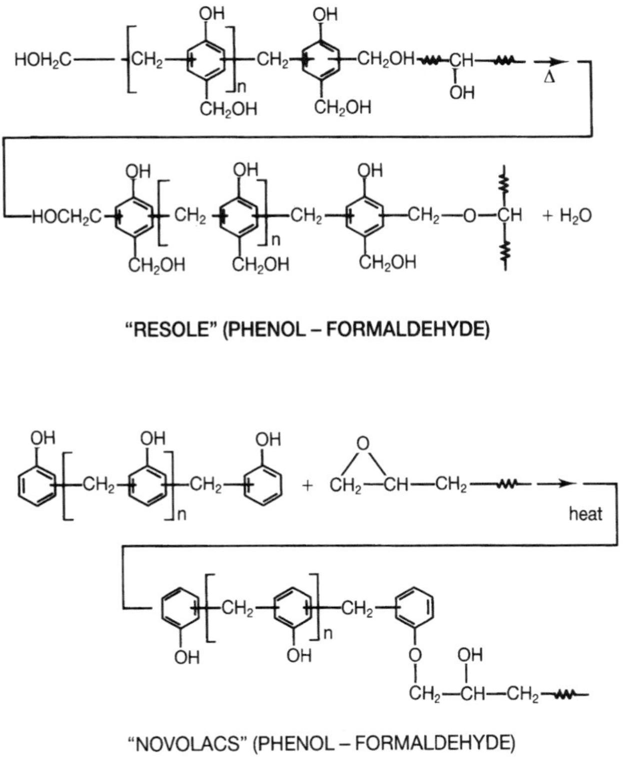

"RESOLE" (PHENOL – FORMALDEHYDE)

"NOVOLACS" (PHENOL – FORMALDEHYDE)

Figure 11-2 Chemical structure of high performance Novolac system.

other circuit materials including the substrate (PCB or ceramic), the silicon IC chips and their solder or wire bond interconnections. This problem is handled in the epoxy systems through the judicious incorporation of appropriate fillers. Using fillers such as fumed silica and crystalline quartz, epoxy systems having TCEs below their T_g in the range of 10-15 ppm/°C are commonly available.

Another issue arising from the properties of epoxy systems is related to their excellent adhesion characteristics coupled with their high modulus. Because of these problems of substrate warping (particularly with ceramic substrates) and encapsulant cracking during low temperature excursions have been encountered. If the area of the substrate covered by the epoxy is small relative to the area of the substrate, warping is minimized. Reducing the TCE, as described above, significantly reduces the incidence of these problems as well. However, these problems have been minimized further through the toughening of the basic epoxy polymer. This has been achieved through the incorporation of elastomeric butadiene or silicone rubber flexibilizing agents into the polymer backbone structure. The resulting epoxy polymers have somewhat reduced moduli along with increased extensibility. Recently, the use of epoxy encapsulants by IBM and Hitachi with an epoxy TCE (~25-28 ppm) closely matched with the flip-chip interconnect solder joint has proven to extend the interconnect temperature cycling results.

Silicones

Polydimethylsiloxanes, polydiphenylsiloxanes and polymethylphenylsiloxanes are generally called silicones. With a repeating unit of alternating silicon-oxygen siloxanes chemical backbone structure, silicones possess excellent thermal stability and flexible mechanical properties that are superior to most other materials. Polydimethylsiloxane, with a free rotating methyl group, provides a very low T_g (glass transition temperature ~ −125°C) material which also is suitable for use at very low temperatures. The basis of commercial production of silicones is that chlorosilanes are readily hydrolyzed to give disilanols which are unstable and condense to form siloxane oligomers and polymers. Depending on the reaction conditions, a mixture of linear polymers and cyclic oligomers is produced. The cyclic components can be ring opened by either acids or bases to become linear polymers and it is these linear polymers that are of commercial importance. The linear polymers are typically liquids of low viscosity and, as such, are not suited for use as encapsulants. These must be cross-linked (or vulcanized) in order to increase the molecular weight to a sufficient level for the properties to be useful. Three methods of cross-linking are used: (1) classified as condensation cures (see Fig. 11-3), (2) addition cure systems (see Fig. 11-4), and (3) peroxide free-radical cure systems (see Fig. 11-5). For electronic applications, only the high purity room temperature vulcanized (RTV) condensation cure silicones which used alkoxide cure systems giving non-corrosive alcohol by-products, and the platinum-catalyzed addition heat cured vinyl and hydride (hydrosilation) silicone systems are suitable for device encapsulation.

Room temperature vulcanized (RTV) silicones are typical condensation cure system materials. The moisture-initiated, catalyst (such as organotitanate, tin dibutyldilaurate, etc.) assisted RTV process generates water or alcohol by-products which could cause outgassing and voids. However, by careful control of the curing process, a very reliable encapsulant can be achieved. Since the silicone has a low surface tension, it tends to creep and run over surfaces to which it is exposed. To control better the rheological properties of the material, a thixotropic agent (such as fumed silica) is usually added to the formulation. The thixotropic agent provides a yield stress, increases the storage modulus (G'), loss modules (G'') and dynamic viscosity (η^*) of the encapsulant. Filler-resin and filler-filler interactions are

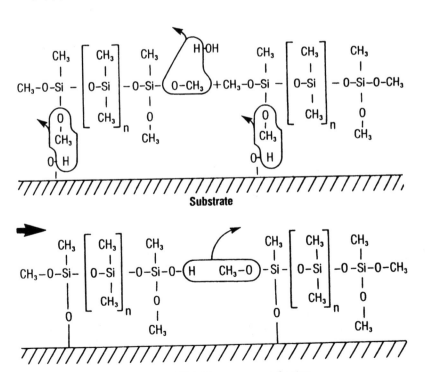

Figure 11-3   RTV silicone cure mechanism.

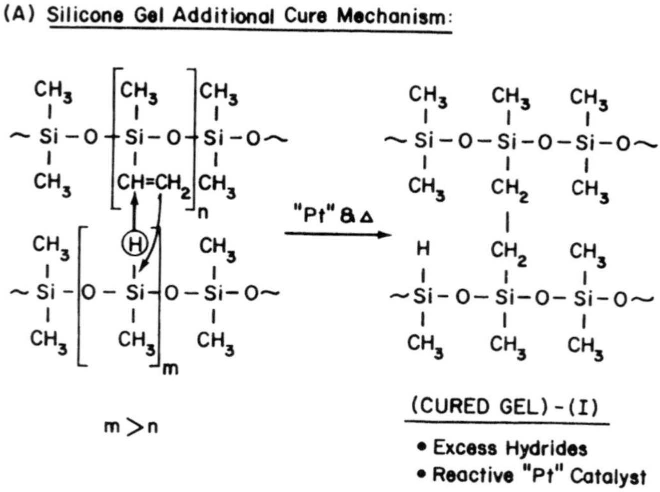

(A) Silicone Gel Additional Cure Mechanism:

$$
\begin{array}{c}
\overset{CH_3}{\underset{CH_3}{\sim Si - O}} \left[\overset{CH_3}{\underset{CH=CH_2}{Si - O}}\right] \overset{CH_3}{\underset{CH_3}{Si - O \sim}}
\end{array}
$$

$$
\begin{array}{c}
\overset{CH_3}{\underset{CH_3}{\sim Si - }} \left[\overset{\textcircled{H}}{\underset{CH_3}{O - Si}}\right]_m \overset{CH_3}{\underset{CH_3}{O - Si - O \sim}}
\end{array}
$$

$$m > n$$

$\xrightarrow{\text{"Pt" & }\Delta}$

$$
\begin{array}{c}
\overset{CH_3}{\underset{CH_3}{\sim Si - O}} - \overset{CH_3}{\underset{CH_2}{Si - O}} - \overset{CH_3}{\underset{CH_3}{Si - O \sim}}
\end{array}
$$

$$
\begin{array}{c}
\overset{H}{\underset{CH_3}{\sim Si - O}} - \overset{CH_2}{\underset{CH_3}{Si - O}} - \overset{CH_3}{\underset{CH_3}{Si - O \sim}}
\end{array}
$$

(CURED GEL) - (I)

• Excess Hydrides
• Reactive "Pt" Catalyst

Figure 11-4 Heat cure silicone mechanism (hydrosilation).

important in obtaining a well-balanced and well-controlled encapsulant. This rheologically controlled material (see Fig. 11-6) tends to flow evenly into each circuit edge, covers all the underchip area, and prevents wicking and run-over of the circuits which is a critical parameter in coating production. In addition, pigments such as carbon-black and titanium dioxide are usually added as opacifier to protect light-sensitive devices. Organic solvents such as xylenes are incorporated into the formulation to reduce the encapsulant viscosity. Coupling with its low viscosity and reactive functional groups that are capable of forming chemical bonds at the encapsulated substrate interfaces, silicone becomes one of the best device encapsulants in achieving superior performance in temperature-humidity-bias (THB) accelerated electrical testing (see Fig. 11-7). Furthermore, the heat curable silicon-hydride and silicon-vinyl addition system provides a fast and deep section cure material which is preferred as an encapsulant for some electronic components.

Heat curable hydrosilation silicone (either elastomer or gel) has become an attractive device encapsulant. Its curing time is much shorter than the RTV-type silicone. With its jelly-like (very low modulus) intrinsic softness, silicone gel is a very attractive encapsulant in wire bonded large chipsize IC devices. The two part heat curable system which consists of the vinyl and hydride reactive functional groups, and the platinum catalyst hydrosilation addition cure system, provides a fast cure system without any by-product (see Fig. 11-4 for cure mechanism). The key to formulating a low modulus silicone is the deliberate undercrosslinking of the silicone system. A few ppm

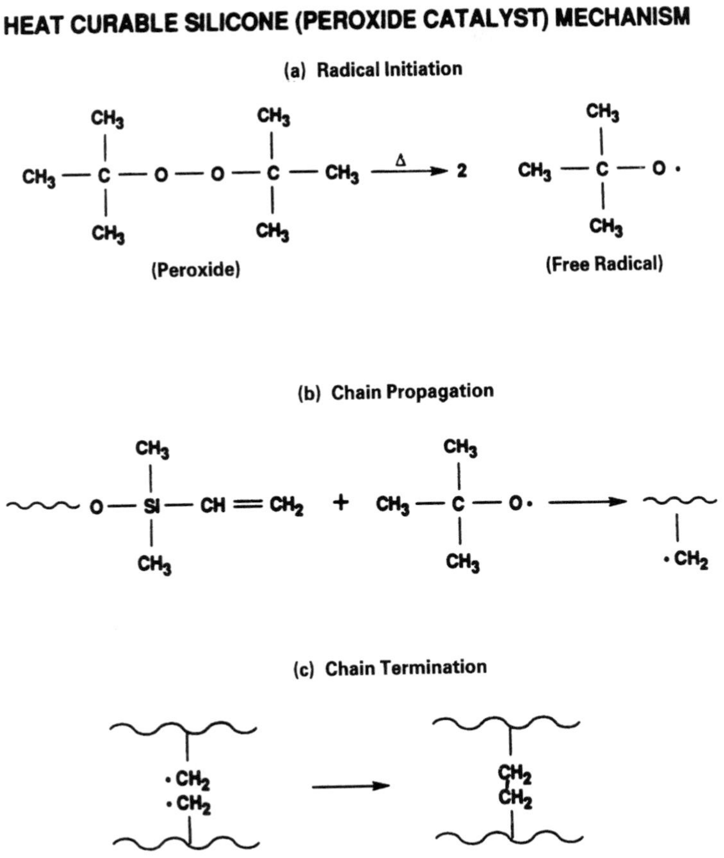

HEAT CURABLE SILICONE (PEROXIDE CATALYST) MECHANISM

(a) Radical Initiation

(Peroxide)   (Free Radical)

(b) Chain Propagation

(c) Chain Termination

Figure 11-5   Peroxide cure silicone mechanism.

platinum catalyst such as chloroplatanic acid or an organoplatinum, is used in this system. This catalyst is usually incorporated in the part A vinyl portion of the resin.This solventless type of heat-curable silicone gel will have increased use in electronic embedding of the dedicated wire bonded or flip-chip bonded electronic components.

Sycar™ (Trademark of Hercules)

A relatively new class of materials based on silicone-carbon resins may also prove useful as an encapsulant. These materials, called Sycar™ (Hercules Chemical) are thermosets with the structure shown in Fig. 11-8.

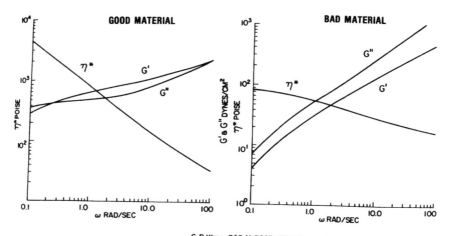

C. P. Wong & D. M. ROSE, IEEE Transactions on CHMT, 6, 485–493 (1983)

Figure 11-6   Rheology of RTV silicones.

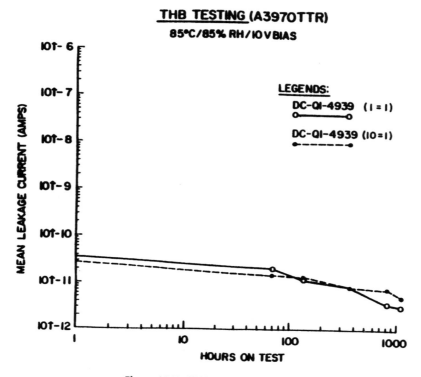

Figure 11-7   THB testing of silicones.

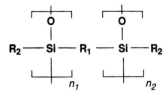

Figure 11-8 Chemical structure of Hercule's Sycar™.

The material is cured through the hydrosilation of silicon hydride and silicon vinyl groups with trace amount of platinum catalyst. The material is a fast curing system (< 15 min. at 180°C) which resulted in a low moisture absorption material that outperforms conventional thermosets such as polyimides and epoxies. Furthermore, the Sycar™ material has excellent mechanical and physical properties that can make it useful as a COB encapsulant.

Polyimides

Polyimides are one of the fastest growing material areas in polymers for electronic applications. They were first developed at Du Pont in the 1950s. During the past couple of decades, there has been tremendous interest in this material system for electronic applications. The superior thermal (up to 500°C), mechanical and electrical properties of polyimides have made their use possible in many high performance applications, from aerospace to microelectronics. In addition, polyimides show very low electrical leakage in surface or bulk. They form excellent interlayer dielectric insulators and also provide excellent step coverage which is important in multilayer IC structures. They have excellent solvent resistance and ease of application. They can be easily spun-on or flow-coated and imaged by a conventional photolithography and etch process.

Most polyimides are aromatic diamine and dianhydride compositions. Polyamic acids are precursors of the polyimides. Thermal cyclization of polyamic acid is a simple curing mechanism for this material (see Fig. 11-9). Siemens of Germany developed the first photodefinable polyimide material. However, Ciba Geigy has recently announced a new type of photodefinable polyimide which does not require a photoinitiator. Both of these photodefinable materials are negative resist type polyimides. A positive resist type polyimide which reduces the processing step in IC fabrication is not yet commercially available. Hitachi has prepared a series of ultra-low thermal coefficient of expansion polyimides. The rod-like, rigid structure of the polyimide backbone structure is the key in preparing the low TCE

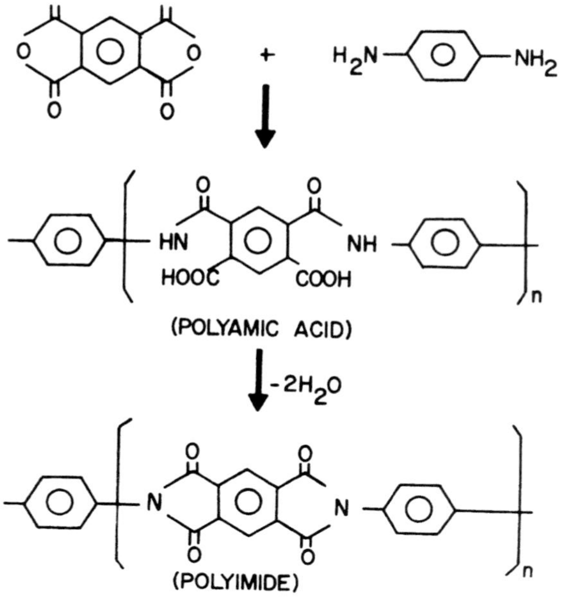

Figure 11-9 Polyimide cure mechanism.

polyimide. By simply blending a high and a low TCE polyimide, it will be possible to achieve a desirable encapsulant which could match the TCE of the substrate, and reduce the thermal stress problem in the encapsulated device during temperature cycling tests. However, their affinity for moisture absorption, high temperature cure, and high cost of the polyimide are drawbacks which have prevented their use in general electronic application. Preimidized polyimides which cure by evaporation of dissolved solvent may reduce the drawback of high temperature cure. Advances in polyimide synthesis have reduced the material moisture absorption and improved its adhesion. For COB application, these polyimides are often loaded with inert fillers such as silica, calcium carbonates, etc., to form a glob-top type encapsulant. These new types of polyimide systems will have significant implications in COB device packaging. A siloxane-polyimide copolymer which combines both the silicone and polyimide properties, has recently gained acceptance as an IC device encapsulant.

Polyurethanes

Polyurethanes are based on reactions of di-isocyanates and diols or polyols. However, recent work is focused on the use of intermediates which are low molecular weight polyethers with reactive functional groups such as hydroxyl or isocyanate groups able to further crosslink, chain extend or branch with other chain extenders to form higher molecular weight polyurethanes. Diamine and diol are chain extended with a prepolymer (either polyester or polyether) to form polyurethanes with urea or urethane linkages, respectively. The morphology of polyurethane is well characterized. Hard and soft segments from di-isocyanates and polyols, respectively, are the key to excellent physical properties of this material (see Fig. 11-10). Bases are more widely used than acids as catalysts for polyurethane polymerization. The catalytic activity is increased with the basicity. Amines such as tertiary alkylamines and organic metal salts, such as tin or lead octoates, promote the reaction of isocyanate and hydroxyl functional groups in the polyurethane system and accelerate the crosslinking. However, the hydrolytic stability of the polyurethane can be affected by the catalyst used. UV stabilizers are usually added to reduce the radiation sensitivity of the material. In addition, polyurethanes generally have high unique strength, high modulus, high hardness, and high elongation. They are some of the toughest elastomers used today. High performance polyurethane elastomers are used in conformal coating, potting and in reactive injection molding of IC devices.

Screening Tests and Analysis

Final qualification for high volume products is an expensive undertaking and should be launched with a high degree of confidence of success. In order to maximize the likelihood of success, screening tests for selecting the encapsulation systems are employed.

Screening tests can take many forms based on an assessment of what failure mechanisms need to be accelerated in time. The Institute for Interconnecting and Packaging Electronic Circuits (IPC) standard ANSI/IPC-SM-789, "Guidelines for Chip on Board Technology Implementation" contains a typical COB acceptance testing program (see Fig. 11-11). Most electronic component producing and using companies have their own "qualification" acceptance program. A good example for Military acceptance is Mil. Spec. 883C. By utilizing a subset of these acceptance tests it is possible to develop a screening test plan. An example of one critical element of an effective screening plan is the use of a "triple track resistor" test vehicle and 85°C/85% RH; 10V bias for 1000 hours THB testing.

The IEEE committee S32.1 has defined the method of testing using a triple track resistor as a test pattern.

Figures 11-12 through 11-15 show representative data for silicone gels tested using THB testing. The results are typical of excellent encapsulation

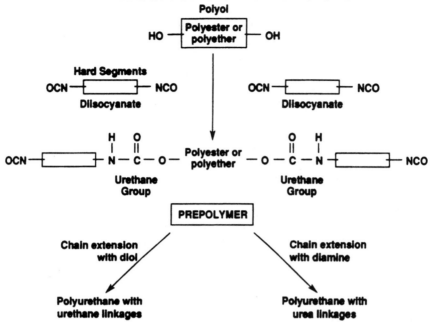

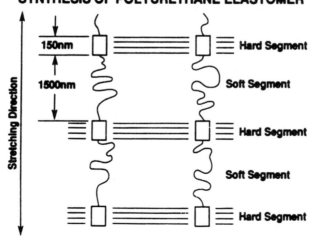

Figure 11-10   Chemical structure of polyurethanes.

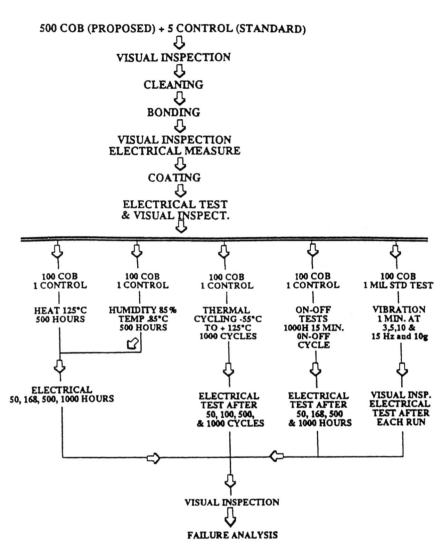

Figure 11-11   Chip-on-board testing programs.

materials. In the triple track resistance measurement, we grounded the two outer tracks and biased the center track. Then we measured the resistance change of the center conductor line. The test process measures the degree of "electro-oxidation" of the center track. Leakage current due to impurities is measured. The change of the triple track resistance in relation to the original resistance increases with time. This is mainly due to the oxidation process.

The smaller the change in resistance and leakage current with time, the better the encapsulant material will be. For the 1:1 vinyl-hydride mixing ratio silicone gel, there is no noticeable change in leakage current (10E-11 amp) after 1700 hours of testing (see Fig. 11-12).

All the silicone gels (samples 1, 2, and 3) with a 1:1 mixing ratio perform relatively the same as the control RTV sample (see Fig. 11-12). However, the leakage current of the corresponding 10:1 vinyl-hydride mixing ratio of silicone gels (samples 4, 5, and 6) shows an order of magnitude higher in leakage current measurement (10E-10 amp vs. RTV leakage current of 10E-11 amp). Furthermore, with leakage current of 10E-10 amp 1700 hours of THB testing, we still consider these silicone gels to be good device encapsulants (see Fig. 11-13). For the (delta R/R) resistance change during the initial 800 hours testing, the 10:1 mixing ratio silicone gels appeared to have an order of magnitude higher resistance change as compared to the RTV silicone control. However, when the THB testing continued to ~1000 hours, the leakage current performances were almost identical to the RTV control (see Fig. 11-14). For the 1:1 mixing ratio of the silicone gel, the delta

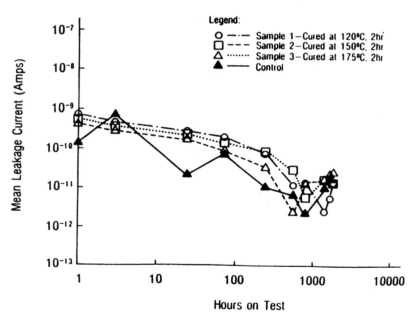

Figure 11-12 THB testing of silicone elastomers (leakage current measurements).

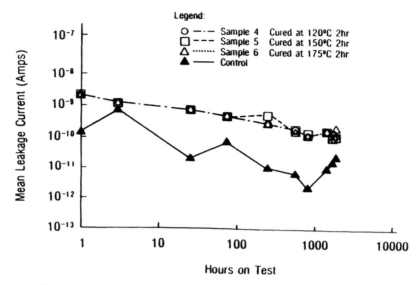

Figure 11-13   THB testing of silicone gels (leakage current measurements).

R/R change of the higher temperature cured samples (such as 150°C and 170°C) perform better than the 120°C cured sample (see Fig. 11-15).

All of the silicone gel samples tested show excellent THB electrical performance as IC device encapsulants, but the softer gels possibly show the lower change in resistance.

In addition to THB testing, other test vehicle or actual product-like prototypes utilizing the following accelerated aging tests should be considered.

The primary acceleration aging tests to evaluate potential encapsulants are the following Life Tests (typical):

Temperature cycle	−40°C to 130°C 600 cycles
Thermal shock	−65°C to 150°C liquid./liquid. 15 cycles
Thermal aging	150°C for 100 hours
Steam bomb	121°C at 2 atm. for 100 hours

Control elements (known good material and results) should be employed in all of these screening tests to validate the experimental results.

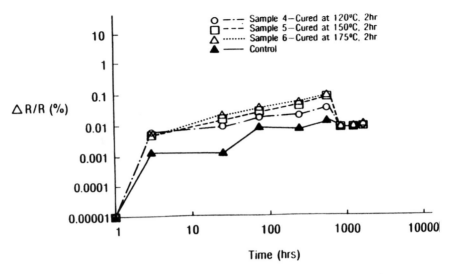

Figure 11-14   THB testing of silicone gels (resistance measurements).

Failure mode analysis (physical examinations) of "defective" and good samples alike are required to understand the effectiveness of the encapsulation systems. This physical examination can take many forms: careful dissection and cross-sectioning and also nondestructive acoustic microscopy. The use of acoustic microscopy is a relatively new tool for evaluation of material adhesion within a multilayer structure. Because it is relatively new, a brief description of this technique is given below.

Acoustic Microscopy

Acoustic microscopy utilizing two particularly useful tools can provide essential information about material adhesion and cohesion after accelerated stress testing. The scanning laser acoustic microscope passes acoustic energy thru a sample and produces an X-ray type of image. Flaw sites (delamination, cracks, voids, etc.) block or significantly reduce acoustic energy transmission while bonded materials transmit the acoustic energy and produce the image. The C mode (Sonoscan) scanning acoustic microscope utilizes focused acoustic energy and reflected energy from

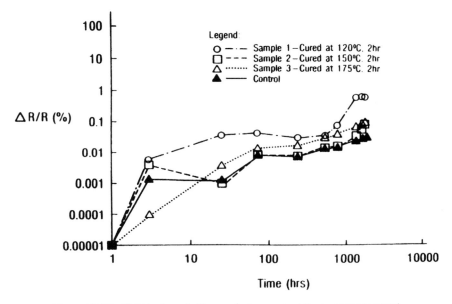

Figure 11-15   THB testing of silicone elastomers (resistance measurements).

dissimilar material interfaces and flaws to produce images. Both instruments provide image information regarding the structure integrity of multimaterial interfaces.

A few concepts are presented to introduce the basic principles of C mode scanning acoustic microscopy. No attempt is made here for completeness, since there are numerous papers and textbooks available.

The audible frequency range extends from about 20 Hz to 20 kHz. Ultrasonic waves are those with a frequency exceeding 20 kHz. Ultrasonic waves of frequencies between 1 and 100 MHz are typically used for flaw detection.

In acoustic microscopy systems, these waves are generated using a piezoelectric crystal. When a high frequency electrical potential is applied, the crystal expands and contracts. The resulting vibration is transmitted, generally using water as the transmission medium, to the sample being examined. The reflected ultrasonic wave from the sample then returns to strike the same crystal which in turn generates an electrical potential which can be read on an oscilloscope.

Just as echoes can be created when sound waves are reflected off a wall, a similar phenomenon occurs as ultrasonic waves travel from one medium to another. When ultrasonic waves, borne by water for example, encounter a different medium, some of the sound is reflected while some is transmitted. A transmission/reflection ratio is a function of the acoustic impedances of the two media. Acoustic impedance, Z, can be defined simply by:

$$Z = \rho C$$

where ρ = density of the material
$\quad\;\; C$ = speed of sound in the material

The ratio of the amplitude of the incident waves to the amplitude of the reflected waves is reflectance. The amount and nature of the reflectance, R, depends on the acoustic impedance of the two materials:

$$R = \frac{Z_2 - Z_1}{Z_2 + Z_1}$$

where Z_1 and Z_2 are the acoustic impedances of the two materials before and after the interface, respectively.

The shape of the echo as recorded by the oscilloscope depends on whether the specific interface of interest involves sound traveling from a material of higher impedance to lower impedance, or vice versa (i.e. whether R is positive or negative). Figure 11-16 shows a schematic cross-section of a hypothetical component with various interfaces, and a simplified view of the oscilloscope trace along three possible paths through the component. The echo return time depends upon the distance from the transducer to the interface or defect. The echo polarity is positive at the boundary of a higher impedance material and negative at the boundary of a lower impedance material.

Although in principle the shape and the relative return times of these echoes appear to be straightforward, the actual oscilloscope trace is generally more complicated. This is because, in addition to reflection, transmission, T, through the interface also occurs, following this relation:

$$T = \frac{2Z_2}{Z_1 + Z_2}$$

As the acoustic energy reaches the next surface, once again transmission and reflection occurs. In this way, the reflected wave can reverberate between a number of interfaces (including the transducer itself). Any portion of this additional acoustic information which makes its way back to the transducer is therefore also seen on the oscilloscope trace.

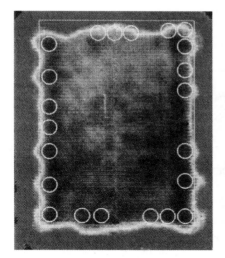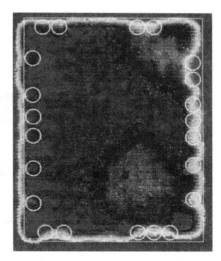

Figure 11-16   An example of a COB encapsulated sample. Left: acoustic image of a flip chip sample showing encapsulation coverage between the substrate and the silicon chip surface. Coverage is complete. The circles represent the reflowed solder joint locations. Right: Acoustic image of a flip chip sample with simulated voiding.

The acoustic image is generated by selecting a plane of interest in the material containing an interface. An electronic gate can be opened and closed around a specific echo to exclude other echoes from other planes. The transducer then scans the sample, and the size of the returning echo determines the brightness on a display. In addition, using a polarity map, the color of the image area can be altered depending on the polarity of the returning echo.

Proper utilization of these tools can greatly facilitate a fundamental understanding of the reliability of the encapsulated structure and plays a key role in screening and qualification testing.

11.3 MANUFACTURING PROCESS FOR ENCAPSULATION OF COB

The manufacturing process for chip-on-board encapsulation, Figures 11-17 and 11-18 shows an example of a COB encapsulated sample which, like that for other encapsulation processes, involves several very important areas. Included among these are pre-encapsulation cleaning, application and

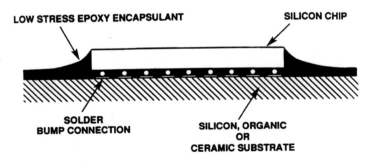

ENCAPSULATED FLIP-CHIP DEVICE

LOW STRESS EPOXY ENCAPSULANT   SILICON CHIP

SOLDER
BUMP CONNECTION

SILICON, ORGANIC
OR
CERAMIC SUBSTRATE

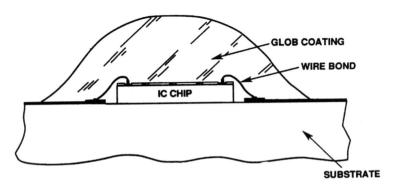

GLOB COATING

WIRE BOND

IC CHIP

SUBSTRATE

DIRECT CHIP ATTACHMENT WITH
PROTECTIVE POLYMER OVERCOAT ("GLOB COATING")

Figure 11-17   Glob top encapsulant.

curing of the encapsulant and repairability issues. Each of these areas is discussed below.

11.3.1 Pre-encapsulation Cleaning

Prior to IC device encapsulation, the pre-encapsulation cleaning is the most critical step to ensure the long-term reliability of the device. Encapsulating a contaminated (dirty) device is a guarantee for device failure. It is imperative to remove even trace amounts of the contaminant from the IC device surface prior to the encapsulation process. There are two main cleaning processes: conventional cleaning and reactive oxygen cleaning.

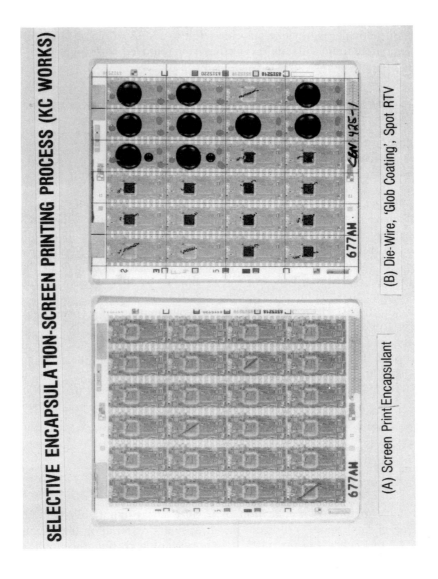

Figure 11-18 Example of chip-on-board encapsulated hybrid IC.

Conventional Cleaning

Conventional cleaning process includes organics, such as detergents and solvents (i.e. chlorofluorohydrocarbon (CFC)-Freons, and chlorohydrocarbon-trichloroethane, methylene chloride, etc.) to remove organic contaminants. Inorganic ionic salts are most effectively removed with high quality aqueous water or polar solvents, such as alcohols.

Reactive Oxygen Cleaning

In addition to the conventional cleaning process, reactive oxygen is very effective in removing low-level organic contaminants. UV-ozone is very effective in removing a few monolayers organics from the substrate surface. However, the device under cleaning must be placed directly under the UV source. Plasma oxygen operates at 13.6 MHz Radio Frequency (RF), is fast and effective in cleaning MOS devices, and preserves the aluminum metallization of the devices. However, the thermal stress associated with the plasma process may damage some device structures. Microwave discharge of oxygen at 2.5 GHz RF is also a powerful technique in the device cleaning process. This process is similar to the oxygen plasma process, except that the microwave frequency is used.

To clean the IC device effectively, the combination of conventional cleaning solvents, highly purified deionized water (with $18 \, M\Omega/cm$ resistivity) and reactive atomic oxygen gas cleaning processes are ideal in providing the thorough cleaning of the IC devices. In addition, the cleaning process should be performed in a clean room (class 1000) or clean hood (class 100) environment where the encapsulation process could proceed as soon as cleaning is completed to minimize contamination of the pre-cleaned devices.

11.3.2 Application of Encapsulants

The primary method of application of a COB encapsulant involves the dispensing of small spots or globs of the encapsulant material on the silicon devices on the board or substrate. It is important here that the applied encapsulant spots or globs provide complete environmental protection of the active surfaces of the silicon devices. In the case of wire or face-up TAB bonded devices, this is relatively easy since the active surface of the chip is facing upward. However, in the case of face-down TAB or solder bumped silicon devices with their active face downward, the task is more demanding.

In this instance, it is necessary for the method of encapsulant application to insure that the encapsulating material flows into the relatively small annular space between the board (substrate) and the underside (the active face) of the silicon device. In either case, a void-free encapsulation is required. This is important since voids at the active surface of the silicon device can represent potential sites for moisture condensation and subsequent circuit failure.

The primary method of applying glob top (see Fig. 11-17) encapsulants is dispensing suitably degassed material from a reservoir through a needle-like nozzle onto the chip. In the case of a single component encapsulant, only one reservoir is necessary. For two-component materials, two reservoirs are used from which the components are fed into a static mixer from which the mixed material is dispensed through a needle-like nozzle. For the single component materials, pot life is an issue which is frequently controlled by the incorporation of a suitable reaction inhibitor in the encapsulant formulation. With a two-component system, pot life is not a problem.

In the application of a glob top encapsulant, the rheological character-istics of the material are extremely important. First, the material must have a rheology that will permit it to be easily dispensed through an application nozzle. This may be characterized as a high shear rate process where the material viscosity needs to be relatively low. On the other hand, once the material is on the circuit, it is usually desirable for it to remain where it is dispensed. That is, excessive flow on the circuit is undesirable for two reasons. First, it may flow into regions of the circuit where its presence may interfere with other circuit assembly operations (e.g., external lead bonding). Secondly, excessive flow on the circuit tends to deplete material from the silicon device it is intended to encapsulate. For a solder bump attached chip, excessive flow may compromise filling the area under the chip. As a result of these considerations, it is generally desirable for the encapsulant to be a non-linear thixotropic material. The use of fillers in the encapsulant helps to achieve the necessary degree of thixotropy.

11.3.3 Curing of Encapsulants

In order to optimize each COB material property, the complete cure of the material is essential. Various analytical methods can be used to determine the degree of cure for a given material. Differential Scanning Calorimetry (DSC), Fourier-Transform-Infrared (FT-IR) and microdielectrometry are examples of methods which provide quantitative measures of the curing process.

Differential Scanning Calorimetry (DSC)

DSC measures the heat flow from the material during its curing process. It measures the heat capacity endothermic and exothermic transitions of the sample, and provides quantitative information regarding the enthalpic changes in the material. A DSC scan plots energy supplied against average programmable temperature. The peak areas could be directly related to the enthalpic changes quantitatively (see Fig. 11-19). T_g can be taken as the temperature at which one-half change in heat capacity, ΔC_p, has occurred. Various kinetic properties could be extracted from the DSC. Eact (activation energy), R (reaction rate), reaction order kinetic information can also be obtained (see Fig. 11-20). When materials are fully cured, there shouldn't be any residual heat capacity remaining in the DSC scan.

Fourier-Transform Infrared (FT-IR)

FT-IR is a very sensitive tool in measuring the vibrational energy of the reactive functional groups of the encapsulant such as heat curable silicones (see Fig. 11-21). The strong absorption of Si-H at $\sim 2100\,cm^{-1}$ shows the presence of this uncured silicone. Upon the hydrosilation cure, the decrease of Si-H absorption at the initial cure stage could be easily measured by its peak height and peak area integration. This provides a useful process tool to quantify the material cure. When the Si-H absorption stabilizes, it is a good indication of the complete cure of the material. Figure 11-22 shows the FT-IR detection of the curing kinetic process. Other encapsulants, such as polyimide are routinely analyzed by this FT-IR method.

Microdielectrometry

The recently developed highly sensitive microdielectrometry, which is capable of measuring from 0.05 Hz to 10 kHz, frequencies of the encapsulant was studied. This microdielectrometry, which utilizes the Micromet Instruments' Eumetric System II microdielectrometer with a miniature IC sensor and a wide range of frequencies (from 0.05 Hz to 10 kHz) to monitor the loss factor (E'') of the encapsulant, is a very sensitive technique to detect the final stage of the material cure study. A thin layer of freshly prepared specimen (~ 20 mil thickness) was coated on the miniature IC sensor and placed inside a programmable oven. The temperature of the oven was set to the predescribed temperature (i.e., 120, 150 or 175°C) and the loss factor (E'') at various frequencies (0.05, 1, 100, 1000, 10,000 Hz) was monitored periodically during the curing time. Results of the microdielectric

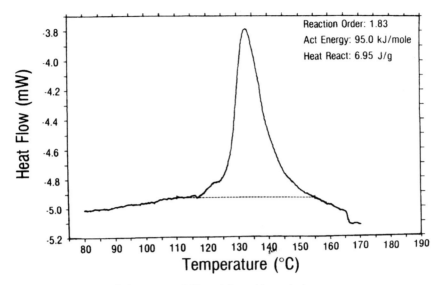

Figure 11-19   Differential scanning calorimetry.

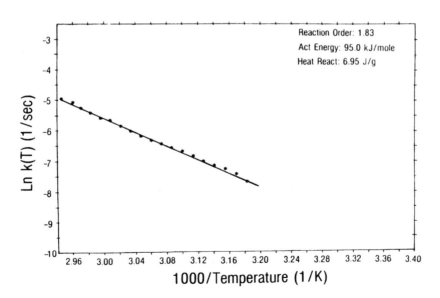

Figure 11-20   Kinetic information obtained from DSC.

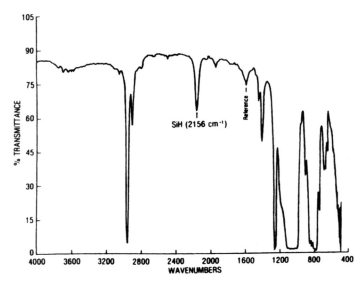

Figure 11-21 FT-IR of Si-H absorption in heat curable (hydrosilation, silicone).

FT-IR CURE STUDY OF SILICONE GEL (150°C)

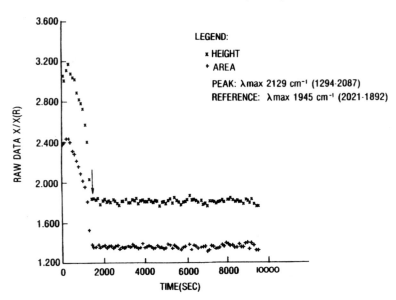

Figure 11-22 FT-IR cure kinetic.

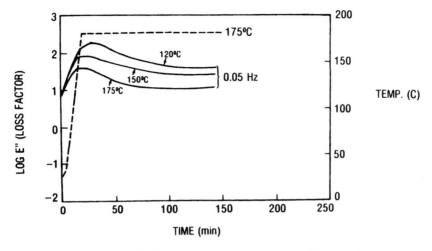

Figure 11-23 Encapsulants cure study by microdielectrometry.

loss factor measurements are shown in Fig. 11-23. When all the E' (dielectric constants) and E'' (loss factors) measurements are stabilized during the isothermal cure, it is a good indication of a complete cure. With the low frequencies sweep experiment of this microdielectrometry, it provides a very sensitive tool in quantifying the material complete cure. In a production environment, a fast cure (SnapCure in a few minutes) system is preferred for the high volume automatic process.

Repairability Issues

Rework and repair for COB depends on the type of encapsulation material used and the robustness of the interconnections/assembly technology used.

Typically, wire bonded chips and "rigid" epoxy-like encapsulated IC are virtually not repairable.

Lower modulus encapsulants are easily removable with appropriate solvents and therefore are more readily repairable. RTV silicones or silicone gels are easily dissolved in strong bases such as tetramethylammonium hydroxide which is often used in production for stripping silicone encapsulated devices for repairs.

However, the best (least expensive) repair process is the one which makes repair unnecessary because extremely high quality robust components, assembly and testing methods minimize or even eliminate the need to rework or repair.

11.4 SUMMARY

In the past two decades, the electronic industry went through an enormous growth, both technological and economical. From the technological point of view, electronic packaging is becoming a bottleneck of electronic system integration. It reduces the high-speed component transmission speed due to propagation delay of the high dielectric material. In order to provide a reliable packaged product, a low-cost but reliable COB material and process are needed. Further technology advancement requires even low dielectric constant (<2) material with high thermal resistance, high thermal conduction, and high breakdown voltage, and a material easy to use in process and repair. It is an area that still needs further R & D.

REFERENCES

1. Wong, C. P., "Recent Advances in IC Passivation and Encapsulation: Process Techniques and Materials," in *Polymers for Electric and Photonic Applications*, C. P. Wong, ed., Chapter 4, pp. 167-214, Academic Press, San Diego, CA, 1992.
2. Wong, C. P., "Integrated Circuit Encapsulants," in *Polymers in Electronics*, Second Edition of *The Encyclopedia of Polymer Science and Engineering*, Vol. 5, p. 638, 1986, John Wiley and Sons Publisher, New York.
3. Wong, C. P., "Can IC Surface Protection Replace Hermetic Ceramic VLSI Packaging?," p. 45, *Program and Extended Abstracts, 5th VLSI Packaging Workshop*, Co-sponsored by IEEE Trans. on Components, Hybrids and Manufacturing Technology Society and National Bureau of Standards, November 17-18, 1986, Paris, France.
4. Wong, C. P., and D. M. Rose, "Modified RTV Silicone as Device Packaging," *33rd Electronic Components Conference Proceedings*, p. 505, 1983.
5. Wong, C. P., "Thermogravimetric Analysis of Silicone Elastomers and IC Device Encapsulants," Chapter 23, in "Polymers in Electronics," *American Chemical Society Symposium Series*, No. 242, 285, 1984, and references therein.
6. Wong, C. P., "Effects of RTV Silicone Cure in Device Packaging," *Polymer Science and Engineering Proceedings*, American Chemical Society, Vol. 55, p. 803, 1986.
7. Martin, J., and L. D. Hanley, "Humidity Test of Premolded Chip Carrier," *IEEE Trans. on Components, Hybrids and Manufacturing Technology*, Vol. 4, No. 2, 210, 1981.
8. Wong, C. P., "IC Encapsulants," p. 12, chapter 6 of "Polymer Materials for Electronic Applications," an intensive short course; University of California at Berkeley, August 1983. M. C. Volk, et al., *Electrical Encapsulations*, Reinhold, N.Y., 1962.
9. Noll, W., *Chemistry and Technology of Silicone*, Academic Press, New York, 1968.

10. Shanefield, D. J., and W. Collins, "Proposed Standard Environmental Test of Component Coatings," *Proceedings of 1st International SAMPE Electronics Conference*, pp. 70-72, 1987.
11. Wong, C. P., "High Performance Silicone Gel as IC Device Chip Protection," Material Research Society Symposium Proceedings, Vol. 108, p. 175, 1988, and reference therein.
12. Sinnadurai, F. N., ISHM Nordic, Stockholm, April 1988, also *Handbook of Microelectronics Packaging and Interconnection Technologies*, Electrochemical Publications Limited, p. 8, 1985.
13. Miller, R., "Encapsulation in Non-Hermetic PGAs," *Proceedings of NEPCON West*, p. 867, 1986.
14. Wong, C. P., "High Performance RTV Silicone for IC Encapsulants," *The International Journal for Hybrids and Microelectronics*, Vol. 4 (2), p. 315, 1981.
15. Wong, C. P., and D. E. Maurer, "Improved RTV Silicone for IC Encapsulants," National Bureau of Standards, Special Publication 400-72, *Semiconductor Moisture Measurement Technology*, p. 275, 1982.
16. Wong, C. P., "Improved Room-Temperature Vulcanized Silicone Elastomers as Integrated Circuit Encapsulants," "Polymer Materials for Electronics Applications," *American Chemical Society Symposium Series*, No. 184, 171, 1982.
17. Danielsson, H., "Will ASIC Technology Demand a New Interconnection Technology Instead of Soldering in Automotive Electronic?," *International Society for Hybrids and Microelectronics Proceedings*, p. 135, 1988.
18. Balde, J. W., "IEEE Gel Testing Task Force, a Report on the Goals and Activities," *IEPS Technical Proceedings*, p. 867, 1986.
19. Balde, J. W., "IEEE Gel Task Force, a Progress Report," *IEPS Technical Proceedings*, p. 949, 1988.
20. White, M. L., "Encapsulation integrated circuits," *Proc. IEEE*, Vol. 27, p. 1610, 1969. J. H. Martin and L. D. Hanley, "Humidity test of premolded chip carriers." *IEEE Trans. Comp., Hybrids, Manuf. Technol.*, Vol. CHMT-4, p. 210, 1981.
21. Mancke, R. G., "A Moisture Protection Screening Test for Hybrid Circuit Encapsulants," *IEEE Trans. Comp., Hybrids, Manuf. Technol.*, Vol. CHMT-4, p. 492, 1981.
22. Sinnardurai, N., "An Evaluation of Plastic Coatings for High Reliability Microcircuits," *Microelectron. J.*, Vol. 12, no. 6, 1981.
23. Wong, C. P., and D. M. Rose, "Alcohol Modified RTV Silicone Encapsulants for IC Device Packaging," *IEEE Trans. Comp., Hybrids, Manuf. Technol.*, Vol. CHMT-6, 1983.
24. Otsuka, K., Y. Shirai, and K. Okutani, "A New Silicone Gel Sealing Mechanism for High Reliability Encapsulation," *IEEE Trans. Comp., Hybrids, Manuf. Technol.*, Vol. CHMT-8, 1985.
25. Wong, C. P., "Can IC Surface Protection Replace Hermetic Ceramic VLSI Packaging," in *Proc. IEEE 5th VLSI Packaging Workshop*, p. 45, 1986.
26. Miller, R., "Encapsulation in Non-Hermetic PGAs," *ISHM J.*, Vol. 4, p. 315, 1986.

27. Otsuka, K., et al., " High Reliability Mechanism of Silicone Gel Sealing in Accelerated Environmental Test," in *IEPS Conf. Proc.*, p. 720, 1986.
28. Shanefield, D. J., and W. Collins, "Proposed Standard Environmental Test of Component Coatings," in *Proc., SAMPE Electron. Conf.*, 70, 1987. K. Otsuka et al., "The Mechanisms that Provide Corrosion Protection for Silicon Gel Encapsulated Chips," *IEEE Trans. Comp., Hybrids, Manuf. Technol.*, Vol. CHMT-12, 1987.
29. Wong, C. P., "High Performance Silicon Gel as IC Device Chip Protection," in *Material Research Soc. Proc.*, Vol. 108, p. 175, 1988.
30. Wong, C. P., J. M. Segelken, and J. W. Balde, "Understanding the Use of Silicon Gels for Non-Hermetic Plastic Packaging," *IEEE Trans. Comp., Hybrids, Manuf. Technol.*, Vol. CHMT-12, p. 421, 1989.
31. Balde, J. W., "The IEEE Gel Task Force: An Early Look at the Final Testing," *IEEE Trans. Comp., Hybrids, Manuf. Technol.*, Vol. 12, pp. 426-429, December 1989.
32. Morgan, B., "Chip-on-Board Grows Despite Video Game Slump," *Electronic Products*, 21, 1984.
33. Kessler, L. W., "Acoustic Microscopy–An Industrial View," in *Proceedings of the 1988 IEEE Ultrasonics Symposium*, Chicago IL, October 2-5, 725, 1988. IEEE Catalog No. 88CH2578-3, Institute of Electrical and Electronic Engineers, Piscataway, NJ.
34. Krautkramer, J., and H. Krautkramer, *Ultrasonic Testing of Materials*, Springer-Verlag, New York, 1983.
35. Balde, J. W., "New Packaging Strategy to Reduce System Costs," *IEEE Trans. Comp., Hybrids, Manuf. Technol.*, Vol. CHMT-7, no. 3, September 1984.
36. Knausenberger, W. H., and L. W. Schaper, "Interconnection Costs of Various Substrates–The Myth of Cheap Wire," *IEEE Trans. Comp., Hybrids, Manuf. Technol.*, Vol. CHMT-7, no. 3, September 1984.
37. Osuka, K., Y. Shirai, and K. Okutani, "A New Silicone Gel Sealing Mechanism for High Reliability Encapsulation," *IEEE Trans. Comp., Hybrids, Manuf. Technol.*, Vol. CHMT-7, no. 3, September, 1984.
38. Marshall, J. F., "New Application of Tape Bonding for High Lead Devices," *Solid State Technology*, 175 (1984).
39. Meyer, D. E., "Tape-Automated Bonding Suits Surface Mount," *Electronic Products*, October 1, p. 71, 1984.
40. Miller, H., "Chip-on-Board and Tape Automated Bonding Markets," *IPC Technical/Marketing Research Council Proceedings*, December, 1984.
41. Pearne, N., "Chip-on-Board and PCB Industry," *IPC Technical/Marketing Research Council Proceedings*, December, 1984.
42. Greer, S. E., "Low Expansivity Organic Substrate for Flip-Chip Bonding," *IEEE Trans. Comp., Hybrids, Manuf. Technol.*, Vol. CHMT-2, March, 140, 1979.
43. Small, D., and N. Sinnadurai, "The Manufacture and Reliability of the EP&IC Chip Carrier," *Proceedings of the Printed Circuit World Convention III*, (IPC), May, 1984.

44. Malhotra, A. K., G. E. Leinbach, J. J. Straw, and G. R. Wagner, "Finstrate: A New Concept in VLSI Packaging," *Hewlett-Packard Journal*, August, p. 24, 1983.
45. Goldman, P. J., "The Plating of Bondable Gold," *P.C. Fab Expo Technical Seminar*, p. 27, 1984.
46. Blackshaw, M. F., F. J. Dance, and P. J. Goldman, "The Design and Manufacture of High Density Printed Circuit Boards for Chip-on-board Assemblies," *1984 Nepcon Proceedings*, p. 745, 1984.
47. Burkhart, A., and M. Bonneau, "Considerations for Choosing Chip-on-board Encapsulants," *Electronics*, September, p. 67, 1985.
48. Dance, F., "Chip-on-Board Has Designs on High-Density Packaging," *Electronic Packaging & Production*, October, p. 75, 1985.
49. Unsigned, "Surface Mounting of Chips on Printed Wiring Boards to Become Common," *Silicon Valley Tech. New*, January 21/February 22, p. 1, 1985.
50. Tuck, J., "Chip-on-Board Technology," *Circuits Manufacturing*, March, p. 78, 1984.
51. Fuchs, E., "Chip-on-Board: An Economical Packaging Solution," *Electronic Packaging & Production*, January, p. 182, 1985.
52. Ginsberg, G., "Chip and Wire Technology: The Ultimate in Surface Mounting," *Electronic Packaging & Production*, August, p. 78, 1985.
53. Keeler, R., "Chip-on-Board Alters the Landscape of PC Boards," *Electronic Packaging & Production*, July, p. 62, 1985.
54. Unsigned editorial, "Let's Get Serious About Chip-on-Board," *Electronics*, March, p. 4, 1985.
55. Brown, C., "Low Cost Pin Grid Array Packages," *Solid State Technology*, May, p. 239, 1985.
56. "Market Trends: TAB and Future Technologies," *SM Trends Newsletter*, Vol. 2, #12.
57. Ginsberg, G., "Chip-on-Board Profits from TAB and Flip-Chip Technology," *Electronic Packaging & Production*, September, p. 140, 1985.
58. Meyer, D., A. Kohli, H. Firth, and H. Reis, "Metallurgy of Ti-W/Au/Cu System for TAB Assembly," *J. Fac. Sci. Technol.*, A3 (3), American Vacuum Society, May/June, p. 173, 1985.
59. Brow, D. B., and M. G. Freedman, "Is there a Future for TAB," *Solid State Technology*, September, p. 173, 1985.
60. Marshall, J., "Encapsulated Chip Packaged on Tape," *Semiconductor International*, August, p. 170, 1985.
61. Koukootsedes, G. J., and R. C. Antonen, "Specialty Silicone Elastomers Cope with Diverse Hybrid Circuit Applications," *Hybrid Circuit Technology*, September, p. 11, 1985.

12

Underfill Encapsulation for Flip Chip Applications

Darbha Suryanarayana and
Donald S. Farquhar

12.1 INTRODUCTION

In this chapter we will discuss the application of underfill encapsulation to flip chip packages. Underfill encapsulation is a technique used to reinforce the solder joints between the chip and the substrate. It is used in manufacturing by dispensing liquid encapsulant and allowing capillary action to draw it between the chip and substrate of the assembled flip chip package. The encapsulant then solidifies upon oven curing and reinforces all the solder joints, typically resulting in a tenfold improvement in fatigue life compared to an unencapsulated package.[1,2]

This relatively new technology has gained rapid acceptance because of the recent development of high performance filled epoxy based materials suitable for use as underfill encapsulants. Underfill encapsulation with these epoxy materials provides dramatic fatigue life enhancement with minimal impact on the manufacturing process flow. The fatigue life improvement obtained by underfill encapsulation has extended the flip chip packaging technology to even larger footprints and to a variety of organic substrate materials.[3,4] Figure 12-1 shows a schematic drawing of a variety of flip chip surface mount applications on an organic printed circuit board.[5]

The discussion of underfill encapsulation in this chapter is organized as follows. First, the role of underfill encapsulation in flip chip technology is discussed, and the candidate materials and the process are introduced. Details of the material requirements and suitable test techniques for underfill encapsulants are then presented. Next, the underlying mechanisms

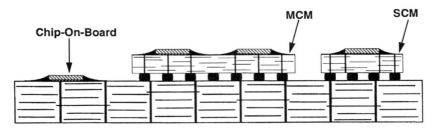

Figure 12-1 Schematic of flip chip packages on organic printed circuit carriers using underfill encapsulation.[5]

of solder joint reliability enhancement are discussed, and the reliability gains that have been obtained in several case studies are reported. Recommendations for manufacturing applications are then presented, and finally the direction of current and future development activities is discussed.

12.1.1 Packaging Strategies for Flip Chips

Although flip chips have been used in packaging applications by IBM since the 1960s[6] the technology has only recently been adapted to bare flip-chip-on-board applications, offering improvements in packaging of circuits, electrical performance, and size and weight considerations.[3,5,7,8] Prior to the advent of underfill encapsulants, flip chips were typically mounted on alumina ceramic (6.7 ppm/°C) substrates which provided a low coefficient of thermal expansion (CTE) to more closely match the CTE of the silicon chip (2.6 ppm/°C).[9] Table 12-1 shows key properties, including the CTE and elastic modulus, of selected packaging materials. Decreasing the CTE mismatch provided less stress on the solder joint interconnections which are typically controlled collapse chip connections, or C4 joints. Ceramic substrate materials were often the materials of choice for first level packaging.[10,11] In addition to a close CTE match, ceramics provide a stable and planar joining surface for C4 attach, and the ability to withstand the high temperature ($\sim 360°C$) joining process of the typical 95%Pb/5%Sn C4 joining metallurgy.[9] Reliability of C4 interconnections is attained primarily by controlling CTE mismatch and limiting the maximum size of the C4 interconnection footprint pattern.[12]

12.1.2 Candidate Underfill Encapsulant Materials

A variety of polymers have been considered as potential underfill encapsulants, including thermosetting molding compounds such as

Table 12-1 Properties of Selected Packaging Materials[11,40]

Material	Density (g/cc)	CTE (ppm/°C)	Elastic Modulus (GPa)	Thermal Conductivity (W/mK)	Dielectric Constant
Non-organic					
Silicon	2.33	2.6	107	150	11.8
GaAs	5.31	5.7	124	58	12.9
Aluminum	2.9	23.0	79	205	
Copper	8.9	17.0	133	393	
Gold	19.3	14.2	82	297	
Pb-5%, Sn	11.3	29.0	53	63	
96% Alumina	3.9	6.6	345	20	9.4
AlN	3.3	3.3		230	8.8
Glass/Ceramics	2.5	3.5	48	5.0	4–8
Silica	2.6	0.5	119	1.01	4.6
Organic					
Epoxy/ceramic reinf.	1.8	36–18	20	0.5	5.2
FR-4 (x-y plane)		15.8	40	0.2	4.7
Polyimide	1.4	45	5	0.2	3.5
Fluorocarbon, PTFE	2.3	200	0.5	0.1	2.2
Fluorocarbon/ceramic	2.5	36–30	4.0	2.8	

silicones, epoxies, polyimides and parylenes.[1,13-23] Table 12-2 lists the properties of a variety of encapsulant materials. As a general statement, thermoplastics can be eliminated from consideration because of their high viscosity in the melt, which prevents them from flowing under a chip. Based on extensive testing, cycloaliphatic epoxy resins with anhydride curing agents have been identified as the material system of choice.[14-16] These are high modulus materials offering certain unique advantages such as high glass transition temperature (Tg), good chemical resistance, low moisture absorption, favorable viscosity for flow and good adhesion strength to the packaging materials.[20] In addition, the coefficient of thermal expansion (CTE) of the epoxies can be tailored with the addition of ceramic fillers (e.g. SiO_2), thus reducing stresses between the substrate and the encapsulant. Figure 12-2 shows a schematic view of an encapsulated flip chip on board. Use of such low CTE epoxies for underfill encapsulation has become an enabling technology for recent advances in flip chip packaging technology. New applications include the use of even larger footprints and direct attach to a variety of thermally mismatched organic substrate materials.[3,4]

Table 12-2 Properties of Semiconductor Encapsulants

Material	T_g (°C)	CTE (ppm/°C)	Elastic Modulus (GPa)	Tensile Strength (MPa)	Elongation (%)	Ref.
Molding compound	175	17.0	14.2	2.0		
Die-attach adhesives	100	50.0	5.0			
Filled epoxies	140–170	20–50	2.8–3.4	56–84	2.7–5.6	20,24,42
Flexible epoxies	117	60	1.4		60	41,43
RTV silicones	<20	100–150		2.0–6.0	100–200	44,45
Polyimide	>400	20–40	1.4–2.4		15–40	46
Parylene		35–69	2.4–3.2	45–75	10–200	17,47
Urethane	<20	>100		1–14	60–450	44

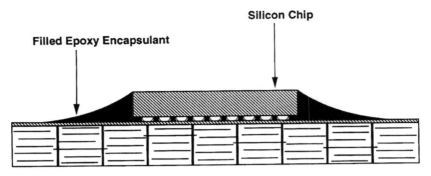

Figure 12-2   Schematic of flip chip package on a printed circuit board with underfill encapsulation.

12.1.3 The Underfill Encapsulation Process

Underfill encapsulation is typically performed as the last step in the assembly process, after the flip chip has been attached to the substrate or board. A measured amount of encapsulant is dispensed at the perimeter of the chip, and is drawn under the chip by capillary action. The gap between the chip and substrate is about 3 mils. Pre-heating of the part may be required to achieve the desired flow rates. Once the underside of the chip is completely filled, encapsulant will continue to flow from the dispense side, and form a fillet around the entire perimeter of the chip. The formation of fillets around the perimeter of the chip is critical in reducing stress concentrations. Following dispensing, the encapsulated parts are cured and the encapsulant develops its final properties. Figure 12-3 contains a micrograph showing a cross-sectional view of the encapsulant fillet geometry at the edge of the chip. The details of the filler distribution can also be seen in the micrograph. Figure 12-4(a) shows encapsulant coverage on three different size flip chips mounted on ceramic substrates, and Figure 12-4(b) shows encapsulated flip chips attached to an FR-4 printed circuit board.

12.2 MATERIAL REQUIREMENTS FOR UNDERFILL ENCAPSULANTS

While filled epoxy based materials have been available as glob-top encapsulants for some time,[22,24] considerable development effort has been expended to obtain suitable rheological and mechanical properties for underfill applications.[19,25] A variety of silica filled, anhydride cured, cycloaliphatic epoxy systems for use as underfill encapsulants have been

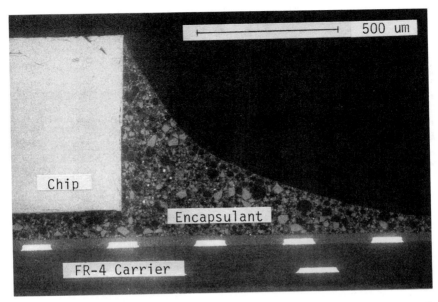

Figure 12-3   Scanning electron micrograph showing the distribution of filler present in the epoxy encapsulant used under the chip.

developed recently.[14-16,20] In this section, material characteristics of such encapsulants and testing methodologies will be examined. In particular the performance criteria that are used to evaluate their suitability, as well as the relevant test methodologies, will be discussed.

Discussion of specific material properties is divided between the properties of the underfill encapsulant in the uncured and cured states. Rheological properties of the uncured encapsulant, and the required curing conditions, are of great importance in determining suitability for manufacturing. On the other hand, the physical properties of the cured encapsulant, such as mechanical, thermal, and electrical properties, are of great importance in determining the degree of reliability enhancement achieved. Requirements for some of the key properties are listed in Table 12-3.

12.2.1 Properties of the Uncured Encapsulant

Flow Characteristics

The flow characteristics of an underfill encapsulant are critical to its performance. One measure of encapsulant flow is viscosity, which can be

(b)

(a)

Figure 12-4 (a) Encapsulated flip chip devices on metallized ceramic modules, showing the epoxy encapsulant coverage at the entrance as well as at the exit sides.[2] (b) Encapsulation of flip chip devices on FR-4 printed circuit board.[4]

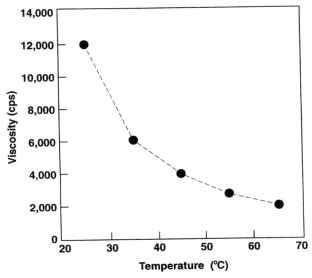

Figure 12-5   Plot of Brookefield viscosity of underfill epoxy encapsulant measured at various temperatures (shear rate 5/sec).

readily determined using conventional techniques. Typical viscosity values are in the range of 10-30 Kcps under low shear rates at room temperature. These values are an order of magnitude lower than the conventional epoxies used in glob-top applications. The encapsulant viscosity can be further reduced by elevating the dispense and flow temperatures. Figure 12-5 shows the viscosity vs. temperature of a typical underfill encapsulant.

While the viscosity of an encapsulant is a useful measurement for relative comparisons, encapsulants are reactive, non-Newtonian, thixotropic fluids with complex flow characteristics. The flow rate of the material in a particular application will depend not only on its viscosity, but also on the C4 pattern, the substrate material, the chip passivation layer, the degree of cure of the encapsulant, and the dispense temperature. Two encapsulants having similar viscosities might reveal significantly different flow characteristics.

Encapsulant flow rates can be determined experimentally using two closely spaced glass slides to form a capillary channel. Typical experimental results showing flow distance versus time for two different encapsulants are shown in Figure 12-6. Similarly, flow experiments can be conducted *in situ* using transparent quartz simulated chips, assembled on the substrates.[19] Using this technique, the flow pattern and fillet formation of the

Table 12-3 Desired Properties of Underfill Materials[19]

Solids	100%
Viscosity	<20 Kcps
CTE	<40 (ppm/°C)
T_g	>125°C
Modulus	>9.0 GPa
Fracture toughness	>1.3 MPa–m$^{1/2}$
Cure temp	<130°C
Filler size	<50 μm
Filler content	<70 wt%
Ionics (Cl^-)	<20 ppm
Extractable chlorinated solvents	<20 ppm
Alpha activity (needed for memory)	<.005 counts/hr/Cm^2
Shelf life (@ −40°C)	>6 months
Pot life	>16 hrs
Electrical resistivity	>1.0 × 10^{12} ohm/cm
Dielectric constant	<4
Good chemical resistance	Against process solvents
Low moisture absorption	Essential
Flow under larger chips	Essential
C4 life improvement	5–10×
T & H reliability	2000 hrs

Note: All values are at 25°C unless specified.

encapsulant can be studied as a function of time and temperature for a particular chip geometry.

Impurities and Volatiles

Volatile components of the uncured epoxy encapsulant may evolve during the high temperature curing cycle, with the resulting possibility of void formation. Because these materials exhibit brittle behavior at their normal service temperatures, they are more prone to cracking if voids occur. Accordingly, an epoxy system that is 100% solids is desirable.

A typical underfill encapsulant has a dielectric constant less than 3.4, and electrical resistivity on the order of 10^{14} ohms/cm.[26] Ionized impurities, such

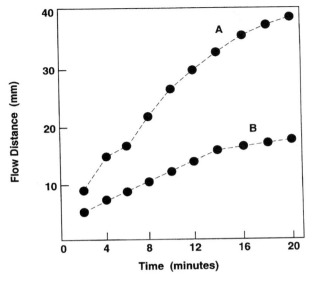

Figure 12-6   Plot of encapsulant flow distance vs. time measured by a flow monitor.

as Na^+, Cl^-, etc., can have potentially adverse effects on the electrical properties of encapsulants.[27] The concentration of these ions and their mobility tends to increase with moisture absorption, leading to insulation resistance breakdown during the temperature and humidity testing. These effects can be minimized by choosing an epoxy system with low levels of ionized impurities, chlorinated solvents, and moisture absorption.

Ionized impurities normally exist in epoxy resin, and in fillers to a lesser degree. This is due to the by-products (chlorine-containing compounds) generated during the production of epoxy resins. In addition, chlorinated solvents may be used in cleaning processing equipment, and the solvent residue can be picked up by the encapsulant. Careful analysis has shown that maximum levels of 20 ppm of ionics and chlorinated solvents are acceptable (see Table 12-3 for details).

An additional concern when packaging memory chips is the presence of trace levels of certain radioactive elements. For example, uranium and thorium may be present in very low concentration (e.g. parts per million) in packaging materials such as Sn/Pb and ceramic. The alpha particles emitted by such elements can cause soft error problems in memory devices.[28,29] In general, such errors can be prevented by using a thin layer of polymer encapsulant. However, the presence of alpha emitters associated with the ceramic fillers in some commercially available encapsulants has been reported.[19]

12.2.2 Properties of the Cured Encapsulant
Thermal Properties

The mechanical and thermal properties of epoxies depend strongly on the extent of cross-linking during the curing process. The extent of cross-linking will in turn depend on chemistries of the base resin, the curing agent, and any other additive such as flexibilizers or accelerators. Moreover, for any particular epoxy system, the temperature and time of curing will also affect the degree of cross-linking. Two properties of principal interest in predicting the performance of underfill encapsulants are the glass transition temperature (Tg) and the coefficient of thermal expansion CTE, both of which depend on the extent of cross-linking in the epoxy. These properties can be readily measured using conventional thermal analysis methods.[20]

The Tg is a second order thermodynamic transition: that is, it occurs when the heat capacity of a polymer changes from one value to another. In addition to a change in heat capacity rate, there is also a change in the rate of thermal expansion. Furthermore, a change from brittle to ductile behavior occurs, corresponding to a 1 to 2 order of magnitude drop in the elastic modulus. Above the Tg, an underfill encapsulant loses the properties which are essential for C4 reinforcement such as low thermal expansion, high moisture resistance, and high elastic modulus.

The underlying mechanism that produces this transition is the increased mobility of the molecular structure as its expands during heating. The Tg coincides with a significant increase in the mobility of some part of the moelcular structure, such as the main backbone, or a large side group. In some materials, primary and secondary transitions occur, but in underfill encapsulants, a single transition is generally well defined. Figure 12-7 shows the variation in CTE as a function of temperature for a typical underfill encapsulant. Above the Tg, the CTE will typically increase by a factor of 2 to 3.

Development studies are often directed at optimizing the properties of the underfill encapsulant without resorting to unnecessarily long times or high temperatures during curing. Figure 12-8 shows the variation of Tg vs. temperature of cure at a fixed time.

Mechanical Properties

Along with the CTE, the mechanical stiffness of an underfill encapsulant is of great importance in determining the level of reinforcement provided to the C4 solder joints. The elastic modulus of the composite depends on both the modulus of the base epoxy and the volume fraction of silica filler. The

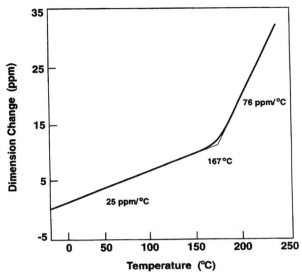

Figure 12-7   Thermal mechanical analysis (TMA) of epoxy encapsulant showing CTE and T_g parameters.

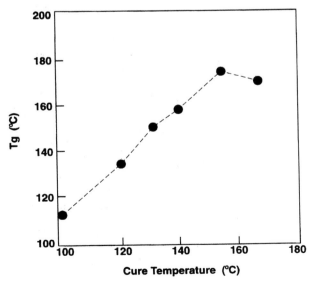

Figure 12-8   Plot of epoxy encapsulant T_g (measured by differential scanning calorimetry) vs. cure temperature.

base epoxies exhibit room temperature elastic moduli on the order of 4 to 5 GPa, while the filled materials are in the range up to 15 GPa. Figure 12-9 shows the variation in modulus with volume fraction of filler.[30]

Whereas the elastic modulus and CTE to a large extent determine the stress level in the solder joints, there is also potential for the underfill encapsulant itself to fail during thermal cycling. The encapsulant to chip and encapsulant to substrate interfaces must be sufficiently strong to prevent adhesive failure. Moreover, the encapsulant must have sufficiently high fracture toughness to resist cohesive cracking. In general, underfill encapsulants bond strongly to the chip and substrate materials, but may be more prone to cohesive cracking initiating in the fillet area.

It has been observed that, in some instances, microcracks or surface flaws of the encapsulants can propagate under repeated loads or a combination of loads and environmental attack which eventually lead to failure. To ensure the mechanical integrity of designs associated with a given encapsulant, the crack resistance or fracture toughness of the encapsulant must be carefully characterized. The fracture toughness is an intrinsic material property, which in the case of an encapsulant will depend on factors such as the epoxy chemistry, the volume fraction and size of the filler, the curing conditions and resulting cross-link density, and environmental factors such as thermal history and humidity exposure.

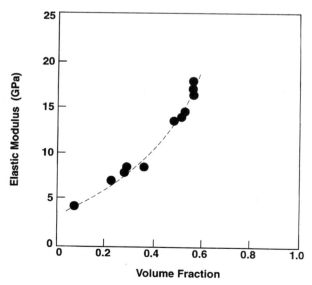

Figure 12-9 Plot of elastic modulus vs. filler volume fraction in epoxy encapsulant.[30]

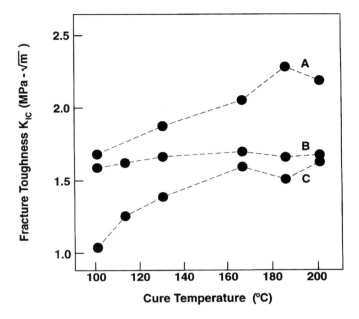

Figure 12-10 Plot of fracture toughness vs. cure temperature of epoxy encapsulant.

The loading of an existing crack by normal tensile forces, denoted as Mode I loading, is of most relevance to underfill encapsulants. There are many accepted methods for measuring the Mode I fracture toughness, and the plane strain compact tension test method is especially suitable for measuring underfill encapsulants.[31,32]

Figure 12-10 shows typical results of how fracture toughness values vary with cure temperature. It is seen that the fracture toughness increases with cure temperature until a critical level. Beyond that temperature, the fracture toughness levels off or starts to decline. A similar dependence of Tg vs. cure temperature is seen in Figure 12-8. This suggests that the crack resistance has a strong dependence on the degree of cross-linking. When cured at extremely high temperatures, crack resistance starts to decrease due to the degradation of epoxy systems or formation of cracks. The optimal cure temperature range for any particular encapsulant must be determined experimentally. Process conditions should be designed to fall in the range that provides the best mechanical strength. Note that regardless of the cure temperature, when exposed to high humidity,[25] the crack resistance of encapsulants declines, indicating that service conditions are also important when selecting an encapsulant.

Environmental Resistance

Underfill encapsulants are sensitive to temperature and humidity exposure in both the uncured and cured states. One of the consequences of high humidity is that it can degrade the encapsulant's physical properties, such as its hardness, elastic modulus, and fracture toughness.[25]

Water can enter the encapsulant by diffusion through exposed edges or by migration along cracks and crazes.[27] The dominant mode of transport is diffusion through the encapsulant, which is governed by Fick's Law of Diffusion[33]

$$\frac{M_t}{M_\infty} = \frac{4}{L} \left(\frac{Dt}{\pi}\right)^{1/2} \tag{12-1}$$

where, M_t is mass uptake of water at time t and M_∞ is the mass uptake at equilibrium, L is film thickness, and D is the diffusion constant. According to this equation, the fractional water uptake varies with the square root of time. The equilibrium water content that a given underfill encapsulant absorbs will depend on temperature and humidity[27]

$$M_\infty = KH^x \tag{12-2}$$

where H represents relative humidity, and K and x are constants.

Testing of moisture absorption can be accomplished by subjecting the underfill encapsulant to superheated steam (125°C, 15 psi) in a laboratory autoclave.[25] Figure 12-11 shows the percentage of weight gain versus time of exposure for several underfill encapsulants used at 125°C, 15 psi. The rate of moisture absorption depends on the cross-linking density of the epoxy encapsulants, where a lower cure temperature results in greater moisture absorption. This is expected since the epoxies have greater free volume as the temperature approaches their Tg, and a lower cure temperature generally results in a lower Tg. The typical diffusion constant for H_2O in epoxies is reported to be $1.5 \times 10^{-13} \, \text{m}^2/\text{sec}$.[33]

12.3 RELIABILITY GAINS AND UNDERLYING MECHANISMS

In this section the mechanics of C4 life enhancement are examined, and the results of finite element analysis and experimental stress analysis studies are reviewed. Then, the basic elements of reliability testing are discussed, and the results of previously published reliability tests are reviewed.

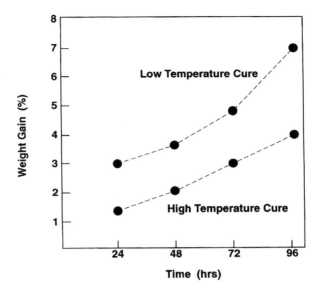

Figure 12.11   Plot of percentage moisture uptake vs. time in epoxy encapsulant when subjected to high humidities using pressure cooker (125°C, 15 psi).

12.3.1 Mechanics of C4 Fatigue Enhancement

It is well established that the fatigue life of metals decreases as the amplitude of cyclic strain and the number of fatigue cycles increases. The fatigue life enhancement of C4 solder joints due to encapsulation can be predicted by the Coffin-Manson equation[34] in the simplified form

$$\frac{N_2}{N_1} = \left(\frac{\varepsilon_1}{\varepsilon_2}\right)^z \tag{12-3}$$

where N_2/N_1 is the ratio of encapsulated to unencapsulated cycles to fail, $\varepsilon_1/\varepsilon_2$ is the ratio of strain unencapsulated to encapsulated, and z is a constant with a value of 1.9 for C4 joints.[35]

While the Coffin-Manson equation shows the influence of a reduction in the thermally induced cycle strain life in a predictable way, calculation of the strain itself is a difficult problem. Analytical solutions to relevant but simplified equations[36] show the common parametric form

$$\varepsilon = \Delta T \times \Delta CTE \times DNP \times G\{E_i, L_i, CTE_i\} \tag{12-4}$$

where ΔT is the temperature cycle, ΔCTE is the thermal mismatch between chip and substrate, and DNP is the distance from the neutral point. With other factors held constant, the thermal strain will increase with a greater CTE mismatch, a larger temperature cycle, and a larger DNP. The unknown function $G\{E_i, L_i, CTE_i\}$ incorporates the influence of these other factors, including the elastic properties E_i, the exact geometry of the structure L_i, and the thermal expansion CTE_i of all of the materials including the chip, the substrate, the solder, and the encapsulant.

Even with the simplifying assumptions of linear elastic behavior and perfectly bonded interfaces, three-dimensional solutions yielding the details of the strain field in the C4 interconnections are intractable. As a result, numerical and experimental stress analysis have been used to study the strain amplitude in the C4 joints as a function of thermal loading, package geometry, and material properties. The results of some recent studies of particular interest will be reviewed in the following paragraphs.

Finite element analysis of encapsulated flip chip packages has been based on the assumption that the encapsulant is perfectly bonded to the chip, substrate, and the C4 joint. Moreover, the encapsulant itself is assumed to be sufficiently strong and tough not to fracture or fail in fatigue. For thermal cycling below the glass transition temperature of the encapsulant, the behavior of the chip, substrate, and encapsulant can be approximated as linear elastic behavior.

The chip attach process is at elevated temperature to accomplish solder reflow, so the stress-free equilibrium state for the package is at the solidification temperature for the solder. Upon further cooling, the substrate tends to shrink more than the chip, thus deforming the C4 joints in a shear-dominant mode as depicted in Figure 12-12(a). The influence of encapsulation on the deformations is shown in exaggerated form in Figure 12-12(b), where the bending of the substrate has reduced the strain in the joints. Finite element analysis of flip chips on both ceramic and FR-4 substrates has yielded similar results.[2,4]

Figure 12-13 shows the variation in maximum principal strain versus DNP for both unencapsulated and encapsulated flip chip modules on ceramic substrates.[2] In the unencapsulated case, the strains increase with DNP, whereas they are essentially uniform for the encapsulated case. Similar results have been obtained for finite analysis of flip chip on organic carriers.[4]

Jackson et al. used a p-type finite element model where a relatively coarse mesh can be used in concert with high order polynomial shape functions.[37] While the details of the strain field in the C4 joint were not determined in their model, the relative displacement of the top and bottom of the joint were computed for a ceramic substrate, and assumed to be directly related to the maximum principal strain in the solder. The influence of the elastic

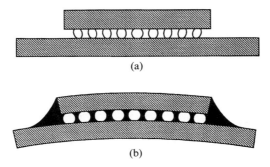

Figure 12-12 Schematic of exaggerated thermal displacements in a flip chip package as predicted by finite element modelling. (a) Shear dominates without encapsulation, (b) bending dominates with encapsulation.[2]

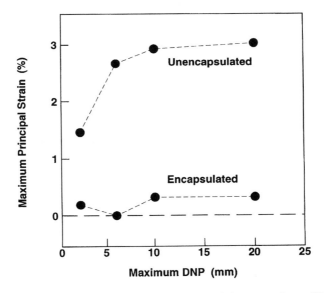

Figure 12-13 Maximum principal strain in C4 joint vs. maximum DNP.[2]

modulus is shown in Figure 12-14(a), and the influence of thermal expansion is shown in Figure 12-14(b). From this analysis, a minimum elastic modulus value of 5 GPa is indicated, while values above 10 GPa provide diminishing returns. This analysis is consistent with the observations that soft materials such as silicones do not protect the C4 joints, and lead to early fatigue failures.[19]

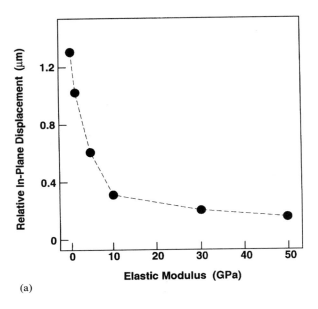

(a)

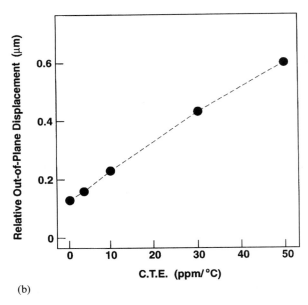

(b)

Figure 12-14 (a) C4 displacements vs. elastic modulus of the encapsulant as determined by finite element analysis. (b) C4 displacements vs. CTE of the encapsulant as determined by finite element analysis.[37]

Guo et al. (1992) used experimental stress analysis to study the influence of underfill encapsulation.[38] Using Moire interferometry the strains throughout the encapsulated package can be studied during a thermal cycle. The frequency of the fringes is proportional to the strain, and shows the deformations associated with the encapsulated package. High sensitivity measurements were also performed to determine the strains in the C4 joints. These experimental results corroborate the conclusions based on the finite element analyses.

Mechanical analysis indicates that the effect of the high modulus underfill encapsulants is to reduce the strains in the C4 joints. In addition, it may be that the presence of the encapsulant also affects the deformation mechanisms in the solder joint, reducing the initiation of fatigue cracks that ultimately lead to failure. Such a behavior may explain the improvement in C4 life that is observed with a thin (3 micron) conformal Parylene coating[17] which in itself has negligible influence on the C4 strains.

12.3.2 Reliability Testing and Case Studies

Fatigue reliability of the C4 joints is routinely assessed using the accelerated thermal cycling (ATC) testing. Extensive ATC reliability studies have been reported recently on encapsulated flip chips mounted on ceramic substrates.[1,2,18,19]

A variety of chip sizes have been tested, representing a wide selection of C4 patterns. One critical measure of the size of a chip is the center to perimeter distance of the C4 pattern, referred to as distance from neutral point. Figure 12-15 shows a schematic view of typical chip foot prints showing the complex C4 patterns and number of C4s present.

The preliminary step in an ATC test is a time-zero (T0) resistance measurement. This establishes a baseline for all future measurements. A failure criterion can be determined by an increase in the electrical resistivity of C4 joints. Typically an excess of 0.2 ohms over the T0 value may be considered as a fail. Following the T0 readout, the test hardware is subjected to ship-shock testing to simulate worst case shipping conditions to which the package may be exposed. Typical ship-shock conditions subject the test parts to 10 thermal cycles from -40 to $+65°C$ at a frequency of one cycle per hour. A post ship-shock readout is then performed.

Once the electrical readout after ship-shock is complete, the samples are put into ATC chambers. Typical thermal excursions used in ATC testing are 0 to 100°C, with three cycles per hour and typical dwell time of five minutes at the extreme temperatures. Some product qualifications require even more severe ATC test conditions, such as cycling from -55 to $+155°C$[1] or -200 to $+24°C$.[2] During ATC, the hardware may be continuously monitored

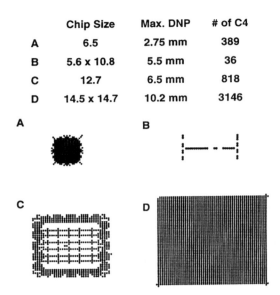

	Chip Size	Max. DNP	# of C4
A	6.5	2.75 mm	389
B	5.6 x 10.8	5.5 mm	36
C	12.7	6.5 mm	818
D	14.5 x 14.7	10.2 mm	3146

Figure 12-15   Schematic of various C4 footprints used in ATC test vehicles, showing the DNP and number of C4s. Footprint (D) is a test chip.[2,3]

electrically for any C4 degradation. Alternatively, electrical readouts may be taken at discrete intervals (e.g. 500 cycles) during the ATC. Upon encapsulation, semiconductor packages generally become much more reliable, so that ATC testing is often continued through 10,000 cycles.

Results of ATC testing on MC modules both with and without encapsulation are summarized in Table 12-4. Control parts without underfill encapsulation showed early fails, with a smaller DNP corresponding to slightly longer ATC life. In one study, the median ATC cycles to failure varied from 3399 for 6.5 mm chips to 233 cycles for 14.7 mm chips, demonstrating the influence of DNP on fatigue life.

Reliability data from ATC testing on MC modules, which passed 10,000 cycles without failures,[2] demonstrate the dramatic reliability enhancement achieved with underfill encapsulation. This corresponds to 5 to 10× improvement in flip chip fatigue life on ceramics with underfill encapsulation. Also, the MC parts thermal cycled under severe conditions (−200°C to +24°C) also yielded 10× improvement in C4 fatigue life. The trend in C4 life improvement is the same even when different types of epoxy encapsulants formulations are used as long as good under chip coverage is achieved. Improvements in flip chip fatigue life have also been reported independently by various groups on both ceramics and organic cards.[1,3,4,18,19]

Table 12-4 ATC Stress Data on Flip Chip Packages

Chip DNP (mm)	Carrier Material	Underfill Material	Encapsulant CTE (ppm/°C)	Life Time (Cycles)	Ratio	Ref.
6.57	Ceramic	None	—	230+	1.0	1
6.57	Ceramic	Epoxy	73	510+	2.2	1
6.57	Ceramic	Epoxy/glass	26	2200+	9.6	1
2.75	Ceramic	None	—	1,686#	1.0	2
6.5	Ceramic	None	—	370#	1.0	2
10.2	Ceramic	None	—	182#	1.0	2
2.75	Ceramic	Epoxy/glass	<40	10,000*	6.0	2
6.5	Ceramic	Epoxy/glass	<40	10,000*	25.0	2
10.2	Ceramic	Epoxy/glass	<40	5,000*	25.0	2
1.09	FR-4 PCB	None	—	100#	1.0	3
1.09	FR-4 PCB	Epoxy/glass	<40	>1200*	12	3
3.78	FR-4 PCB	Epoxy/glass	<40	>1200*	12	3
5.52	FR-4 PCB	Epoxy/glass	<40	>1250*	13	3
5.52	FR-4 PCB	None	—	260 @	1.0	4
5.52	FR-4 PCB	Epoxy/glass	<40	>1727*	6.0	4

* No fails observed during ATC testing
\# Typical first fail occurred during ATC
\+ About 50% parts failed during ATC
@ All parts failed during ATC

ATC stress conditions used were: (−55 to 155°C);[1] (0 to 100°C);[2,3] (−25 to 115°C).[4]

Besides ceramics, ATC reliability test results of flip chip packages on organic printed circuit boards (FR-4) have also been reported.[3,4] Unlike ceramics, the flip chip attachment (FCA) to FR-4 cards uses eutectic solder (e.g. 63%Sn/37%Pb), which is first deposited on the copper pads. Chips having standard solder bumps (e.g. 97%Pb/3%Sn) are bonded directly to the eutectic solder on the PCB, by reflowing the solder at the eutectic melt temperature ($< 200°C$). The resulting solder joint in FCA therefore has a different alloy and structure compared to a conventional C4.

Since the thermal mismatch problem in the organic FCA packages is much more severe than in ceramics, the unencapsulated C4 joints in FCA are expected to yield very limited fatigue life during thermal cycling. Hence, encapsulation is an essential element of the FCA technology. Reliability testing on encapsulated FCA parts has shown significant reliability gains, passing 1200 ATC cycles without fails.[3] Similar results have also been reported on low CTE PCB laminates as well as FR-4 cards.[4] The fatigue life improvement has also been demonstrated on mechanically reworked and encapsulated FCA packages.[39]

While the reliability results published in the literature show marked improvement, during the development phase a variety of failure mechanisms may be observed prior to optimizing the encapsulation technique. Such problems may be due to both the encapsulation process and the materials characteristics, including factors such as flow, fillet formation, voiding, moisture absorption, cracking, adhesion and others. In particular, flow irregularities that result in void formation can cause a variety of problems. For example, since the encapsulants are brittle, cracks tend to initiate from voids and inclusions. In the case of low temperature C4 joints on FCA, if the compressive stresses during thermal cycling are sufficiently high, solder can be extruded into adjacent voids and can lead to joint failure.[3]

12.4 RECOMMENDATIONS FOR MANUFACTURING

Underfill encapsulation has been proven to offer tremendous reliability gains to flip chip packages with minimal impact on the manufacturing process flow. However, the encapsulants are highly engineered, reactive mixtures with high volume fractions of suspended particles and complex flow properties. Accordingly, a number of factors relating to the material and process must be carefully controlled to ensure package reliability. These factors include, for example, storage and handling of the encapsulant, temperature and humidity control during encapsulation, and the dispense and curing processes.

Underfill encapsulants are premixed at the supplier, packaged in plastic syringes (5 to 100 cc), and then frozen at $-40°C$ to prevent curing. Shipping the materials requires special handling to maintain the low temperatures continuously, and subsequent storage requires uninterrupted temperature control. As received at the manufacturing site, receiving inspection is necessary to verify that the properties meet the material specification put in place during the development phase. The material specification would include a set of tests to verify properties (see Table 12-3) such as viscosity, filler content, filler size, ionics, residual chlorinated solvents, Tg and CTE.

Maintained at $-40°C$, underfill encapsulants may have a storage life of approximately one year. The extent of degradation of the materials after longer storage times can be studied by repeating the receiving inspections and comparing with the original material specification. As the material is needed, it can be removed from the freezer and thawed at room temperature. Depending on the size of the syringe, this may take an hour or more. Once thawed, the encapsulant has a plot life of four to eight hours, depending on the particular material and the room temperature and humidity.

Dispensing the encapsulant through a syringe can be accomplished with either air pressure or a screw-driven pump. The motion of the dispense tip is controlled by programming an automated three axes control tool, and coordinated with the sequencing of pump to achieve the desired dispense pattern. The temperature of the encapsulant will affect its viscosity, and in turn affect the amount of material dispensed. The weight of material dispensed must be maintained within a range which provides proper fillet formation.

Generally, a properly formulated encapsulant can be dispensed at a single point along the perimeter of the chip, and it will flow sufficiently to surround the C4s, and also form fillets on the other three sides. After the encapsulant is dispensed, the temperature of the flip chip package may be maintained at some intermediate temperature that lowers its viscosity but limits the rate of curing. Once exit fillets are completely formed, the encapsulant may be gelled and cured. The gelling step, where the encapsulant has solidified but is not completely cured, is necessary when the encapsulation on the opposite side of the card is required.

A generic process flow showing the sequence of events required when encapsulating a two-sided FR-4 circuit board with direct flip chip attach is shown in Figure 12-16. The underfill encapsulation materials are sensitive to environmental factors such as temperature and humidity, both in the uncured and cured state. Temperature and humidity need to be controlled at various stages of the process, including dispensing, gelling, and curing. Studies have shown that the idling time between the dispense and cure plays a role in the encapsulant performance. Moreover, moisture contained in the FR-4 cards may affect the properties of the cured encapsulant, in which case pre-baking of the cards is required.

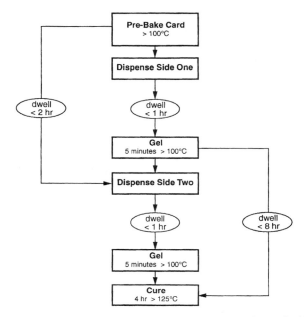

Figure 12-16   Typical process flow in manufacturing for encapsulating double-sided card.

12.5 SUMMARY AND FUTURE DIRECTIONS

Underfill encapsulation of flip chips is a relatively new technique for reinforcing the solder interconnections. It has found applications in both ceramic packages and FCA and organic carriers, and has opened the possibility of packaging even larger DNP chips. Two of the key properties required of an underfill encapsulant are low CTE and high elastic modulus to reduce the cyclic strain in the C4 during thermal loading. The best materials presently available are epoxy resins filled with low thermal expansion ceramic particles.

While having minimal impact on the existing chip assembly process flow, underfill encapsulation with these epoxy-based materials has the disadvantage of not being reworkable. This lack of reworkability is a significant trade-off in applying encapsulation to highly integrated packages such MCM, and is certainly a challenge for future research and development efforts. Other material advancements of immediate practical importance to manufacturing would include reduced time and temperature for cure and improved shelf life. Furthermore, underfill encapsulation may find application in additional package configurations as a result of improvements in its fracture toughness, adhesion and flow properties.

ACKNOWLEDGMENTS

Thanks are due to the following contributors: Jack McCreary, Dave Wang, Kostas Papathomas, Jim Clementi, Jeff Palomaki, Gary Hill, Doug Powell, Don Gould, Barry Bonitz, Jack Varcoe, Jimmy Ellerson, Kris Nair, T. M. Niu, Tien Wu, Doug Thorne, and Bob Jackson.

REFERENCES

1. Nakano, F., T. Soga, and S. Amagi, "Resin-Insertion effect on Thermal Cycle Resistivity of Flip-Chip Mounted LSI Devices," *ISHM '87 Proceedings*, pp. 536-541.
2. Clementi, J., J. McCreary, T. M. Niu, J. Palomaki, J. Varcoe, and G. Hill, "Flip Chip Encapsulation on Ceramic Substrates," *Proc. of 43rd Electronic Components & Technology Conference*, Orlando, FL, June 1993, p. 175.
3. Powell, D. O., and A. K. Trivedi, "Flip-Chip on FR-4 Integrated Circuit Packaging," *Proc. of 43rd Electronic Components & Technology Conference*, Orlando, FL, June 1993, p. 182.
4. Tsukada, Y., Y. Mashimoto, T. Nishio, and N. Mil, "Reliability and Stress Analysis of Encapsulated Flip Chip Joint on Epoxy Based Printed Circuit Packaging," *Proc. of the 1992 Joint ASME/JSME Conf. on Electronic Packaging*, Vol. 2, Materials, Process, Reliability, Quality Control and NDE, p. 827, 1992.
5. Boyko, C., T. Bocek, V. Markovich, and D. Mayo, "Film Redistribution Layer Technology," *Proc. of 43rd Electronic Components & Technology Conference*, Orlando, FL, June 1993, pp. 302-305.
6. Totta, P. A., and R. P. Sopher, "SLT Device Metallurgy and Its Monolithic Extension," *IBM J. Res. Develop.*, 5, pp. 226-238, May 1969.
7. Milkovich, C., M. A. Gaynes, and J. S. Parkins, "Double Sided Flexible Carrier With Discretes and Thermally Enhanced FCA/COF," *Proc. of 43rd Electronic Components & Technology Conference*, Orlando, FL, June 1993, p. 16.
8. Lyn, R. J., "Encapsulation for PCMCIA Assemblies–An Overview," *NEPCON West*, Anaheim, CA, pp. 743-751, 1992.
9. Koopman, N. G., T. C. Reiley, and P. A. Totta, "Chip-to-Package Interconnections," *Microelectronics Packaging Handbook*, ed. R. R. Tummala, and E. J. Rymaszewski, Van Nostrand Reinhold, New York, pp. 361-453, 1989.
10. Schwartz, B., "Review of Multilayer Ceramics for Microelectronic Packaging," *J. Phys. Chem. Solids*, 45 (10), 1051-1068, 1984.
11. Tummala, R. R., and E. J. Rymaszewski (eds), *Microelectronics Packaging Handbook*, New York, Van Nostrand Reinhold, 1989.
12. Goldmann, L. S., "Geometric Optimization of Controlled Collapse Interconnections," *IBM J. Res. Develop.*, pp. 251-265, May 1969.
13. Beckham, K., et al., "Solder Interconnection Structure for Joining Semiconductor Devices to Substrates that have Improved Fatigue Life, and Process for Making," U.S. Patent No. 4,604,644, Aug. 5, 1986.

14. Christie, F. R., K. I. Papathomas, and D. W. Wang, "Solder Interconnection Structure and Process for Making," U.S. Pat. 4,999,699, March 12, 1991.
15. Christie, F. R., K. I. Papathomas, and D. W. Wang, "Solder Interconnection Structure and Process for Making," U.S. Pat. 5,089,440, February 2, 1992.
16. Christie, F. R., K. I. Papathomas, M. D. Poliks, and D. W. Wang,"Dielectric Composition and Solder Interconnection Structure for Its Use," U.S. Pat. 5,194,930, March 16, 1993.
17. Tong, H. M., L. Mok, K. R. Grebe, H. L. Yeh, K. K. Srivastava, and J. T. Coffin, "Parylene Encapsulation of Ceramic Packages for Liquid Nitrogen Applications," *Proc. of 40th Electronic Components & Technology Conference,* Las Vegas, pp. 345-350, 1990.
18. Gabrykewicz, T., D. Sengupta, and T. Thuruthumaly, "Glob top material selection for flip chip devices," in *Proc. 1986 Int. Symp. on Microelectronics,* Reston, VA, pp. 707-713, 1986.
19. Suryanarayana, D., R. Hsiao, T. P. Gall, and J. M. McCreary, "Enhancement of Flip-Chip Fatigue Life by Encapsulation," *IEEE Transactions on Components and Hybrids, and Manufacturing,* Vol. 14, No. 1, pp. 218-223, 1991.
20. Wang, D., and K. I. Papathomas, "Encapsulant For Fatigue Life Enhancement of Controlled Collapse Chip Connection (C4)," *Proc. of 43rd Electronic Components & Technology Conference,* Orlando, FL, June 1993, p. 780.
21. Suhir, E., "Application of an Epoxy Cap in a Flip-Chip Package Design," *Trans. ASME, J. of Electronic Packaging,* vol. 111, 1989, no. 1, pp. 16-20.
22. Hunadi, R., and N. Bilow, "Low Expansion Blob Top Encapsulants—A New Generation of Materials for Chip-on-Board and Hybrid Packaging," *Proc. of the First Intl. SAMPE Electronics Conf.,* June 23-25, 1987, pp. 397-410.
23. Ichikawa, K., T. Kubota, and M. Suzuki, "Flip-Chip Joining Technique for 400DPI Thermal Printheads," *OKI Technical Reviews (3),* vol. 55, pp. 18-24, September 1988.
24. Emerson, J. A., J. J. Sparapany, A. R. Martin, M. R. Bonneau, and D. A. Burkhart, "Robust Encapsulation of Hybrid Devices," *Proc. of 40th Electronic Components & Technology Conference,* Las Vegas, NV, May 21-23, 1990, pp. 600-605.
25. Suryanarayana, D., T. Y. Wu, and J. A. Varcoe, "Encapsulants Used in Flip-Chip Packages,' *Proc. of 43rd Electronic Components & Technology Conference,* Orlando, FL, June 1993, p. 193.
26. Burkhart, A., "Hysol FP4510 Improves Heat Resistance of Flip Chip Devices," Private communication, 1992.
27. Anderson, J. E., V. Markovac, and P. R. Troyk, "Polymer Encapsulants for Microelectronics: Mechanisms for Protection and Failure," *IEEE Transactions on Components and Hybrids, and Manufacturing,* Vol. 11, No. 1, pp. 152-158, 1988.
28. May, T. C., and M. H. Woods, "A New Physical Mechanism for Soft Errors in Dynamic Memories," *Proc. Int. Rel. Phys. Symp.,* pp. 33-40, April 1978.
29. Howard, R., "Characterization of Low Alpha-Particle Emissions from Semiconductor Memory Materials," *J. Electronic Materials,* 10 (4), p. 747, July 1981.

30. Farquhar, D. S., private communication, 1993.
31. Stawley, J. E., and W. F. Brown, "Fracture Toughness Testing Methods," *ASTM STP* 381, pp. 133-145, 1965.
32. Broek, D., *Elementary Engineering Fracture Mechanics*, Martinus Nijhoff Pub., pp. 170-176, 1982.
33. Kinloch, A. J. (ed.), *Durability of Structural Adhesives*, Applied Science Publishers Ltd., New York, 1983.
34. Manson, S. S., *Thermal Stress and Low-Cycle Fatigue*, Reprint Edition, Malabar, FL, Robert E. Kreiger Publishing Co., 1981.
35. Norris, K. C., and A. H. Landzberg, "Reliability of Controlled Collapse Interconnections," *IBM J. Res. Develop.*, 13 (3), pp. 266-271, May 1969.
36. Chen, W., and C. Nelson, "Thermal Stress in Bonded Joints," *IBM Journal of Research and Development*, 23, no. 2, pp. 179-180, 1979.
37. Jackson, R. L., P. Carnevali, and J. D. Frazier, "Finite-element Modelling of Encapsulated Flip-Chip Packaging Assemblies," *ISHM 1991 Proceedings*, pp. 82-85.
38. Guo, Y., W. T. Chen, and C. K. Lim, "Experimental Determinations of Thermal Strains in Semiconductor Packaging using Moire Interferometry," *Proceedings of ASME San Jose Conference on Electronic Packaging*, 1992, pp. 779-783.
39. Tsukada, Y., Y. Mashimoto, and N. Watanuki, "A Novel Chip Replacement Method For Encapsulated Flip Chip Bonding," *Proc. of 43rd Electronic Components & Technology Conference*, Orlando, FL, June 1993, pp. 199-204.
40. Lewis, C. F., (ed.), *Materials Engineering*, Materials Selector, A Penton Publication, 1991.
41. Edwards, M., "Flexible Epoxies," *Electronics Materials Handbook*, Volume 1, Packaging, Technical Chairman, M. L. Minges, ASM International, Materials Park, OH, 1989.
42. May, C. A., "Epoxy Materials," *Electronics Materials Handbook*, Volume 1, Packaging, Technical Chairman, M. L. Minges, ASM International, Materials Park, OH, 1989.
43. Zwiers, R. J. M., H. J. L. Bressers, B. Ouwehand, and D. Baumann, "Development of a New Low-Stress Hyperred LED Encapsulant," *IEEE Trans. on Hybrids, and Manufacturing Technology*, Vol. 12, No. 3, 1989, pp. 387-392.
44. Castall, Inc., Weymouth Industrial Park, East Weymouth, MA, 1989.
45. Suhir, E., and J. M. Segelken, "Mechanical Behavior of Flip-Chip Encapsulants," *J. Electronic Packaging*, Vol. 112, pp. 327-332, 1990.
46. Craig, J. D., "Polyimide Coatings," *Electronics Materials Handbook*, Volume 1, Packaging, Technical Chairman, M. L. Minges, ASM International Handbook Committee, Materials Park, OH, 1989.
47. Beach, W. F., "Parylene Coatings," *Electronics Materials Handbook*, Volume 1, Packaging, Technical Chairman, M. L. Minges, ASM International Handbook Committee, Materials Park, OH, 1989.

Authors' Biographies

John H. Lau is a senior engineer of Hewlett-Packard Company located in Palo Alto, California. His research activities cover a broad range of electronics packaging and manufacturing technology.

He received a BE degree in civil engineering from the National Taiwan University, an MASc degree in structural engineering from the University of British Columbia, an MS degree in engineering mechanics from the University of Wisconsin, an MS degree in management science from the Fairleigh Dickinson University, and a PhD in theoretical and applied mechanics from the University of Illinois.

Prior to joining Hewlett-Packard Laboratories in 1984, he worked for Exxon Production and Research Company and Sandia National Laboratory. He has more than 24 years of research and development experience in applying the principles of engineering and science to the electronic, petroleum, nuclear, and defense industries. He has authored and coauthored over 100 technical publications in these areas and is the author and editor of the books, *Solder Joint Reliability: Theory and Applications, Handbook of Tape Automated Bonding, Thermal Stress and Strain in Microelectronics Packaging, Handbook of Fine Pitch Surface Mount Technology*, and *The Mechanics of Solder Alloy Interconnects*.

He is an associate editor of the *ASME Transactions, Journal of Electronic Packaging*, the *IEEE Transactions on Components, Packaging, and Manufacturing Technology*, and the *International Journal of Microelectronics Packaging*. He has also served as session chairman, program chairman, general chairman, and invited speaker of several IEEE, ASME,

ASM, ISHM, and NEPCON conferences and symposiums. He is an IEEE fellow, a registered professional engineer, and is listed in the American Men and Women of Science.

Charles F. Hawkins has worked with I_{DDQ} testing and CMOS IC defect electrical characterization since 1984. He is a Professor in the Electrical and Computer Engineering Department at the University of New Mexico and works as a contract consultant to Sandia National Labs Multichip and Custom Products department.

He is active in teaching testability short courses for industry and teaches VLSI Design and Test at the university. He is also the Program Chair for the 1993 International Test Conference. He and Jerry Soden have given over a dozen one day I_{DDQ} seminars to semiconductor companies over the last three years. Chuck is a coeditor of the special November 1992 issue on IDDQ testing in the *Journal of Electronic Testing: Theory and Applications (JETTA)*.

David W. Palmer manages the Integrated Circuit Packaging Technology Department at Sandia National Laboratories in Albuquerque, New Mexico. He received his PhD for performing solid state applied physics experiments at CalTech in 1975.

Since then, at Sandia, he has been involved in the development of high reliability hybrid microcircuits including systems that functioned in 300°C environments. In addition, he has studied SAW devices, converted cable/connector design to an expert system, helped establish the computer hardware and software for VLSI design, and improved the yield and reliability of packaging production.

His current interests consist of developing a complete set of test chips for evaluation of packaging technology and manufacturing, developing enablers so that multichip module technology is the cost effective approach to the next generation system development, understanding the IC fabrication rules that will minimize bond pad cratering, and optimizing productivity with desktop computer software and networking. Dave has been active in the Components, Hybrids, and Manufacturing Society of the IEEE including producing the Society Newsletter with desktop publishing technology.

R. Keith Treece joined Sandia National Laboratories in 1962 where he has been involved in the design and development of electronic test systems, high-field varistor components, radiation hardened integrated circuits, and most recently, multichip modules. Keith received undergraduate degrees in mathematics and electrical engineering in 1962 from the University of Missouri and Carnegie Mellon University, respectively, and an MSEE degree from the University of New Mexico in 1964. His current interests include known good die testing and direct chip-attach techniques.

Harry K. Charles, Jr received a BSEE degree from Drexel University, Philadelphia, Pennsylvania, and a PhD degree in electrical engineering from The Johns Hopkins University, Baltimore, Maryland, in 1967 and 1972, respectively.

In 1973, following a post-doctoral appointment in the Research Center of The Johns Hopkins University Applied Physics Laboratory (APL), he joined the Microelectronics Group at APL where he has held various engineering and managerial positions including Group Supervisor. Since January 1989, Dr Charles has been leading the Engineering and Fabrication Branch which is responsible for all electronic and mechanical hardware design and fabrication at APL.

Dr Charles has been active in several fields of engineering research and development including: the physics of crystalline and amorphous semi-conductors; electronic energy conversion devices; microelectronic circuit development and processing; and electronic packaging, interconnection, and reliability. His current research interests include the modeling and experimental analysis of high speed interconnection and packaging structures for multichip module applications. He has published over 120 papers in these and other fields during the last 22 years. Many of these papers have appeared in prestigious peer reviewed journals ranging from Physics Review Letters to various Transactions of the IEEE. To date, Dr Charles has received six Best of Session Paper Awards and one Best of Conference Paper Award. In addition to his publications he holds one patent with several applications pending. In 1989, he received the Distinguished Young Engineer Award from the State of Maryland for his "outstanding contributions in the miniaturization of electronic instrumentation and microcircuits for space, underwater, and biomedical applications".

Dr Charles is currently a Senior Member of the IEEE and has previously served on the Administrative Committee of the IEEE Components, Hybrids and Manufacturing Technology (CHMT) Society for seven years. He was Chairman of both the Standards Committee and the Nominations Committee for the CHMT. Dr Charles is a member of the Electronic Components and Technology Conference (ECTC) Reliability Subcommittee and has served the ECTC as both session chair and co-chair as well as a contributing author. Other professional society memberships include: the American Physical Society, the International Solar Energy Society, The Microbeam Analysis Society, ASM International, the Society for Advanced Materials and Processing Engineering, and the International Society for Hybrid Microelectronics (ISHM). Dr Charles has been an active ISHM participant, both at the local chapter level and on the national committee. In

1987, he received ISHM's Technical Achievement Award for "his contributions in thick and thin film assembly techniques, semiconductor processing and advanced microelectronic packaging applications". In 1989, he served as Technical Program Chair for the International Microelectronics Symposium. From 1990-1992 he was ISHM's Technical Vice President. Dr Charles has been active in both the IEEE and ISHM educational activities. He has taught in The Johns Hopkins University Continuing Engineering Education Program for the last thirteen years developing seven new courses to expand and enrich the engineering program. In 1990, Dr Charles received the Continuing Professional Programs Outstanding Instructor Award from The Johns Hopkins University's G. W. C. Whiting School of Engineering.

Dr Charles is married and lives with his wife and eleven year old daughter in Howard County, Maryland. His hobbies include: stamp collecting, woodworking, and gardening.

Sung K. Kang received his PhD in Materials Science from the University of Pennsylvania, Philadelphia, in 1973, and his BS in Metallurgical Engineering from the Seoul National University, Seoul, Korea, in 1969.

He was a research scientist at the Center for Joining of Materials, Carnegie Mellon University, Pittsburgh, 1973-76, and also at the Nova Scotia Technical University, Halifax, Canada, 1976-77. Then he taught at the Stevens Institute of Technology, Hoboken, N. J., as an Assistant Professor in the Department of Materials Science, 1977-80, and as an Adjunct Professor, 1980-84. Later, he became a Senior Scientist at the International Nickel Company, Research & Development Center, Sterling Forest, N.Y., 1980-84.

In 1984, he joined IBM T. J. Watson Research Center, Advanced Packaging Technology Laboratory, Yorktown Heights, N.Y., where he has been working on chip interconnection technology (TAB, C4, wire bonding), surface mount technology, solder joint reliability, and multichip module packaging.

He was the Chairman of Microelectronic Interconnections and Assembly Committee, ASM International Electronic Materials & Processes Division, Materials Park, Ohio, 1989-91. He has taught a graduate course on "Electronic Packaging Technology", at the Polytechnic University of NY, Spring, 1990, and at the Inter-University Semiconductor Research Center of SNU, Korea, June 1991.

William T. Chen is a Senior Technical Staff member of the IBM Microelectronics Division. Dr Chen received a BS degree in mechanical engineering from Queen Mary College, University of London; an MS from Brown University; and a PhD from Cornell University in theoretical and applied mechanics. He joined IBM in 1963.

Dr Chen has been active in various aspects of electronic packaging product development and manufacturing, including designs, materials, processes, and reliability. His interest has been in applying knowledge of mechanics, materials, and physics to engineering practices of design and manufacturing for leading-edge electronic packaging products.

Dr Chen is a Fellow of the American Society of Mechanical Engineers; he has been elected to the IBM Academy of Technology and is a technical editor of *ASME Transactions, Journal of Electronic Packaging*.

Richard B. Hammer is currently at IBM Corporation in Endicott, New York. In his present position, he is an Engineering Manager of the Precision Flex Line in IBM Microelectronics. He has been involved with various aspects of First Level Device/Module Packaging including: Ceramics, TAB, Area Arrays, BGA's, Tape and Materials. Prior to joining IBM, he was a Research Manager with ITT. Dr Hammer received his BS degree in 1966 from Alfred University and his PhD in Physical Organic Chemistry in 1972 from Syracuse University. He has 10 US Patents and numerous publications. He is an active member of the American Chemical Society and has served as Chairman of the International TAB Society.

Frank E. Andros has been with IBM Corporation since 1968. During this time he has had several technical and managerial assignments in Electronics Packaging. Management experience includes Packaging Manager, Technical Support (Thermal, Electrical, Reliability) for Semiconductor Packages; Product Development Manager for Ceramic Substrates/Packages and TAB Product Development Manager. He is currently a Senior Engineer in IBM Microelectronics, Precision Flex Operations, located in Endicott, New York.

He received a BS from Michigan Technological University in Mechanical Engineering and MS and PhD degrees from Arizona State University in Mechanical Engineering. Dr Andros is a member of the American Society of Mechanical Engineers.

Lewis S. Goldmann is a Senior Engineer at IBM's General Technology Division. During most of his career at IBM, he has been involved with the mechanical modeling and testing of electronic packages, chips and materials. He has published numerous papers, including several on flip-chip joining.

He holds a BS degree in Pre-Engineering and Liberal Arts from Queens College, CUNY; a BSME from Columbia University, and an MSME from MIT. He is a member of Phi Beta Kappa and Tau Beta Pi Honor Societies, and is currently a member of ASME and an Associate Editor of their *Journal of Electronic Packaging*.

Prior to IBM, Mr Goldmann worked in the Electro-Mechanical Switching Laboratory at Bell Labs.

Paul A. Totta is an IBM Fellow at the Technology Products East Fishkill site. Before his appointment in 1987 he was Project Manager of Interconnection Technology in the Laboratory for many years. He is presently a technical assistant to the Director of the Semiconductor Research and Development Center. His areas of responsibility have been the development of multilevel thin film wiring and dielectrics on the surface of semiconductor devices including the ohmic and Schottky contacts, packaging materials, and chip solder bump connections.

He is a metallurgical engineer by training, and a graduate of RPI. He is a member of Tau Beta Pi, Phi Lambda Upsilon and Sigma Xi. Recently he was elected a Fellow of ASM, International. In his IBM career he has achieved the ninth invention plateau and has several outstanding contribution and invention awards. He has also received the recognition award from ISHM for his development of solder bump metallurgy and thin film interconnections.

His areas of expertise are thin film metal and insulator technology for electronic devices, metallurgy of joining, physical metallurgy of corrosion and magnetic materials. Prior to his joining IBM he also worked as a metallurgist for Central Electric and Handy and Harman.

Goran Matijasevic is a Post-Doctoral Researcher at the University of California, Irvine, Calif. Dr Matijasevic has been involved in work on Microelectronic Packaging. His other research interests include nondestructive testing techniques, especially acoustic microscopy, as well as failure analysis and reliability issues. He has presented his work at the International Reliability Physics Symposium, the Electronic Components and Technology Conference, and the International Electronic Packaging Conference, and has co-authored 15 papers. He received a BEng degree in applied physics from the Faculty of Electrical Engineering of the University of Belgrade, Yugoslavia in 1984 and MS and PhD degrees in Electrical and Computer Engineering from University of California, Irvine in 1985 and 1991 respectively. Dr Matijasevic is a member of the IEEE, MRS, IEPS, and Eta Kappa Nu.

Chin C. Lee is an Associate Professor at the University of California, Irvine, Calif. Dr Lee received BE and MS degrees in electronics from National Chiao-Tung University, Hsinchu, Taiwan, in 1970 and 1973 respectively, and a PhD degree in electrical engineering from Carnegie Mellon University, Pittsburgh, Pa., in 1979. From 1979 to 1980, he was a research associate with the Electrical Engineering Department of Carnegie

Mellon University. From 1980 to 1983, he was with the Electrical Engineering Department of the University of California, Irvine, as a research specialist. In 1984, he joined the same department as an Assistant Professor and later became Associate Professor of Electrical and Computer Engineering in 1990.

His research interests include thermal design and measurement of integrated circuit devices, electromagnetic theory, packaging technologies of semiconductor devices, scanning acoustic microscopy and its application to fiber optics. He is the co-author of two book chapters and 80 papers in the subject areas mentioned above. Dr Lee is a member of International Society for Boundary Elements and Tau Beta Pi.

Chen Y. Wang is a Post-Doctoral Researcher at the University of California, Irvine, Calif. He received a BS degree in physics from the National Taiwan University, Taiwan, in 1986, and the MS and PhD degrees in Electrical Engineering from the University of California at Irvine in 1989 and 1992 respectively. He has worked as a Teaching Assistant and Research Assistant at the University of California, Irvine, since 1988. His current research interests include bonding technologies of electronic devices, finite element analysis of thermal stress, failure analysis and reliability issues. He is a co-author of two book chapters and 13 papers in the subject areas mentioned above. Dr Wang is a member of the IEEE and MRS.

Robert A. Christiansen is a graduate of the University of Missouri with a BS Chemical Engineering degree. Mr Christiansen has over 30 years of experience in materials and process engineering, primarily in electronics packaging. He was manager of Materials Engineering for IBM's Space Systems Center during the design, development, build and launch of the Saturn V and Skylab on-board computer systems. He was also responsible for sealing and passivation systems for military applications of thick film hybrid microelectronic assemblies. At IBM/Lexington, Kentucky, primary assignments have been the necessary research, development and test to implement microelectronics packaging into competitively priced, reliable desk-top business equipment such as typewriters, printers and keyboards. This included the development of chip on board materials and processes for both flex and rigid card assemblies and to intermix this technology with SMT and PTP packaging.

Mr. Christiansen is currently employed by Lexmark (now an independent company formed from the sale of the Lexington facility by IBM to Clayton, Dubilier and Rice). He holds a number of patents in electronic packaging. Publications on the topic of chip on board include:

"The Extension of Chip on Board to Highly Reliable System Assemblies", given at the IEEE/CHMT Conference, September 1990.

"Chip on Board: Alternative Electronic Packaging Technique", guest speaker on the National Technological University satellite network as part of the series "Contemporary Issues in Electronic Manufacturing: Beyond SMT", air date May 10, 1991.

Tom C. Chung is Tandem Computers' Visiting Representative at Microelectronics and Computer Technology Corporation (MCC) in Austin, Texas. Dr Chung has more than 14 years of experience in the field of advanced computer packaging and interconnect technologies, including surface mount, tape automated bonding, flip-chip, advanced single chip packaging, multichip module technologies, etc. He worked on semiconductor assembly, test, packaging and interconnect related technology developments for United Technologies Corporation and MCC for five years before joining Tandem Computers Inc. in 1989 as a Development Engineering Manager in the advanced computer packaging and interconnect department. Since early 1993, he has taken the responsibility as Tandem's technical representative at MCC to lead a group to develop flip-chip based MCM technology for the next generation of computer workstations. He has a BSE from NCKU in Taiwan, an MSME from Texas Tech University, and an MS in engineering management, and a doctorate degree in engineering from the Southern Methodist University, Dallas, Texas. He owns four US patents and has published more than 20 technical papers. He is also a co-author of the *Handbook of Tape Automated Bonding* published in 1991.

James Chang is a Senior Research Fellow at GINTIC Institute of Manufacturing Technology (GIMT) in Singapore. Dr Chang received his PhD in Engineering Mechanics from the University of Texas at Austin, an MS in Mechanical Engineering from the Southern Methodist University, and a BS in Mechanical Engineering from NCKU in Taiwan. He has extensive experience in microelectronics packaging and manufacturing technology. Prior to joining GIMT in 1993, he worked for Tandem Computers in Cupertino, California and Microelectronics and Computer Technology Corporation (MCC) in Austin, Texas as a Development Engineer and Member of Technical Staff respectively. He has more than 10 years of industry experience in mechanical design, electronic package design and manufacturing, and reliability. He owns two US patents and has published more than 10 technical papers.

Ali Emamjomeh is a Staff Design Engineer at Tandem Computers, Cupertino, California. He has been involved in many aspects of Electronic

Packaging, his specialty has been in the application of TAB technology to MCMs; design, modeling, and assembly process optimization. His current interests are thermal and mechanical finite element modeling at chip, module and board level. Prior to joining Tandem, he was employed by NARA Technologies Corporation and the National Semiconductor Corporation as TAB Engineering Manager. He is a registered Professional Engineer in the state of California. He received his BS degree in Mechanical Engineering in 1983 from California State University, Fresno and an MS degree in Engineering Mechanics in 1988 from Santa Clara University in California. Mr. Emamjomeh holds three patents and has published four technical papers. He is a member of ASME, IEEE, ASM International and Society for Experimental Mechanics.

Yutaka Tsukada is a Senior Technical Staff member of Yasu Technology Application Laboratory, IBM Japan. He is the manager of Packaging Technology Development and is responsible for new technology introduction to electronic packagings.

He joined IBM in the Circuit Packaging area in Yasu plant in 1970. He has experience in electronic packaging from semiconductor through large system as a manufacturing engineer. He has also worked on machine vision and automated tooling design.

He has moved from manufacturing to development in 1988 when Yasu laboratory was established. He has worked on developments of QFP, TAB and Surface Laminar Circuit and flip-chip attach technologies. He received a BME and an MSME from Hokkaido University in Japan and is a member of JSME, SHM and JPCA. He has several patents and has published approximately 40 technical papers in the field of semiconductor chip packaging and printed circuit boards.

Kenzo Hatada is a Senior Staff Engineer at Matsushita. Dr Hatada has a PhD in electrical engineering from Kobe University, Kobe, Japan. Dr Hatada joined the VLSI Technology Research Laboratory, Semiconductor Research Center, Matsushita Electric Industrial Co., Ltd, in 1967, where he was engaged in the research and development of photolithography technology for LSI. In 1977, Dr Hatada became involved in the research and development of assembly technology for LSI. Dr Hatada pioneered many Packaging Technology innovations for LSI, for example the Transferred Bump TAB Method and the Micron Bump Bonding Method.

Dr Hatada is Director of the Society for Hybrid Microelectronics of Japan and a member of the Japan Society of Applied Physics, the Institute of Electronics, Information and Communication Engineers, and ISHM of USA.

C. P. Wong received a BS degree in chemistry from Purdue University, and a PhD degree in organic and inorganic chemistry from the Pennsylvania State University. After his doctoral study, he was awarded two years as a postdoctoral scholar with Nobel Laureate Professor Henry Taube at Stanford University, where he conducted studies on electron transfer and reaction mechanism of metallocomplexes. He was the first person to synthesize the first known lanthanide and actinide porphyrin complexes which represents a breakthrough in metalloporphyrin chemistry.

He joined AT&T Bell Laboratories in 1977 as a member of the technical staff. He has been involved with the research and development of the polymeric materials (inorganic and organic) for electronic applications. He became a senior member of the technical staff in 1982, a distinguished member of the technical staff in 1987 and an AT&T Bell Laboratories Fellow in 1992. His research interests lie in the fields of polymeric materials, high Tc ceramics, materials reaction mechanism, IC encapsulation in particular, hermetic equivalent plastic packaging, and electronic manufacturing packaging processes. He is one of the pioneers who demonstrated the use of silicone gel as a device encapsulant to achieve reliability without hermeticity in plastic IC packaging. He holds over 28 U. S. patents, numerous international patents, has published over 80 technical papers and 90 presentations in the related area. He is the editor and author of the Academic Press text book on "Polymers for Electronic and Photonic Applications" in 1993.

Dr Wong is a Fellow of the IEEE, an AT&T Bell Labs Fellow, and member of Phi Lambda Upsilon, the National Honorary Chemical Society, and the Materials Research Society. He was the program chairman of the 39th Electronic Components Conference in 1989, and the general chairman of the 41st Electronic Components and Technology Conference in 1991. He was elected to the Board of Governors of the IEEE-CHMT Society from 1987-1989, served as the IEEE-Components, Hybrids and Manufacturing Technology Society technical vice president (1990 & 1991), and currently serves as the president of the IEEE-CHMT Society (1992 & 1993).

John M. Segelken received a BSME degree from the University of Maryland, College Park, and an MSME degree from Purdue University, West Lafayette, Ind.

He joined AT&T Bell Laboratories as a Member of Technical Staff and was responsible for various physical design/system engineering aspects of public telephones, residential telephones, and business terminals. He became a Supervisor in the Switching Apparatus Laboratory in 1979, responsible for connector applications and development. In 1982, he was a Supervisor in the Interconnection Technology Laboratory, responsible for connector applications and planning. In 1984 he was a supervisor responsible for

Physical Design of Advanced VLSI Packaging. From 1989 to 1993 he was a Supervisor in the research area of AT&T Bell Laboratories, Murray Hill, N. J., investigating the use of polymeric materials for electronic/optic packaging applications and has been awarded numerous patents pertaining to electronic/optic packaging. Currently, he is a Technical Manager in Network Wireless Systems, AT&T Bell Laboratories, Whippany, N. J., responsible for Component Engineering and Reliability. He is also currently active in numerous professional activities regarding single chip, multichip packaging, and hermetic equivalent packaging. He has received best paper awards, chaired and organized many technical sessions, and served as Guest Editor and Associate Editor of the Components, Packaging (Hybrids), and Manufacturing Technology transactions of the IEEE. He served as a chairman of the Packaging Program Committee and currently Assistant Program Chair for the Electronic Components and Technology Conference and was elected to the Board of Governors of the CHMT for two consecutive 3-year terms.

Mr Segelken is a member of Tau Beta Pi and Pi Tau Sigma; he is an ASME fellow and a registered Professional Engineer in the State of New Jersey.

Courtland N. Robinson is Manager of the Encapsulation and Materials Group at AT&T Bell Laboratories at North Andover, Mass. He is responsible for the R&D and manufacturing implementation of encapsulation and packaging systems for hybrid integrated circuits.

Dr Robinson received his BS degree in chemistry from Virginia Union University in 1956 and his PhD in polymer science from the University of Utah in 1973. In addition to his work at AT&T Bell Laboratories, Dr Robinson also served as an Adjunct Professor of Chemistry in the polymer program at Lehigh University.

Prior to joining AT&T Bell Laboratories in 1973, he spent nearly 15 years as a chemical scientist and technical manager in the aerospace industry working on the development of solid propellants for missiles and rockets.

Dr Robinson is a member of the Board of Governors of the Components, Hybrids and Manufacturing Technology Society, the ECTC Program Committee and is or has been a member of several professional societies including IEEE, ISHM, the American Chemical Society, the Society of Plastic Engineers and the Sigma Xi Scientific Honorary Society.

Darbha Suryanarayana is an Advisory Engineer at Endicott Electronic Packaging, IBM Microelectronics Division, Endicott, New York. He joined IBM in 1982 at East Fishkill Facility, N. Y. and later moved to Endicott in 1984. His R&D activities include underfill encapsulants for flip-chip mounted packages, advanced ceramic substrate materials and sol-gel materials.

Prior to IBM, he was a Research Associate at: University of Houston, Houston, Texas (1980-82), Oakland University, Rochester, Michigan (1977-80), University of Saskatchewan, Saskatoon (Canada) (1975-77). He received a PhD degree in Physics (1975), from the Indian Institute of Technology, Madras (India); and a BSc degree from Andhra University, Waltair (India). His academic research activities include polymer chain dynamics, structure of polymer gels, defects in crystalline solids and free radical study using electron spin resonance and electron spin echo modulation spectroscopic techniques.

He has published over 50 scientific research papers in several international journals, and is the author of a number of industrial technical reports, US patents and inventions. He was listed in *Who is Who in Technology Today* and is a fellow of American Institute of Chemists.

Donald S. Farquhar joined Endicott Electronic Packaging, IBM Microelectronics, in 1992 as a Staff Development Member. His area of specialization is the design and fabrication of composite materials, in which he has a Doctorate in Engineering from Cornell University. He is currently working in process development for IBM's high performance packaging technology. Dr Farquhar was previously employed at the US Navy David Taylor Research Center and at the Texaco Bellaire Research Center.

Index

Accelerated aging wafer test structures, 102
Accelerated thermal cycling (ATC), 523–6
Acoustic microscopy, 488–91
Activation energy, 150, 175
Ag-In alloys, 266
Ag-Sn alloys, 266
Al-Au intermetallics, 128, 160, 175–7
Al-Cu chip metallizations, 298
Al_2Cu intermetallics, 297
Al-Cu-Si intermetallics, 297–8
Al-Mg wire, 156
Al-Si alloy wire, 156, 313, 323
Al-Si alloys, 154, 160
Alloys 42 lead frames, 157
Aluminum alloy wire, 131
Aluminum bonding wire, 313
Aluminum bump, 193–5
Aluminum metallization, 153–4, 170
Aluminum pad metallization, 296
Aluminum spikes, 153–4
Aluminum wedge/ultrasonic wire bonding,
    308–12
Aluminum wire, 129, 131, 155
AMKOR plastic MCCM, 33
Analog test standard definition, 117
Anisotropic conductive adhesive and epoxy,
    400

Anisotropic conductive adhesive films
    (ACAFs), 258
Antitarnish coatings, 254
Applications specific IC (ASIC), 20–1,
    109–10, 115, 116
Aqueous cleaning, 383
Area array chip, number of possible I/Os on,
    14
Area array packages, 33
Area array pads, physically possible number
    of, 12
Area Array TAB (ATAB) package, 219, 222,
    223
Assembly technology, 1
ASTAP (Advanced Statistical Analysis
    Program), 440
ASTM Test Method, 168
AT&T high-density thin film package, 32
Au_5Al_2-Au_4Al interface, 177
Au-Al diffusion-induced strains, 178
Au-Al-Si contact system, 177
Au-Cu alloy system, 159
Au-Ge alloy, 261
Au-In alloy, 262
Au-In equilibrium phase diagram, 263
Au-In multilayer bonding, 263
Au-Si eutectic alloys, 261
Au-Sn alloys, 261, 266

Au-Sn eutectic alloy, 262
Auger electron spectroscopy (AES), 153
Automatic wire bonding, 134, 140

Backplane Module Test and Maintenance
   (MTM) Bus Protocol, 117
Ball bond shear strength, 140
Ball bond shear test, 151
Ball bonding cycle, 130
Ball-limiting metallurgy (BLM), 230, 233–4
Ball shear failure modes, 172
Ball shear strength data, 170
Ball shear test, 138, 161, 162, 170–1
Ball-wedge bonds, 127
Balltape, 188–9, 200–1
Bare chip testing, alternative approaches,
   101–2
Biphenyldiaminine-phenyldiamine (BPDA-
   PDA), 232
Board level DFT tools, 115
Bond joints, inspection of, 383–6
Bonding material requirements, 256–66
Boundary scan (BSCAN), 102, 109–10, 115,
   116
Boundary scan masters (BSMs), 116
Boundary scan test of inter-chip connections,
   21
BPSG, 178, 179
Built-in current (BIC) monitors, 114, 119
Built-in self-test (BIST), 21, 109, 111–12
Bump deformation, 214
Bump heating, zero contact resistance, 208
Bump structures, 189
Bumped-TAB (BTAB) tape, 197–8
Bumped tape, 188–9
Bumping processes, 187

Ceramic ball grid array (CBGA), 30
Ceramic DIPs (CERDIPs), 157
Ceramic single chip packages, 31
Ceramic substrates, 160
Characteristic impedance, 42–7, 47
   shielding stripline in homogeneous
      medium, 45–7
   stripline in homogeneous medium, 43–5
Chemical vapor deposition (CVD), 155
Chip-and-wire method, 252
Chip attachment, 251–74
   bonding materials requirements, 256–66
   quality, 267–8
   types of COB attach, 255–6

Chip burn-in, 19–21
Chip carriers, 20, 29
Chip interconnection methods, current size
   and performance criteria, 133
Chip level interconnects, 25–9
   solder bumped flip chip, 228–50
   wafer bumping and inner lead bonding,
      186–227
   wire bonding for MCMs, 124–85
Chip metallizations and dielectrics, 153–5
Chip metallurgy, 296–8
Chip on board (COB), 33, 57–80, 275–342
   acceptance testing program, 483
   advantages of, 277
   applications, 278–81
   assembly, evalution matrix, 397
   assembly testing, 329
   circuitization, 287–8
   comparison to MCM-L, 35–7
   comparison to tape automated bonding,
      281
   defect analysis, 118–19
   description of process, 276
   design, 107
   design for diagnostics, 108–9
   design for testability tools, 109–18
   design, materials, and process sets, 281–5
   design targets, 278–81
   die attachment, 301–7
   early products, 275
   electrical system diagnostics, 102
   electrical testing, 47, 328–30
   encapsulation. See Encapsulation
   environmental stress testing, 338
   layouts, 288–90
   manufacturing process flow sequence, 338
   product development, 275–6
   research and development, 21
   rework of, 21
   substrate cleanliness, 298–300
   substrate materials, 48
   substrate options, 286–7
   substrate plating, 291–6
   substrate testing, 47
   TAB technology for, 395–6
   testing, 101–23
   TOB, 61–4, 395–6
   yield, 15–19
   see also Encapsulation; Wire bonding
Chip-on-flex, 287
Chip on glass (COG) technology, 84

Chip on tape (COT) devices, 345, 365, 366, 382
  assembly equipment selection, 393
  E&F'd defects, 367
  excised and formed (E&F'd), 366–70
  removal of defective, 388–9
  removal with die-attach adhesive, 389
  removal without die-attach adhesive, 389
  TAB-related applications, 396–8
  technology, 344
Chip passivation, 231–2
  inorganics, 231–2
  polymers, 232
Chip preparation, 254
Chips, 3–24
  fracture toughness measurement, 22–4
  key elements, 9
  number of possible I/Os (pads) on, 12–13
  number of required I/Os (pads) on, 13–15
  number of required I/Os with different
    circuit designs, 15
  number per wafer, 3–4
Chlorofluorocarbons (CFCs), 365
Circuit scan, 109, 109–10
Cleaning process, 389–91
  conventional, 494
  pre-encapsulation, 492
  see also under specific processes
Cleaning solvents, 364–5
CMOS defects, 113, 118
CMOS devices, 9, 473
CMOS gate arrays, 14
CMOS VLSI logic, 435
Coefficient of thermal expansion (COT), 52,
    155, 267, 314, 475, 476, 482, 505, 506,
    514
  mismatch, 52–3, 421, 424
  of SLC, 423
Coffin-Manson equation, 76, 519
Complementary metal oxide semiconductor.
    See CMOS
Composite solders, 260
Concurrent engineering, 119–20, 253
Condensation cures, 477
Conductive adhesives, 258
Contamination-interface cleaning study, 151
Contamination removal, 254, 298–300, 383
Continuity test, 47
Controlled collapse chip connection (C4)
    bumps, 255
  fatigue enhancement, 519–23

  technology, 29, 195
Copper bump, 193
Copper-clad stainless steel, 157
Copper conductor leads, 361–2
Copper, high ductility, 53–4
Copper metallization, 287
Corner tie bar to hot bar clearance, 354
Cratering, 150–1, 161, 177–9
Cu-Sn alloys, 266
Cu-Sn-Au alloys, 204
Cu-Sn-Au compound, 205–6
Curing agents, 301
Curing mechanism, 481
Curing stress and time, temperature effect on,
    59
Curing systems, 477, 495–9
  peroxide free-radical, 477

Dage-Precima MCT 20 microtester, 140
Delamination, 258
Design for testability (DFT) circuit, 109–18
Destructive ball bond shearing, 170–1
Destructive wire bond pull test, 148, 149, 161,
    164–8, 173
Die attachment, 301–7
  adhesives, 303, 389–90
    material curing, 392
    replenishment methods, 391–2
  dispensing process, 303–6
  formulation, 301
  materials, 157, 301–3, 363–4
  quality control, 307
Diebond pad size, 290
Differential scanning calorimetry (DSC), 495,
    496
Direct chip interconnect (DCI), 258
Dissipation factor, 48–9
Double torsion specimen, 22
Dual in-line package (DIP), 30
Duktilomat hydraulic deep-drawing test, 54
Dwell time and wire tensile strength, 145–7
Dynamic random access memory (DRAM)
    devices, 473

Electric discharge machining (EDM), 355
Electrical overstress/electrostatic damage
    (EOS/ESD), 113–14, 118, 119
Electrical testing, 328–30, 391
Electroless nickel/immersion gold plating
    process, 294
Electroless nickel plating, 295

Electroless plating, 415
Electronic packaging
    delay, 12
    hierarchy of, 1
    levels of, 2
    major functions of, 2
Electro-oxidation, 485
Electroplated gold/nickel process
    full additive, 291–2
    semi-additive, 292–4
Electroplating process, 244, 254
Electrostatic boundary value problem, 43
Emitter coupled logic (ECL), 9
Encapsulation, 299, 330–7
    application, 494–5
    chip on board, 470–503
    curing systems, 495–9
    dispense equipment, 337
    dry preform systems, 334–6
    manufacturing process, 491–9
    material requirements, 474
    polymeric materials, 474–6
    pre-encapsulation cleaning, 492
    process details, 336–7
    purposes and goals, 472–4
    reliability, 499
    screening tests and analysis, 483–8
    SLC, 423–5
    see also Underfill encapsulation
Energy dispersive X-ray (EDX), 265
Environmental stress testing, COB, 338
Epoxies
    encapsulation, 474–6
    mechanical and thermal properties of, 514
Epoxy encapsulant systems, 331–4
Epoxy off-gas contamination, 151
Epoxy resins, 301
Evaporation process, 243–4

Fiducial marks, 290
Fillers, 302
Fine pitch developments, 289
Finite elements analysis (FEM), 178, 422
First level packages, 29–37
Flatpack, 30
Flexural rigidity measurement, 55–6
    cantilever beam method, 56
    simply supported beam method, 56
Flip bare chips on board (COB), 2
Flip chip attach (FCA), 73, 252, 256, 410–11,
    418–30, 526

eutectic carrier bump build, 427
    manufacturing process, 427–30
    stress analysis, 422–6
    see also SLC/FCA package
Flip chip bonding, 418–22
Flip chip packages, underfill encapsulation.
    See Underfill encapsulation
Flip chip packaging strategies, 505
Flip chip rework capability, 102
Flip chip solder bumps. See Solder bumped
    flip chip
Flip-TAB, 365
Fluorine plasma, 261
Flux cleaning, 382–3
Flux materials, 364
Flux removal, 261
Fluxes, 389
Fluxless soldering, 261
Fourier-Transform-Infrared (FT-IR), 495,
    496
FPSG film, 178
Fracture toughness
    double torsion specimen, 22
    indentation crack specimen, 22–4
    measurement of chip, 22–4
    three-point loaded bend specimen, 24

Gallium arsenide, 154, 160–1
Gang thermode bonding, force application
    mechanism, 374–5
Garofalo-Arrhenius steady-state creep, 68
Gate delays, 9
Gate integration, 9
Glass adhesives, 259
Glass transition temperature, 49–52, 253, 506,
    514
Gold bump material, 188
Gold bumping process, 189–91
Gold bumps, 206
Gold-gold weld, 128
Gold metallization, 168, 171, 173
Gold metallized ceramic substrates, 167, 169
Gold pad, 128
Gold plating, 254
Gold thermosonic ball and stitch, 308
Gold thermosonic ball wire bonding, 312
Gold thin film metallization, 170
Gold wire, 128, 131, 132, 149, 156
Gold wire bond testing, 166
Gull-wing pull test set-up, 384

Hard solders, 261–2
HCFC solvents, 300
HDI process for wafer level burn-in, 105–6
Heat sinks, 438
Hermetic packages, 157
High Density Interconnect (HDI), 102
Hostile environments, 474
Hot air solder leveling (HASL), 254
Hot air thermode (HAT), 216
Hot bar location, 356
Hot bar selection, 354–6
  folded metal, 354
  solid ceramic or ceramic coated metal, 356
  solid metal, 355

IBM MCCM, 33
IEEE 1149.1 standard, 102, 115
IEEE P1149.4 standard, 102, 117
IEEE P1149.5 standard, 102, 117
Imidazole, 254
Indentation crack specimen, 22–4
Inner lead bonding (ILB), 187
  bonding process window, 214
  equipment manufacturers, 203
  plating material and bonding options, 362
  residual stress, 366
  single-point laser, 368
  solder attach processes, 218–22
  solder joining tool-hot air thermode,
    216–18
see also Tape automated bonding (TAB)
Input/output (I/O) system, 1
  density, 124
  number of pads required, 13–15
  physically possible number of pads, 12
  required number on chip with different
    circuit designs, 15
Inspection and testing, 391
Inspection of bond joints, 383–6
Institute for Defense Analysis (IDA) Report
    R-338, 1988, 120
Integrated circuit (IC)
  chip, 1
  designs, 9
  gang bonding, 461–2
Interconnection density. See Wireability
Intermetallic failure mechanisms, 175–7
Inverted chip solder reflow, 137
Isolation test, 47
Isopropyl alcohol (IPA) cleaning, 383

Keeper bar design, 353
Known good dies (KGD), 19–21, 102–7, 255
  objective of, and business needs, 20
  testing at wafer level, 103–4
Known good substrates (KGS), 47
Kulicke-Soffa Model 1419 automatic wire
    bonder, 140

LANs, 82
Laser joining process, 203
Laser pantography, 136
LCD modules by COG technology, 84
Lead frames
  gold-plated, 159
  materials, 157–9
  packages, 187
  wire bonding to, 159
Lead heights, 212
Lead thickness, 212
Leadform design, 349–52
Leadless chip carrier (LCC), 30
LED array module
  constitution of, 453–4
  construction of, 453–7
  electrical characteristics, 456–7
  fabrication process, 454–6
Line delay of high-end and low-end computer
    applications, 9
Linear ball shear model, 148
Linear feedback shift registers (LFSRs),
    111–12
Logic performance, 12

Masking process, 240–3
MCM-C, 32, 34, 125, 470, 471
MCM-D, 32, 34, 125, 344, 345, 439–42, 470,
    471
  assembly considerations, 365–95
  die attachment materials, 363–4
  material considerations, 358–65
  outer lead bonding, 371–82
  substrate materials, 359–61
  TAB, 401–2
    design considerations, 346–58
MCM-L, 32, 33, 34, 125, 251, 435–42, 470
  advantages of, 437
  comparison with COB, 35–7
  competitive advantages, 471
  cross-section, 435
  electrical tests, 47
  heat sink, 438

substrate materials, 48
Mech-E1 Model 827 thermosonic ball
    bonding machine, 139
Metal masks, 240–1
Metal-oxide semiconductor field-effect
    transistors (MOSFETs), 155
Microdielectrometry, 495, 496
Micron bump bonding (MBB), 444–69
    applications, 453–68
    characteristics of, 445–6
    electrical characteristics, 449–51
    insulation resistance between adjacent
        electrodes, 451
    outline and futures of new technology,
        444–52
    principle of method, 447–9
    process method, 446–7
    relationship between compression stress
        and thermal stress, 448
    reliability, 451–2
    schematic cross-section, 444
Mil-Std-883/5011, 299, 303
Mil-Std-883C, 134
Mil-Std-883D, 325
Miniature products, packaging technology
    for, 80–4
Mixed signal test problems, 117
MODEMs, 82
Moiré interferometry, 73
Moisture absorption, 54–5
Multichip carrier modules (MCCM), 33–5
Multichip modules (MCM), 31–3, 124, 186,
        275, 470
    average wire length, 38
    bare die, 103
    categories of, 32
    construction of, 463–8
    cross-section, 464
    defect level vs fault coverage, 18, 19
    DFT tools, 117
    formal definition, 32
    generic structure with wire bonds added,
        127
    generic structure without wire bonds, 125
    materials, 153–61
    parameters, 126
    process flow, 466–7
    reliability, 467–8
    research and development, 21
    rework of, 21
    signal delay, 41, 42

substrate criteria, 34
substrate materials, 48
substrate testing, 47
transferred bumps on electrodes of LSI
    chip, 465
types I, II, III, 164–8
wire bonding, 124–85, 180
wireability, 39
yield, 15–19
    vs number of chips, 16
    vs test fault coverage, 17, 18
    see also MCM-C, MCM-D, MCM-L
Multilayer ceramic (MLC), 430
Murphy's law, 4

Ni-B alloy plating, 295–6
Ni-P electroless plating, 295
Ni-P electroplating, 296
Nitrides as insulating layers, 155
N-methyl-pyrrolidine 2 (NMP), 232
Non-destructive ball shear (NDBS), 148, 149,
    161, 171
Non-destructive pull test (NDPT), 161, 162,
    168, 174
Non-hermetic packages, 157
Novolac system, 475

Organic adhesives, 257
Organic solvents, 299
Outer lead bonding (OLB)
    anisotropic conductive adhesive and epoxy,
        400
    bonding parameters, 375–80
    comparison of methods, 374
    demountable, 371, 400–1
    fine-pitch bonders, 393
    gang bonding, 382
    interconnection, 186
    metallurgical, 398
    permanent, 371
    placement accuracy, 380–2
    plating material and bonding options, 362
    process, 222–3
    residual stress, 366
    single-point laser, 368
    solder defects, 384
    solder reflow, 364
    thermode, 374
        gimbaling options, 375
        operation sequence and force applied
            mechanism, 375

window, 349–52
Oxides as insulating layers, 155
Oxygen plasma cleaning methods, 170, 180, 300

Packaging density, 37
Packaging of integrated circuits, 157
Packaging strategy
    factors in, 471
    flip chips, 505
Packaging technology for miniature products, 80–4
Pad preparation, 254
Particulate contamination, 300
Pb-In solder joints, 239
Pb-In-Au alloys, 266
Pb-Sn alloys, 266
Pb-Sn solders, 234–6, 259
PCMCIA IC card, 435
PCMCIA PC card, 81–4, 186
Peripheral array chip, number of possible I/Os on, 13
Peripheral packages, 33
Peroxide free-radical cure system, 477
Phosphorus pentoxide, 155
Phosphosilicate glass (PSG), 155, 178, 179
Photolithographic masks, 241–3
Photolithographic process sequence, 245
Photothermosetting resin, 84, 86
Pin grid array (PGA), 30, 430
Planar tape, 188–9
Plasma Enhanced Chemical Vapor Deposition (PECVD), 231–2
Plastic ball grid array (PBGA), 30
Plastic leaded chip carrier (PLCC), 30
Plastic quad flat pack (PQFP), 30, 33, 187
Plating thickness, purity and visual requirements, 296
Poisson's yield formula, 6
Polydimethylsiloxanes, 476
Polydiphenylsiloxanes, 476
Polyimides, 232, 481–2
Polyimidesiloxanes (SIM), 258
Polymer bonding, 256–8
Polymer degradation, 258
Polymerization stress, 59
    resin layer thickness effect on, 59
Polymethylphenylsiloxanes, 476
Polyurethanes, 482–3
    chemical structure, 484
    synthesis of, 484

Power-dwell-temperature effects, 143–5
Printed circuit boards (PCBs), 37–57
    dimensional stability, 54–5
    electrical tests, 47
    flexural ridigity, 55–7
    high copper ductility laminate materials, 53–4
    high glass transition temperature and thermal conductivity laminate materials, 49–52
    in-plane elongation, 50
    layouts, 288–90
    load-deflection curve, 57
    low coefficient of thermal expansion laminate materials, 52–3
    low moisture absorption laminate materials, 54–5
    low relative dielectric constant and dissipation factor laminate materials, 48–9
    materials, 47–55, 252
    out-of-plane elongation, 51
    solderability, 254
    standard features, 47–8
    Young's modulus, 55–6
Printed wiring board (PWB). See Printed circuit boards (PCBs)
Processing performance (MIPS), 6–12
    factors affecting, 9
Propagation delay, 40
Pseudo-random test vectors, 111
Pyromellitic dianhydride-oxydianiline (PMDA-ODA), 232

Quad flat pack (QFP), 430
Quality assurance, 171–4
Quality control, die attachment, 307
Quasi module concept in SLC, 417–18
Quiescent current monitoring (I_{DDQ}), 104, 109, 112–16

Rams, 171
Raw card testing, 329
Reactive oxygen cleaning, 494
Rectangular quad flat pack (RQFP), 30
Relative dielectric constant, 40, 43, 48–9
Reliability of wire bonds, 174–9
Reliability problems, particulate contamination, 300
Remounting, 392
Repair process, 386–7, 392–3

Residual stress, 267
Resin layer thickness effect on polymerization
    stress, 59
Reversion, 258
Rework process, 324–5, 386–7, 392
RISC system/6000, 435
Room temperature vulcanized (RTV)
    silicones, 477
Rosin mildly activated (RMA) solder pastes,
    259

Scanning acoustic microscopy (SAM), 265
Schottky barrier height, 154
SCOAP, 116
Screen printing, 304
Screening tests, encapsulation systems, 483
Secondary ion mass spectrometry (SIMS),
    153
Seeds' formula, 6
Semi-aqueous cleaning, 383
Semiconductor, 9
Shielded stripline in homogeneous medium,
    45–7
Shipped Product Quality Level (SPQL), 279
Signal delay, 48
    time, 40
    vs relative dielectric constant, 41
Silane, 155
Silicon chips
    substrate materials, 32
    toughness of, 22
Silicon nitride, 155, 231
Silicon on silicon (SOS) package, 439–42
Silicon substrate, 32
Silicon wafer, 3
Silicone gels, 486
    THB testing, 487, 488
Silicones, 476–9
    encapsulant systems, 330–1
    rheology, 478
    temperature-humidity-bias (THB)
        accelerated electrical testing, 478
    THB testing, 486
Single chip modules (SCMs), 29–30, 186,
    439–42
Single expose double develop (SEDD)
    process, 197
Single-point bonding, 202, 203, 368
SiO$_2$ films, 155
SiO$_2$ sputtering, 231
SLC/FCA package

advantages of, 430
applications, 434–42
Sloan Dektak IIA stylus profilometer, 177
Small outline integrated circuit (SOIC), 30
Sodium chloride, 299, 473
Soft solders, 259–61
SOIC with J-leads (SOJ), 30
Solder Attach Tape Technology (SATT), 196,
    216, 219
Solder bumped flip chip, 25, 27–9, 195–6,
    228–50
    accumulated effective plastic strain
        distribution, 76
    alternative materials, 239
    basic process description, 230–1
    creep curves, 66
    creep responses of solder interconnects, 70,
        71
    effective steady-state creep strain rate, 70
    effective stress, 70
    effective stress in underfill encapsulant,
        78–9
    finite element model, 75
    fringe patterns of U and V displacement
        fields, 73–4
    high-temperature, 72–9
    load-displacement curve, 79
    low-temperature, 79–80
    process sequences, 245
    reliability, 236
    solder deposition processes, 243–5
    solder joint geometry, 236–9
    solder systems
        materials, 234–6
        processes, 240–5
    solders used, 66–71
    steady-state creep curves, 68
    terminal metals
        materials, 233–4
        process, 240–5
    thermal stresses and strains, 75
    underfill epoxy encapsulants, 72
    uniaxial steady-state creep strain rate, 70
    universal stress-strain rate curve, 71
Solder bumped flip chip on board, 65–80
Solder injection tool, 427–9
Solder mask over bar copper (SMOBC), 254
Solders
    bonding below eutectic point, 262–2
    composite, 260
    hard, 261–2

pastes, 259–60
replenishment methods, 391–2
soft, 259–61
Solid-liquid interdiffusion (SLID), 263, 265
Solid-state diffusion, 262
Solvent cleaning, 151
Sputter cleaning, 240
Sputtering process, 231, 244
Square quad flat pack (SQFP), 30
Stamping, 305
Standard single chip module, 33
Static random access memory (SRAM) chips, 20–1
Statistical process control (SPC), 320
Stress concentrations, 268
Stripline in homogeneous medium, 43–5
Substrate, 286–7
    bonding parameters, 147–8
    cleanliness, 298–300
    criteria, 34
    keep clear areas, 358
    materials, 48, 159–61, 359–61
    metallization, 159–61
    options, 286–7
    pad coverage by lead, 381
    pad design, 357–8
    plating, 291–6
    testing, 47
Surface laminar circuits (SLC), 73, 234, 410–11, 439–42
    CTE of, 423
    encapsulation, 423–5
    eutectic solder bump, 420
    fan-out capability, 418
    manufacturing process, 413–17
    material properties, 75, 415, 423
    quasi module concept in, 417–18
    structure, 411–13
    wiring density, 413
    see also SLC/FCA package
Surface mount component (SMC), 413
Surface mount technology (SMT), 33
Sycar™, 479–81

TAB. See Tape automated bonding (TAB)
TAB on board (TOB), applications, 398–401
Tape automated bonding (TAB), 2, 25, 27, 102, 106–7, 134–6, 186, 252, 343–409
    advantages, 61
    applications, 396–402
    assembly considerations, 365–95

carrier and test/burn-in socket selection, 357
COB, 61–3
    application, 395–6
    process flowchart, 64
comparison of face-up and flip, 399
comparison to COB, 281
conductor leads, 361
configurations, 345–6
cost analysis, 63–4
design considerations on board and on MCM-D, 346–58
design flow chart, 346
flat-mount, 365
flip-TAB, 365
future trends, 222–3, 402–5
history, 343
ILB bonding cycle
    gang bonding bump forging model, 209–14
    mechanics of, 206–14
    thermal and dynamic models, 207–9
ILB gang bonding, 206
ILB inner lead bond bump options, 188–9
ILB pitch, 187
ILB processes and tools, 201–5
inner lead thermocompression bonding process, 205–16
key technology elements, 344
manufacturing flow using pre-encapsulated chips, 345
manufacturing process flow in ceramic package, 345
material considerations, 358–65
MCM-D, 401–2
mount options, 365
package overall dimensions with and without keeper bar, 352
reliability, 62
tape classification, 361–2
tape structure, 188–9
typical dimensions, 351
UBMs for, 62
Tape bumping, 187, 197–201
Tape lead force component and boundary conditions, 385
Tape on board (TOB), outer lead bonding, 371–82
TAPEPAK, 345
Temperature effect on curing stress and time, 59

Temperature-humidity-bias (THB)
    accelerated electrical testing, 478
Temperature shock test, 338
Temporary packaging, 102, 105
Test access port (TAP), 115
Test clock (TCK), 115
Test data input (TDI), 115
Test data output (TDO), 115
Test mode select (TMS), 115–16
Test reset (TRST), 115
Test strategies, 118–19
Thermal coefficient of expansion. See
    Coefficient of thermal expansion
Thermal conductivity, 49–52, 157, 314
Thermal expansion mismatch. See Coefficient
    of thermal expansion
Thermal exposure studies, 150
Thermal head module, 457–63
    bonding process, 459
    characteristics of, 459–61
    construction of, 457–8
    cross-section of, 459
    gang bonding of ICs, 461–2
    mass-production of 400 DPI, 462
    reliability, 463
Thermal stresses, 267, 314
Thermocompression bonding, 127, 128–31,
    203, 205–6, 262
    temperature range, 130
Thermocompression bonding mechanical
    process window, 214–16
Thermode gang bonding, 368
Thermomechanical analysis (TMA), 50
Thermomechanical stress, 59
Thermosonic ball bond shear strength vs.
    dwell time, 146
Thermosonic ball bonding parameters, 137
Thermosonic bonding, 127, 128, 132
Thermosonic gold wire bond testing, 165
Thermosonic single-point bonding, 203, 204
Thermosonically bonded gold wire, 167, 169,
    170, 173
Thin quad flat pack (TQFP), 29
Thin small outline package (TSOP), 29, 186
Thin tape carrier package (TCP), 29
Three-point loaded bend specimen, 24
Token Ring LAN Adapter card, 435
Total quality management (TQM), 119–20
Transfer-bumped tape, 188–9, 198–9
Transfer printing, 305
Transient liquid phase solding, 266

Transmission line theory, 40
Triazole, 254
Type I errors, 108, 109, 120
Type II errors, 108

Ultra thin TCP, 30
Ultrasonic bonding, 127, 128, 131–2, 180
Ultrasonic wedge-wedge bonding, 128, 129,
    131
Under-Bump Metallurgy (UBM), 230
Underfill encapsulation, 504–31
    C4 fatigue enhancement, 519–23
    candidate materials, 505–6
    case studies, 523–6
    cured encapsulant properties, 514–18
    environmental resistance, 518
    epoxy, 72
    flow characteristics, 509–13
    future directions, 528
    impurities, 512–13
    material requirements, 508–18
    process, 508
    recommendations for manufacturing,
        526–7
    reliability, 518–26, 523–6
    uncured encapsulant properties, 509–12
    volatile components, 512–13
Universal test sites (UTS), 285, 329, 338
Unknown bad dies (UBD), 21
Unknown bad substrates (UBS), 47
UTHE 10E ultrasonic generator, 139
UV-ozone cleaning method, 153, 170, 180
UV-visible (UV-VIS) light radiation, 473–4

Vendor Material Safety Data Sheets
    (MSDSs), 336
Very small outline package (VSOP), 30
Very small quad flat pack (VSQP), 30
Vias, 290
VLSI technology, 473
Voids, 267–8
von Mises criterion, 69, 71
von Mises stress distribution, 78
VPSG (Via's Per Square Grid), 413

Wafer, 3
    bumping, 187, 189–96
    cells, 3
    level probe tests, 102
    level sampling, 102
    level testing, 103–4

burn-in, 104
number of chips on, 3–4
number of yielded chips on, 4–6
scale integration (WSI), 136
step plan, 3
testing, 328
Water soluble solder pastes (WSP), 259
Wire bond geometry, 164
Wire bond pull strength, 140, 164
  effects of bond geometry, 169
  effects of bond length on, 167
Wire bond testing, 161–71, 296
  equipment, 162–4
  parametric analysis, 151–3
  prior influences, 148–50
  temperature effects, 150–1
Wire bonding, 2, 25–7, 307–25
  bond dimensions, 321–2
  burn-out current, 315
  chips on board, 58–61
    see also Chip on board (COB)
  cost analysis, 63–4
  current practice, 187
  design, 180
  design maximum current, 315
  electrical current capabilities, 314–15
  equipment requirements, 315–19
  fabrication, 139–40
  in-process monitoring, 320–1
  multichip modules, 124–85
  options, 308–12
  parameter experiments, 141–2
  parameters, 137–40

  principles of, 125–37
  process, 137–53
  process details, 315–19
  process optimization, 141–53
  process set-up, 319–20
  rework, 324–5
  spacing limitations, 135
  tests and test samples, 138–40
  thermal expansion coefficient, 314
  visual inspection of wire bonds, 321–4
  wire metallurgy/controls, 313–14
Wire bonding chips on board
  process flowchart, 64
  reliability, 61
  thermal management, 61
  see also Chip on board (COB)
Wire bonding process, wire types and
    properties, 155–6
Wire length, 187
Wire-like interconnects, 134–7
Wire tensile strength and dwell time, 145–7
Wireability, 37–40
Wirebond pad
  length, 289
  orientation, 289
Wirebond, span, 289

X-ray inspection, 429

Yielded chips, number per wafer, 4–6
Young's modulus, measurement of, 55–6
Yt-Al-garnet (YAG) laser, 200, 204